Functional Integration
and Quantum Physics

This is a volume in
PURE AND APPLIED MATHEMATICS

A Series of Monographs and Textbooks

A list of recent titles in this series appears at the end of this volume.

Functional Integration and Quantum Physics

BARRY SIMON

Departments of Mathematics and Physics
Princeton University
Princeton, New Jersey

ACADEMIC PRESS New York San Francisco London 1979

A Subsidiary of Harcourt Brace Jovanovich, Publishers

ACADEMIC PRESS, INC.
111 Fifth Avenue, New York, New York 10003

United Kingdom Edition published by
ACADEMIC PRESS, INC. (LONDON) LTD.
24/28 Oval Road, London NW1 7DX

Library of Congress Cataloging in Publication Data

Simon, Barry.
Functional integration and quantum physics.

(Pure and applied mathematics, a series of mono-graphs and textbooks)
Based on lectures given by the author at a con-sortium in Lausanne, Switzerland, 1977.
Bibliography: p.
1. Integration, Functional. 2. Quantum theory.
I. Title. II. Series.
QA3.P8 [QC20.7.F85] 510'.8s [530.1'5] 78-22535
ISBN 0-12-644250-9

AMS (MOS) 1970 Subject Classifications: 60J65, 80-00

PRINTED IN THE UNITED STATES OF AMERICA

79 80 81 82   9 8 7 6 5 4 3 2 1

Contents

Preface

In the summer of 1977 I was invited to lecture in the Troisième Cycle de la Suisse Romande, a consortium of four universities in the French-speaking part of Switzerland. There was some discussion of the topic about which I might speak. Since I seem fated to be the apostle of probability to Swiss physics (see [258]), we agreed on the general topic of "path integral techniques." I decided to limit myself to the well-defined Wiener integral rather than the often ill-defined Feynman integral. In preparing my lectures I was struck by the mathematical beauty of the material, especially some of the ideas about which I had previously been unfamiliar. I was also struck by the dearth of "expository" literature on the connection between Wiener integral techniques and their application to rather detailed questions in differential equations, especially those of quantum physics; it seemed that path integrals were an extremely powerful tool used as a kind of secret weapon by a small group of mathematical physicists. My purpose here is to rectify this situation. I hope not only to have made available new tools to practicing mathematical physicists but also to have opened up new areas of research to probabilists.

I am pleased to be able to thank some of my colleagues who aided me in the preparation of this book. During the period of the lectures on which the book is based, I was a guest of the Physics Department of the University of Geneva. I am grateful to M. Guenin, the departmental chairman, and most especially to J. P. Eckmann for making my visit possible. The lectures were given at the EPFL in Lausanne; P. Choquard was a most gracious host there. I should like to thank the Secretariat Centrale of the University of Geneva Physics Department and Mrs. G. Anderson of the Princeton Physics Department for typing the first and second drafts, respectively, of the manuscript. I am also grateful to Y. Kannai for the hospitality of the Weizmann Institute Pure Mathematics Department where Sections 20–24 were written.

Finally, I owe a debt to a number of people for scientific contributions: M. Donsker and M. Kac made various valuable suggestions about what topics might be included as well as offering help on technical questions; L. van Hemmen gave his permission to use an unpublished argument of his; I had valuable discussions with M. Aizenman, R. Carmona, P. Deift, J. P. Eckmann, J. Fröhlich, C. Gruber, E. Lieb, A. Sokal, M. Taylor, A. Truman, and S. R. S. Varadhan; the careful reading of the complete manuscript by R. Carmona was especially valuable; finally, M. Klaus, A. Kupiainen, and K. Miller helped in the proofreading. I am glad to be able to thank all these individuals for their help.

List of Symbols

	page
a_ν (Eq. (9.8))	96
$b(s)$, $\mathbf{b}(s)$; Brownian motion	33, 36
$\mathbf{C}_\theta$; Hölder continuous functions	264
$C(K)$; capacity	84
db, $d\mathbf{b}$, dq, $d\alpha$; stochastic differentials	151, 154, 170
Db, Dq, $D\alpha$; measures for basic processes	38
$E(A)$, $E(f)$, $E(f; A)$; expectations	8
$E(f \mid \Sigma)$; $E(f \mid g_1, \ldots, g_n)$; conditional expectation	21
$G_{D;S}$; Dirichlet Green's function	69
$h(\mathbf{y}, k)$; hitting probability	82
$H_\hbar$	195
$H_{D;S}$; Dirichlet Laplacian	69
$H(\mathbf{a}, V)$; Schrödinger operator	159
$H_0(\mathbf{a})$	161
K_γ; Birman–Schwinger kernel	89
$\overline{\lim}\, A_n$	18
$p(J)$, $P_\Lambda(z, \beta)$; pressure	200, 246
$P_{D;S}$; Dirichlet propagator	69
$P_t(x, y)$	35
$q(s)$; oscillator and also $P(\phi)_1$ process	34, 57

	page
$s(\rho)$; entropy	200
S_n; Schwinger function	253
$u_n(X_1, \ldots, X_n) \equiv \langle X_1, \ldots, X_n \rangle_T$; Ursell function	129
$W_\gamma(t)$, $W_\gamma(a, c)(\mathbf{b})$; Wiener sausage	209, 236
α; Brownian bridge	40
μ_B; conditional measure	68
$d\mu_{0,\mathbf{a},\mathbf{c};t}$; conditional Wiener measure	39
$\rho_\Lambda^{(n)}(\mathbf{x}; \varepsilon, z)$; correlation functions	246
ω; Wiener path	38
$\Omega^\pm$; wave operators	226
Ξ; grand partition function	246
$\mathscr{C}$; continuous functions on $C[0, 1]$	176
$\mathscr{E}_M^{(\alpha)}$; Thomas–Fermi functional	98
$\mathscr{S}$; Schwartz space of tempered functions	12
$\lvert\cdot\rvert$; Lebesque measure of	2
$A \,\Delta\, B$; symmetric difference	8
$:-:$; Wick ordering	27
$\sim$; asymptotic series	212
$\lVert\cdot\rVert_{L_u^p}$; uniform local norm	260

Functional Integration and Quantum Physics

I

Introduction

1. Introduction

It is fairly well known that one of Hilbert's famous list of problems is that of developing an axiomatic theory of mathematical probability theory (this problem could be said to have been solved by Khintchine, Kolmogorov, and Lévy), and also among the list is the "axiomatization of physics." What is not so well known is that these are two parts of one and the same problem, namely, the sixth [287], and that the axiomatics of probability are discussed in the context of the foundations of statistical mechanics. Although Hilbert could not have known it when he formulated his problems, probability theory is also central to the foundations of quantum theory. In this book, I wish to describe a very different interface between probability and mathematical physics, namely, the use of certain notions of integration in function spaces as technical tools in quantum physics. Although Nelson [190, 191] has proposed some connection between these notions and foundational questions, we shall deal solely with their use to answer a variety of questions in conventional quantum theory; some typical problems which we shall solve using functional integration are the following:

(1) Consider a potential V in three dimensions and let $N(V)$ be the number of bound states of $-\Delta + V$ (i.e., the dimension of its spectral projection for $(-\infty, 0)$). In a semiclassical approximation, this is expected to be $(2\pi)^{-3}|\{(p, x) \mid p^2 + V(x) < 0\}|$ (the $(2\pi)^{-3}$ comes from the fact that $\hbar = 2m = 1$ so $h = 2\pi$); i.e.,

$$N_{sc}(V) = (2\pi)^{-3} \int |\min(0, V(x))|^{3/2} \left(\frac{4\pi}{3}\right) d^3x$$

Throughout this book, $|\{\cdots\}|$ denotes the Lebesgue measure of $\{\cdots\}$. It is clear that, in general, $N(V)$ can be zero even though $N_{sc}(V)$ is quite large: for a shallow square well will not bind states, so if V is a sum of 10^6 such wells, each 10^6 light years apart, $N(V)$ will be zero but $N_{sc}(V)$ will be very large. On the other hand, one might hope that $N(V)$ cannot be large without $N_{sc}(V)$ being about as large; i.e., one might hope to prove that

$$N(V) \le cN_{sc}(V)$$

for some suitable universal constant c (see Section 9).

(2) Let $E_n(\beta)$ be the nth eigenvalue of $-d^2/dx^2 + x^2 + \beta x^4$. Since $E_n(0) = (2n + 1)$, we have that $E_3(0) - E_2(0) = E_2(0) - E_1(0)$. A moment's reflection convinces us that the left-hand side of this equality should increase faster with β than the right-hand side. How do we prove that

$$E_3(\beta) - E_2(\beta) \ge E_2(\beta) - E_1(\beta)$$

(see Section 12)? We note that the more general conjecture that "$E_n(\beta)$ is convex in n" is open.

(3) Fix a positive continuous function V on $\mathbb{R}^3$ and $\mathbf{a}$, a C^1-function on $\mathbb{R}^3$ with values in $\mathbb{R}^3$. Then (see, e.g., [132, 215, 229, 242])

$$H(\mathbf{a}) = (-i\nabla - \mathbf{a})^2 + V$$

is essentially self-adjoint on $C_0^\infty(\mathbb{R}^3)$. $H(\mathbf{a})$ is, of course, the Hamiltonian of a particle in a magnetic field $\nabla \times \mathbf{a}$ and electric field $-\nabla V$. Think of fixing V and varying $\mathbf{a}$. Let $E(\mathbf{a}) = \inf \operatorname{spec}(H(\mathbf{a}))$. In [250], Simon showed that $E(\mathbf{a}) \ge E(\mathbf{0})$ by the following elementary argument: Let us compute

$$\begin{aligned} \nabla(u^*u) &= [(\nabla - i\mathbf{a})u]^*u + u^*[(\nabla - i\mathbf{a})u] \\ &= 2\operatorname{Re}(u^*(\nabla - ia)u) \end{aligned}$$

Thus, using $\nabla(u^*u) = 2|u|\nabla|u|$

$$|\nabla|u|| \le |u|^{-1}|\operatorname{Re}[u^*(\nabla - i\mathbf{a})u]| \le |(\nabla - i\mathbf{a})u|$$

Squaring and integrating over x we find

$$(u, H(\mathbf{a})u) \ge (|u|, H(\mathbf{0})|u|)$$

so that

$$E(\mathbf{a}) \ge E(\mathbf{0}) \tag{1.1}$$

by the variational principle. (1.1) says that the energy of spinless Bosons goes up when a magnetic field is turned on. Let V go to infinity as $x \to \infty$ so fast that $Z(\mathbf{a}) = \mathrm{Tr}(e^{-\beta H(\mathbf{a})})$ is finite. How can one prove the finite temperature analog of (1.1):

$$Z(\mathbf{a}) \leq Z(\mathbf{0})$$

(see Section 15)?

(4) Fix V and W, even functions on $(-\infty, \infty)$ with $dW/dx \geq 0$ for $x \geq 0$. Suppose that $-d^2/dx^2 + V$ and $-d^2/dx^2 + V + W$ both have normalized eigenfunctions, Ω_V and Ω_{V+W}, respectively, at the bottom of their spectrum. Since W is "attractive," adding it should "pull the ground state in"; i.e., one expects that for any $a > 0$

$$\int_{-a}^{a} |\Omega_{V+W}(x)|^2 \, dx \geq \int_{-a}^{a} |\Omega_V(x)|^2 \, dx$$

The difficulty with proving this lies in the facts that Ω is only given implicitly, that the energy shift must be taken into account, and finally that the normalization condition must be taken into account (see Section 12).

(5) The Hamiltonian of a hydrogen atom in a constant magnetic field $(0, 0, cB)$ is (units with $2\mu = \hbar = |e| = 1$):

$$H = \left[\left(i\frac{\partial}{\partial x} + \frac{By}{2}\right)^2 + \left(i\frac{\partial}{\partial y} - \frac{Bx}{2}\right)^2 - \frac{\partial^2}{\partial z^2}\right] - \frac{1}{r} \tag{1.2}$$

It is not hard to show that H commutes with $L_z = i[y(\partial/\partial x) - x(\partial/\partial y)]$, and one expects that the ground state ψ of H has $m = 0$; i.e., $L_z\psi = 0$. The usual proof [217] that ground states of systems without statistics have $l = m = 0$ (when rotationally invariant) breaks down since e^{-tH} is no longer positivity preserving; indeed, it is not even reality preserving (see Section 12).

(6) Let V be a positive potential with compact support but also with some rather severe singularities (e.g., $r^{-\alpha}$ with α large). One's intuition is that particles should just "bounce off," so that the basic existence and completeness questions of scattering theory for the pair $(-\Delta, -\Delta + V)$ should be solvable. However, since V and $-V$ have a very different status, the usual perturbation methods [199], will not be applicable (see Section 21).

(7) One can ask to what extent the ground state (lowest eigenfunction), $\psi(x)$, of $-d^2/dx^2 + V(x)$ mirrors properties of V. Suppose that V is even and $\psi(x) = e^{-f(x)}$. Then modulo technical hypotheses we shall prove (see Sections 12 and 13) that $V' \geq 0$ (respectively, $V'' \geq 0$, $V''' \geq 0$) on $(0, \infty)$ implies that $f' \geq 0$ (respectively, $f'' \geq 0$, $f''' \geq 0$) on $(0, \infty)$. We note that the analog for four derivatives is false; see the example at the end of Section 12.

(8) Consider the ground state energy of H_2^+ in the Born–Oppenheimer approximation; i.e., for $R \in [0, \infty)$, let

$$H(R) = -\Delta - |\mathbf{x}|^{-1} - |\mathbf{x} - (R, 0, 0)|^{-1}$$

as an operator on $L^2(\mathbb{R}^3)$. Let $e(R) = \inf \operatorname{spec}(H(R))$. Then $E(R) = e(R) + R^{-1}$ is the Born–Oppenheimer energy curve. It is reasonable to suppose that $e(R)$ is monotone increasing as R increases but how does one prove this? (See Sections 12 and 13.)

It is not true that these problems all *require* functional integration for their solution (although, *at the present moment*, some of them have only been solved with such methods), but they all share the property of being problems with "obvious" answers and with elegant, conceptually "simple" solutions in terms of the tools we shall develop here. Once the reader has understood these methods and solutions, he will probably have little trouble giving a "word-by-word translation" into a solution that never makes mention of functional integration but rather exploits the Trotter product formula (Theorems 1.1 and 1.2 below) and the fact that $e^{t\Delta}$ is an integral operator with a positive kernel. That is, there is a sense, somewhat analogous to the sense in which the Riemann integral is a systematized limit of sums, in which the Feynman–Kac formula is a systematic expression of the Trotter product formula and positivity of $e^{t\Delta}$. In part, the point of functional integration is a less cumbersome notation, but there is a larger point: like any other successful language, its existence tends to lead us to different and very special ways of thinking.

* * *

Basic to a mathematical elucidation of path integration in quantum theory is Trotter's extension [279] to infinite dimensions of a result of Lie. Nelson [189] has isolated a special case (which is the one we mainly need) with an especially simple proof:

Theorem 1.1 (special case of Trotter's product formula [279]) Let A and B be self-adjoint operators on a separable Hilbert space so that $A + B$, defined on $D(A) \cap D(B)$, is self-adjoint. Then

$$e^{it(A+B)} = \operatorname*{s-lim}_{n\to\infty}(e^{itA/n}e^{itB/n})^n \tag{1.3}$$

If, moreover, A and B are bounded from below, then

$$e^{-t(A+B)} = \operatorname*{s-lim}_{n\to\infty}(e^{-tA/n}e^{-tB/n})^n \tag{1.4}$$

Remark The theorem remains true if $A + B$ is merely supposed to be essentially self-adjoint (see [279] or [34]) but the proof below does not extend to this case.

Proof ([189]) Let $S_t = e^{it(A+B)}$, $V_t = e^{itA}$, $W_t = e^{itB}$, $U_t = V_t W_t$, and let $\psi_t = S_t\psi$ for some ψ in $\mathscr{H}$, the underlying Hilbert space. Then

$$\|(S_t - U_{t/n}^n)\psi\| = \left\| \sum_{j=0}^{n-1} U_{t/n}^j(S_{t/n} - U_{t/n})S_{t/n}^{n-j-1}\psi \right\|$$

$$\leq n \sup_{0 \leq s \leq t} \|(S_{t/n} - U_{t/n})\psi_s\| \tag{1.5}$$

Let $\phi \in D(A) \cap D(B)$. Then $s^{-1}(S_s - 1)\phi \to i(A + B)\phi$ as $s \downarrow 0$ and

$$\begin{aligned} s^{-1}(U_s - 1)\phi &= V_s(iB\phi) + V_s[s^{-1}(W_s - 1) - iB]\phi + s^{-1}(V_s - 1)\phi \\ &\to iB\phi + 0 + iA\phi \end{aligned}$$

so

$$\lim_{n\to\infty} [n\|(S_{t/n} - U_{t/n})\phi\|] \to 0; \qquad \text{each} \quad \phi \in D(A) \cap D(B) \tag{1.6}$$

Let D denote $D(A) \cap D(B)$ with the norm $\|(A + B)\phi\| + \|\phi\| \equiv |||\phi|||$. By hypothesis, D is a Banach space. By the above calculations, $\{n(S_{t/n} - U_{t/n})\}$ is a family of bounded operators from D to $\mathscr{H}$ with $\sup_n\{\|n(S_{t/n} - U_{t/n})\phi\|\} < \infty$ for each ϕ. As a result, the uniform boundedness principle implies that

$$\|n(S_{t/n} - U_{t/n})\phi\| \leq C|||\phi||| \tag{1.7}$$

for some C. (1.7) implies that the limit in (1.6) is uniform over compact subsets of D. Let $\psi \in D$. Then $s \to \psi_s$ is a continuous map from $[0, t]$ into D, so that $\{\psi_s | 0 \leq s \leq t\}$ is compact in D. Thus the right-hand side of (1.5) goes to zero as $n \to \infty$.

The proof of (1.4) is similar. ∎

Recently, Kato [154] has found the ultimate version of (1.4), which we will give without proof, primarily as a "cultural aside." Recall [152, 214] that there is a one–one correspondence between positive self-adjoint operators and positive closed quadratic forms given by $a(\phi, \psi) = (A^{1/2}\phi, A^{1/2}\psi)$. Let us extend the notion of positive self-adjoint operator:

Definition Let A be an operator on a subspace D of $\mathscr{H}$ which is symmetric and positive and which is self-adjoint as an operator on $\bar{D}$ (which may not be $\mathscr{H}$). We call A a **generalized positive operator** and set e^{-tA} equal to zero on $D^\perp$ and define it in the obvious way on $\bar{D}$. (Note: e^{-0A} may not equal one.)

With this definition, the above one–one correspondence is between positive closed quadratic forms (not necessarily densily defined) and generalized positive operators. The point is that if a and b are closed quadratic forms,

$a + b$, defined on $D(a) \cap D(b)$, is always a closed quadratic form, but it may not be densily defined, even if a and b are. Given generalized positive operators A and B, with associated forms a and b, we define $A \dotplus B$ to be the generalized operator associated to $a + b$.

Theorem 1.2 (Kato [154]) For any two generalized positive operators A and B, $\text{s-lim}_{n\to\infty} (e^{-tA/n}e^{-tB/n})^n$ exists and equals $e^{-t(A \dotplus B)}$ for any $t > 0$.

Remark 1. An amusing and illuminating example of Kato's theorem is the following: Let U and W be two closed subspaces of $\mathscr{H}$ and let P, Q be the corresponding self-adjoint projections. Let A have domain U with $A = 0$ on U so that $e^{-tA} = P$ for any t. Similarly define B so that $e^{-tB} = Q$ for any t. Then Kato's theorem says that $(PQ)^n$ converges strongly to the projection onto $U \cap W$, a well-known result, but not one previously realized to be a case of a Trotter-type product formula.

2. For the extension to more than two operators, see Kato–Masuda [155a].

Following Nelson [189] we can use the Trotter formula to give sense to Feynman's path integral [83, 84]: Let V be a smooth potential on $\mathbb{R}^3$. Then $e^{-itH} = \text{s-lim}(e^{-itH_0/n}e^{-itV/n})^n$, where $H_0 = -\frac{1}{2}\Delta$ and $H = H_0 + V$. Since e^{-itH_0} is an operator with integral kernel (see, e.g., [215]):

$$K_0(x, y; t) = (2\pi it)^{-3/2} \exp\left(i\frac{|x - y|^2}{2t}\right)$$

we see that $(e^{-itH_0/n}e^{-itV/n})^n$ has an integral kernel $K^{(n)}(x, y; t)$, where

$$K^{(n)}(x_0, x_n; t) = N_n^{-1} \int \exp(iS(x_0, x_1, \ldots, x_n; t))\, dx_1 \cdots dx_{n-1}$$

where N_n is the "normalization factor," $(2\pi it/n)^{3n/2}$, and

$$S(x_0, \ldots, x_n; t) = \sum_{i=1}^{n} \frac{1}{2}|x_{i-1} - x_i|^2\left(\frac{t}{n}\right)^{-1} - \sum_{i=1}^{n} V(x_i)\left(\frac{t}{n}\right)$$

which is an approximation to the action

$$S(\omega) = \frac{1}{2}\int_0^t \dot{\omega}^2(s)\, ds - \int_0^t V(\omega(s))\, ds$$

for a polygonal path going through x_j at time jt/n and linearly in between (the $\frac{1}{2}$ in S is $m/2$ corresponding to the $\frac{1}{2}$ in H_0 being $1/2m$; i.e., $m = 1$ is chosen). Thus, in the limit, we obtain the *formal* expression

$$K(x, y; t) = \int \exp(iS(\omega)) \text{“}d\omega\text{”} \tag{1.8}$$

for the integral kernel for e^{-itH}. In (1.8), the integral is over all paths between x and y in time t. A particularly beautiful aspect of (1.8) is that if we reinsert the $\hbar$'s then $\exp(iS)$ becomes $\exp(iS/\hbar)$ and a *formal* application of stationary phase in the semiclassical $\hbar \to 0$ limit gives the principle of stationary action of classical mechanics. See the end of Section 18.

The “$d\omega$” in (1.8) is certainly not a positive measure because of the normalization and indeed, general arguments [29] imply that it cannot be chosen as a signed measure even if one attempts to absorb the free action into it. (By extending the notion of measure, various attempts at defining the right-hand side of (1.8) have been made; see, e.g., C. Dewitt-Morette [47–50a], Albeverio–Hoegh-Krohn [2] and Truman [280a–f]. These methods, while useful computationally and for formal heuristics, have not thus far turned out to be analytically powerful; we will not discuss them further.) Kac made the fundamental discovery that when one follows the above procedure for e^{-tH}, e^{-tV}, e^{-tH_0} one can make sense out of the combined quantity $\exp(-\frac{1}{2}\int \dot{\omega}^2\, ds)$ “$d\omega$” and get a finite positive measure $d\mu(\omega)$; indeed, this had essentially been done already by Wiener [286]. The resulting analog of (1.8) is now known as the Feynman–Kac formula; see Section 6.

Of course, using the Wiener–Kac approach we obtain information about e^{-tH} and not e^{-itH}, the fundamental object of quantum dynamics. However, if we want to study eigenfunctions, it is hard to claim that e^{-itH} is any more basic than e^{-tH}. Indeed, it is often easier to study the ground state using e^{-tH} since

$$\inf \operatorname{spec}(H) = -\lim_{t\to\infty} t^{-1} \ln(\psi, e^{-tH}\psi) \tag{1.9}$$

for any vector ψ whose spectral measure $d\mu_\psi$ obeys $d\mu_\psi(a, a+\varepsilon) > 0$, where $a = \inf \operatorname{spec}(H)$. Moreover, if $a = \inf \operatorname{spec}(H)$ is a nondegenerate eigenvalue, $H\eta = a\eta$ and $(\psi, \eta) > 0$, then

$$\eta = \lim_{t\to\infty} (\psi, e^{-2tH}\psi)^{-1/2} e^{-tH}\psi \tag{1.10}$$

(1.9) is easy to prove and (1.10) follows by noting that $e^{-t(H-a)}$ converges strongly to the projection $(\eta, \cdot)\eta$ as $t \to \infty$. In fact, the reader of the recent literature in both constructive and “particle-theoretic” field theory could well conclude that we are living in an imaginary time era.

2. Construction of Gaussian Processes

We begin with some probabilistic notions; see Breiman [23] or Feller [82] for further discussion. A **probability measure space** is a triple $(X, \mathscr{F}, \mu)$ of a set X, a σ-field $\mathscr{F}$ of subsets of X, and a positive measure μ on X with $\mu(X) = 1$. (Such a measure is called a probability measure.) A real-valued measurable function on X is called a **random variable**. Given a random variable f, the measure $d\mu_f$ on $(-\infty, \infty)$ defined by $\mu_f(A) = \mu(f^{-1}[A])$ is called the **probability distribution** of f (to be distinguished from the "distribution function," $F(t) = \mu_f(-\infty, t]$ which is common in the literature and which we shall not use). Given n random variables $f_1, \ldots, f_n$ we define $f_1 \otimes \cdots \otimes f_n \colon X \to \mathbb{R}^n$ by $(f_1 \otimes \cdots \otimes f_n)(x) = (f_1(x), f_2(x), \ldots, f_n(x))$ and the **joint probability distribution** $\mu_{f_1, \ldots, f_n}(A) \equiv \mu([f_1 \otimes \cdots \otimes f_n]^{-1}[A])$, a probability measure on $\mathbb{R}^n$. We use $E(A)$ for $\mu(A)$, $E(f)$ for $\int f\, d\mu$, and $E(f; A)$ to denote the integral of f over the set A.

There is a reason for introducing the term "random variable" for the equivalent "measurable function," namely, an implied change of viewpoint. For suppose that one has a "random function on $(-\infty, \infty)$," i.e., for each $t \in \mathbb{R}$ a random variable f_t (with some measurability in t); a functional analyst would most naturally think about $f_t(\cdot)$ for each t or perhaps the function $f_t(x)$ of two variables. The probabilist's language leads to the consideration of the functions $f_.(x)$ for each x.

As is usual, one does not distinguish random variables which are equal almost everywhere. Typically, one goes even further and, at least formally, removes the points of X from consideration: two sets in $\mathscr{F}$, A and B, are called equivalent if and only if $\mu(A \triangle B) = 0$ where $A \triangle B = (A \backslash B) \cup (B \backslash A)$. The equivalence classes are called **events** and the family of events, $\mathscr{F}/\mathscr{I}_\mu$, inherits the notions of intersection and union. A random variable is, more properly, a map (f^{-1}) from $\mathscr{B}$, the Borel subsets of $\mathbb{R}$, to $\mathscr{F}/\mathscr{I}_\mu$ preserving countable unions and intersections and μ_f is the composition of $f^{-1} \colon \mathscr{B} \to \mathscr{F}/\mathscr{I}_\mu$ and $\mu \colon \mathscr{F}/\mathscr{I}_\mu \to [0, \infty)$. An **isomorphism** of two probability measure spaces $(X, \mathscr{F}, \mu)$ and $(X', \mathscr{F}', \mu')$ is a map $T \colon \mathscr{F}/\mathscr{I}_\mu \to \mathscr{F}'/\mathscr{I}_{\mu'}$ which is a bijection respecting countable unions and intersections and with $\mu' \circ T = \mu$. Random variables f on X and f' on X' are said to **correspond** under T if and only if $T \circ f^{-1} = (f')^{-1}$. Having pointed out the need for the above general abstract framework, we will usually be colloquial and talk about points, about random variables as functions, etc.

Let $\mathscr{I}$ be an index set for a family of random variables. If $I \subset \mathscr{I}$ is a set with $n \equiv \#(I) < \infty$, we have a joint probability distribution μ_I on $\mathbb{R}^n$ for $\{f_\alpha\}_{\alpha \in I}$. The measures μ_I are **consistent** in the sense that if $I \subset I'$, then μ_I can be obtained from $\mu_{I'}$ by "integrating out" the variables in $I'\backslash I$; i.e., if $\{f_1, \ldots f_n\} \equiv \{f_\alpha\}_{\alpha\in I}$ and $\{f_1, \ldots, f_m\} = \{f_\alpha\}_{\alpha\in I'}$ $(m \geq n)$, then $\mu_I(A) = \mu_{I'}(A \times \mathbb{R}^{m-n})$. A fundamental result in probability theory is:

Theorem 2.1 (Kolmogorov's theorem) Let $\mathscr{I}$ be a countable set and let a probability measure μ_I on $\mathbb{R}^{\#(I)}$ be given for each finite set $I \subset \mathscr{I}$ so that the family of μ_I's is consistent. Then, there is a probability measure space $(X, \mathscr{F}, \mu)$ and random variables $\{f_\alpha\}_{\alpha\in\mathscr{I}}$ so that μ_I is the joint probability distribution of $\{f_\alpha\}_{\alpha\in I}$. Moreover, this space is unique in the sense that if $(X', \mathscr{F}', \mu')$ and $\{f'_\alpha\}_{\alpha\in\mathscr{I}}$ also have these properties and if $\mathscr{F}$ (and respectively, $\mathscr{F}'$) is the smallest σ-field with respect to which the f_α (respectively, f'_α) are measurable, then there is an isomorphism of the probability measure spaces under which each f_α corresponds to f'_α.

Proof The existence and uniqueness aspects are quite distinct. To prove existence we will take $X \doteq \dot{\mathbb{R}}^{\mathscr{I}}$ where $\dot{\mathbb{R}} = \mathbb{R} \cup \{\infty\}$ is the one-point compactification of $\mathbb{R}$. Since X is compact, we can use the Reisz–Markov theorem to construct Baire measures (this finesses the proof of countable additivity, or more properly, places it on the shoulders of the proof of the Reisz–Markov theorem; see, e.g., Berberian [11] for this proof). Let $C_{\text{fin}}(X)$ denote the family of continuous functions of finitely many coordinates $\{x_\alpha\}_{\alpha\in I}$ (as I runs through all finite subsets) and let

$$\ell(f) = \int f(x_\alpha)\, d\mu_I(x_\alpha)$$

if f is a function of $\{x_\alpha\}_{\alpha\in I}$. ℓ is a well-defined positive linear functional on $C_{\text{fin}}(X)$ because of the consistency conditions and clearly $|\ell(f)| \leq \|f\|_\infty$. By the Stone–Weierstrass theorem, $C_{\text{fin}}(X)$ is dense in $C(X)$, so ℓ extends uniquely to a positive linear functional on $C(X)$. Therefore, by the Reisz–Markov theorem, there is a Baire probability measure, μ, on X with

$$\ell(f) = \int f(x)\, d\mu(x)$$

Define $f_\alpha(x) = x_\alpha$ if $x_\alpha \neq \infty$ and zero if $x_\alpha = \infty$. Then clearly the $d\mu_I$ are the joint distribution of the $\{f_\alpha\}_{\alpha\in I}$. This proves existence.

To prove uniqueness, we first show that the condition that the f_α's generate $\mathscr{F}$ implies that the bounded functions of finitely many f_α's are dense in $L^2(X, d\mu)$. For let $\mathscr{H}$ be the closure of these functions in L^2 and let χ_A in $\mathscr{H}$ be a characteristic function of some set A. Then we claim that there are A_n whose characteristic functions, χ_n, depend on only finitely many f_α's with $\chi_n \xrightarrow{L^2} \chi_A$. For let $g_n \xrightarrow{L^2} \chi_A$ and by passing to a subsequence, suppose that $g_n \to \chi_A$ pointwise almost everywhere. Let $A_n = \{x \mid \frac{1}{2} < g_n(x) < 2\}$. Then

$\chi_n \to \chi_A$ pointwise almost everywhere and so in L^2 by the dominated convergence theorem. Since $\chi_n \to \chi_A$, the set of A's with χ_A in $\mathscr{H}$ is closed under finite intersections since this is true of the generating sets. Clearly it is closed under complements and it is closed under countable unions by the monotone convergence theorem. Thus, the set of A with χ_A in $\mathscr{H}$ is a σ-field, and so it is $\mathscr{F}$. Thus $\mathscr{H} = L^2(X, d\mu)$.

Given two models $(X, \mathscr{F}, \mu)$, $\{f_\alpha\}$ and $(X', \mathscr{F}', \mu')$, $\{f'_\alpha\}$, define $U: L^2(X, d\mu) \to L^2(X', d\mu')$ by $U[F(f_\alpha)] = F(f'_\alpha)$, which is well defined and unitary and extends to L^2 by the above density result. It is not hard to see that U takes characteristic functions into characteristic functions (use $\chi_n \to \chi_A$) so that T defined by $U(\chi_A) = \chi_{T(A)}$ defines a map of $\mathscr{F}/\mathscr{I}_\mu$ to $\mathscr{F}'/\mathscr{I}_{\mu'}$. This T is easily seen to be an isomorphism under which f_α corresponds to f'_α. ∎

$(X, \mathscr{F}, \mu)$ is called a **model** for the μ_I. If we take $\mathscr{I} = \{0, 1, 2, \ldots\}$, then by the above $\mathbb{R}^{\mathscr{I}}$ can be taken as a model. (In the above, $\dot{\mathbb{R}}^{\mathscr{I}}$ was used a priori, but one notes that $\mu\{x \mid \text{some } x_\alpha = \infty\} \le \sum_\alpha \mu\{x \mid x_\alpha = \infty\} = 0$.) It is often useful to know if some nice subset of $\mathbb{R}^{\mathscr{I}}$ has measure one. Two particularly useful subsets are $\mathscr{s}$ and $\mathscr{s}'$ defined as follows: for $m \in \mathbb{Z}$,

$$\mathscr{s}_m = \left\{x \,\middle|\, \sum_n (1 + n^2)^m |x_n|^2 \equiv \|x\|_m^2 < \infty\right\}$$

$\mathscr{s} \equiv \bigcap_m \mathscr{s}_m$ with the Fréchet topology induced by the $\|\cdot\|_m$ and $\mathscr{s}' \equiv \bigcup_m \mathscr{s}_m$. $\mathscr{s}'$ is the topological dual of $\mathscr{s}$ if $x \in \mathscr{s}'$ is associated to the linear functional

$$L_x(y) = \sum x_n y_n$$

The **cylinder sets** σ-field on $\mathscr{s}'$ is the smallest σ-field with respect to which the functionals $x \mapsto L_x(y)$ are measurable for each $y \in \mathscr{s}$. It is easy to see that this is identical to the σ-field generated by viewing $\mathscr{s}'$ as a subset of $\mathbb{R}^{\mathscr{I}}$ with its natural σ-field. Given a probability measure, μ, on $\mathscr{s}'$ with the cylinder set σ-field, we can define the function Φ on $\mathscr{s}$ by

$$\Phi(y) = \int \exp(iL_x(y))\, d\mu(x) \tag{2.1}$$

called the **Fourier transform of μ**. Often this is called the characteristic function of μ. To avoid confusion with the characteristic function of a set, we only use the former name in this book. Φ clearly has the following three properties [(c) follows from the dominated convergence theorem and the continuity of $L_x(\cdot)$]:

(a) $\Phi(0) = 1$.
(b) Φ is **positive definite**; i.e., given $z_1, \ldots, z_n \in \mathbb{C}$ and $y_1, \ldots, y_n \in \mathscr{s}$,

$$\sum_{i,j=1}^{n} \bar{z}_i z_j \Phi(y_i - y_j) \ge 0$$

(c) Φ is continuous when $\mathscr{s}$ is given its Fréchet topology.

The following infinite-dimensional analog of Bochner's theorem (see [215] for a proof of Bochner's theorem) is a special case of a theorem of Minlos [187]. The beautiful proof we use we learned from van Hemmen [283] (although the literature on Minlos' theorem is so large, it may well be older):

Theorem 2.2 (Minlos' theorem for $\mathscr{S}'$) A necessary and sufficient condition for a function Φ on $\mathscr{S}$ to be the Fourier transform of a probability measure on $\mathscr{S}'$ is that it obey (a)–(c).

Proof Necessity has already been discussed, so suppose that (a)–(c) hold. By Bochner's theorem, for any finite I, there is a measure μ_I on $\mathbb{R}^I$ with

$$\Phi((y_\alpha)_{\alpha \in I}) = \int_{x \in \mathbb{R}^I} \exp(i\mathbf{x} \cdot \mathbf{y})\, d\mu_I(\mathbf{x})$$

By the uniqueness part of Bochner's theorem, the μ's are consistent, so by our proof of Kolmogorov's theorem, there is a measure μ on $\mathbb{R}^{\mathscr{I}}$ so that (2.1) holds for all $y \in \mathscr{S}$ with y_α nonzero for only finitely many α's. Each $\mathscr{S}_m$ and thus $\mathscr{S}'$ is measurable in $\mathbb{R}^{\mathscr{I}}$. If we show that $\mu(\mathscr{S}') = 1$, then μ may be restricted to $\mathscr{S}'$ and (2.1) will extend to all of $\mathscr{S}$.

The proof is thus reduced to using (c) to show that $\mu(\mathscr{S}') = 1$. Given ε, we can find m and δ so that $\|y\|_m \leq \delta$ implies that $|\Phi(y) - 1| \leq \varepsilon$. We claim that

$$\operatorname{Re}\Phi(y) \geq 1 - \varepsilon - 2\delta^{-2}\|y\|_m^2 \tag{2.2}$$

for all $y \in \mathscr{S}$. For (2.2) holds if $\|y\|_m^2 \leq \delta^2$, since $|1 - \Phi(y)| \leq \varepsilon$ in that case, and it holds if $\|y\|_m^2 > \delta^2$, since $\operatorname{Re}\Phi(y) \geq -1$ for all y [we use here the fact that condition (b) implies that $|\Phi(y)| \leq \Phi(0)$].

Fix a sequence $\{q_n\}$ and for α and N let $d\sigma_{\alpha,N}$ be the measure on $\mathbb{R}^{N+1}$:

$$d\sigma_{\alpha,N}(y) = \prod_{n=0}^{N} (2\pi\alpha q_n)^{-1/2} \exp\left(\frac{-y_n^2}{2\alpha q_n}\right) dy_n$$

Notice that (for all integrals over $\mathbb{R}^{N+1}$):

$$\int d\sigma_{\alpha,N} = 1, \qquad \int y_i y_j \, d\sigma_{\alpha,N} = \alpha q_i \delta_{ij} \tag{2.3a}$$

$$\int e^{i\mathbf{x}\cdot\mathbf{y}}\, d\sigma_{\alpha,N}(\mathbf{y}) = \exp\left(-\frac{\alpha}{2}\sum_{n=0}^{N} q_n x_n^2\right) \tag{2.3b}$$

Integrating (2.2) with respect to $d\sigma_{\alpha, N}$ and using (2.3), we find that

$$\int_{\mathbb{R}^{\mathscr{I}}} \exp\left(-\frac{\alpha}{2}\sum_{n=0}^{N} q_n x_n^2\right) d\mu \geq 1 - \varepsilon - 2\delta^{-2}\alpha \sum_{n=0}^{N} (1 + n^2)^m q_n \tag{2.4}$$

Choose q_n so that $\sum_{n=0}^{\infty} q_n(1 + n^2)^m \equiv K < \infty$ and take $N \to \infty$, using the monotone convergence theorem:

$$\int_{\mathbb{R}^{\mathscr{I}}} \exp\left(-\frac{\alpha}{2}\sum_{n=0}^{\infty} q_n x_n^2\right) d\mu \geq 1 - \varepsilon - 2\delta^{-2}\alpha K$$

Now take α to zero, using the monotone convergence theorem again to obtain

$$\mu\left\{x \,\middle|\, \sum_{n=0}^{\infty} q_n x_n^2 < \infty\right\} \geq 1 - \varepsilon$$

Choosing $q_n = (1 + n^2)^{-m-1}$, we see that

$$\mu(\mathscr{s}') \geq \mu(\mathscr{s}_{-m-1}) \geq 1 - \varepsilon$$

so that, if we take ε to zero, we see that $\mu(\mathscr{s}') = 1$. ∎

The above theorem only depends on the structure of $\mathscr{s}$ and $\mathscr{s}'$ as a particular topological vector space and its dual. But $\mathscr{s}$ is topologically isomorphic to $\mathscr{S}(\mathbb{R})$ under the Hermite expansion $f \in \mathscr{S} \mapsto \{f_n\}$ with $f_n = \int \phi_n(x) f(x)\, dx$ where $\phi_n(x) = (2^n n!)^{-1/2}(-1)^n \pi^{-1/4} e^{x^2/2}(d/dx)^n e^{-x^2}$, the nth harmonic oscillator wave function (see [240] or [214, Appendix to Section V.3]) and also to $\mathscr{S}(\mathbb{R}^\nu)$ for each ν. As a result, Theorem 2.2 immediately extends to the following.

Definition A **cylinder set measure** on $\mathscr{S}'(\mathbb{R}^\nu)$ is a measure on the σ-field generated by the functions $T \to T(\phi)$ as ϕ runs through all of $\mathscr{S}(\mathbb{R}^\nu)$.

Theorem 2.3 (Minlos' theorem for $\mathscr{S}'$) A necessary and sufficient condition for a function $\Phi(\cdot)$ on $\mathscr{S}(\mathbb{R}^\nu)$ to be the Fourier transform of a cylinder set probability measure on $\mathscr{S}'(\mathbb{R}^\nu)$

$$\Phi(\phi) = \int \exp(iT(\phi))\, d\mu(T)$$

is that $\Phi(0) = 1$, Φ be positive definite, and Φ be continuous in the Fréchet topology on $\mathscr{S}(\mathbb{R}^\nu)$.

Remark The above proof shows a little more; namely, if Φ is continuous in the norm associated to $\mathscr{s}_m$, then μ is concentrated on any set of the form ($\varepsilon > 0$),

$$\left\{ x \,\middle|\, \sum_{n=0}^{\infty} (1 + n^2)^{-m-1/2-\varepsilon} x_n^2 < \infty \right\}$$

For example, these methods imply that if $\omega(t)$ is a Wiener path and $\phi \in C_0^\infty$, then $\phi\omega$ is in the domain of $|d/dx|^{1/2-\varepsilon}$. This is just short of implying continuity (any function in $|d/dx|^{1/2+\varepsilon}$ is continuous), so we will need a subtler argument to get continuity (see Section 5).

* * *

In the above, we considered countably many random variables, but much of the discussion is applicable to arbitrary families, e.g., families indexed by $t \in (a, b) \subset \mathbb{R}$ ("stochastic processes"). In particular, Kolmogorov's theorem, *in the form we give it* (and its proof) extends to arbitrary families. We can use $\dot{\mathbb{R}}^{\mathscr{I}}$ for X; however, the use of $\mathbb{R}^{\mathscr{I}}$ does not extend since $\{x \mid x_\alpha = \infty \text{ for some } \alpha\}$ is no longer even measurable!

There is one subtlety ("the problem of versions") associated with this extension that we shall discuss now and henceforth generally brush under the rug. The subtlety can be illustrated in the following trivial example:

Example Let $(X, \mathscr{F}, \mu)$ be $[0, 1]$ with its Borel sets and Lebesgue measure. For each $t \in [0, 1]$, let $q(t)$ be the random variable on X which is identically one and let $\tilde{q}(t)$ be defined by

$$\begin{aligned} \tilde{q}(t)(x) &= 1, \qquad t - x \text{ nonrational} \\ &= 0, \qquad t - x \text{ rational} \end{aligned}$$

Then for each fixed t, q and $\tilde{q}$ are equal almost everywhere so that, from the Kolmogorov theorem point of view, q and $\tilde{q}$ are identical families. But notice that for every $x \in X$, the map $t \mapsto q(t)(x)$ is continuous while $t \mapsto \tilde{q}(t)(x)$ is discontinuous! This example illustrates dramatically that if

$$\{q(t) \mid t \in [0, 1]\}$$

is a family of random variables on $(X, \mathscr{F}, \mu)$ and $\mathscr{F}$ is the minimal σ-field, then $\{x \mid t \to q(t)(x) \text{ is continuous}\}$ may not be measurable.

This example is especially disturbing because one of the most celebrated results in the development of the Wiener process is that the "paths," $q(t)(x)$ are continuous in t for almost every x. One can, for example, define the Wiener process for $t \leq 1$ a priori on $\dot{\mathbb{R}}^{[0, 1]}$ with the Baire field, in which case it does not strictly make sense to say that $t \to q(t)$ is pointwise almost everywhere continuous. (It does make sense to say that $t \to q(t)$ is continuous in L^2-norm and this will be trivial to prove.) There are a variety of ways around this problem:

(a) (The one we will use) Suppose we show that for almost every $x \in X$, there is a C_x so that

$$|q(t) - q(s)| \leq C_x |t - s|^\delta \tag{2.5}$$

for some fixed δ and all *rational* t, s in $[0, 1]$. Since only countably many t and s are involved, (2.5) is "version independent." Choose any explicit X for the process and consider the $q(t)$ for t rational and let $X_0 = \{x \in X \mid (2.5) \text{ holds}\}$. For $x \in X_0$, $q(t)(x)$ defined for t rational extends to a unique continuous function $\tilde{q}(t)(x)$ for all $t \in [0, 1]$ and thereby, we can define random variables $\tilde{q}(t)$ on X_0 and extend them to be zero on $X \backslash X_0$. Clearly $t \to \tilde{q}(t)(x)$ is continuous for each $x \in X$. Moreover, the joint distributions of $(q(t_1), \ldots, q(t_n))$ agree with those for $(\tilde{q}(t_1), \ldots, \tilde{q}(t_n))$. This is obvious for t_i rational and extends to all t by the pointwise continuity of $\tilde{q}$ and the L^2 continuity of q. We thus have a "**version** of q" with continuous "paths." This point of view is further discussed in [39].

(b) One directly constructs a probability measure on the space of continuous functions and defines this to be the Wiener process; the above philosophy is not then directly relevant. One disadvantage of this method is that a detailed proof of countable additivity can get somewhat more involved than in our discussion. This procedure is used in [14]; see also Section 17.

(c) Use Minlos' theorem to construct a measure on $\mathscr{S}'(\mathbb{R})$ so that the joint distribution of $T(f_1), \ldots, T(f_n)$ agrees with that of $\int q(t) f_1(t)\, dt, \ldots, \int q(t) f_n(t)\, dt$. Then find in $\mathscr{S}'$ a measurable family, F, of distributions equal (as distributions) to continuous functions with $\mu(\mathscr{S}' \backslash F) = 0$. The Lévy [168]–Ciesielski [35] proof (described, e.g., in [183]) can be interpreted in this way.

(d) Use the compact model of $\dot{\mathbb{R}}^{[0,1]}$ but use the Borel field rather than the Baire field (see, e.g., [214] for the distinction) in which case the Hölder continuous functions are Borel measurable. The construction we used in Theorem 2.2 only defines a Baire measure which will have many Borel extensions, but precisely one regular Borel extension. In this extension, the Hölder continuous functions have measure one. This is a point of view advocated by Nelson [188].

While the words leading to the final result are different in the above arguments, the end definition of Wiener measure *as a measure on the space of continuous functions is the same.*

* * *

A particularly important class of abstract processes are the Gaussian processes. It is easiest to describe them by using the Fourier transforms of the joint probability distributions; i.e.,

$$C_{f_1,\ldots,f_n}(t_1,\ldots,t_n) = \int e^{i\sum x_j t_j}\, d\mu_{f_1,\ldots,f_n}(x_1,\ldots,x_n)$$

$$= \int \exp(i \textstyle\sum t_j f_j(y))\, d\mu(y)$$

One reason for this is that $d\mu_{f_1,\ldots,f_m}$ is consistent with $d\mu_{f_1,\ldots,f_n}$ $(m \geq n)$ if and only if

$$C_{f_1,\ldots,f_n}(t_1,\ldots,t_n) = C_{f_1,\ldots,f_m}(t_1,\ldots,t_n,0,0,\ldots,0).$$

We call f a **Gaussian random variable of variance** a if and only if

$$C_f(t) = \exp(-\tfrac{1}{2}at^2)$$

Equivalently,

$$\begin{aligned} d\mu_f(x) &= (2\pi a)^{-1/2}\exp(-\tfrac{1}{2}x^2/a)\,dx \qquad && (a \neq 0) \\ &= \delta(x)\,dx && (a = 0) \end{aligned}$$

We call $f_1,\ldots,f_n$ **jointly Gaussian with covariance** $\{a_{ij}\}$ $(a_{ij} = a_{ji})$ if and only if

$$C_{f_1,\ldots,f_n}(t_1,\ldots,t_n) = \exp(-\tfrac{1}{2}\textstyle\sum a_{ij}t_it_j) \tag{2.6}$$

a_{ij} is only dependent on f_i and f_j, since

$$a_{ij} = \int f_i f_j\, d\mu$$

If $\{a_{ij}\}_{1\leq i<j\leq n}$ is a nonsingular matrix with inverse b, then

$$d\mu_{f_1,\ldots,f_n}(x) = (2\pi)^{-n/2}(\det a)^{-1/2}\exp(-\tfrac{1}{2}\textstyle\sum b_{ij}x_ix_j) \tag{2.6$'$}$$

Given an $n \times n$ real symmetric matrix a_{ij}, it will be the covariance of some jointly Gaussian random variables if and only if a is positive semidefinite. For, in that case, (2.6) defines a positive definite function and so, by Bochner's theorem, a measure via Fourier transformation.

Occasionally (but not here), one discusses Gaussian random variables with mean m_i and covariance a_{ij} in which case those described above are Gaussian random variables of mean zero. The right-hand side of (2.6) is then replaced by

$$\exp(-\tfrac{1}{2}\textstyle\sum a_{ij}t_it_j + i\sum m_jt_j)$$

Throughout this book, the phrase "f and g are Gaussian random variables" will be used to indicate they are *jointly* Gaussian with mean zero.

Theorem 2.3A Let $\mathscr{H}$ be a separable *real* Hilbert space. Then there exists a probability measure space $(X, \mathscr{F}, \mu)$ and for each $v \in \mathscr{H}$ a random variable, $\phi(v)$, so that $v \mapsto \phi(v)$ is linear and so that for any $v_1, \ldots, v_n \in \mathscr{H}$, $(\phi(v_1), \ldots, \phi(v_n))$ are jointly Gaussian with covariance $\langle v_i, v_j \rangle$ ($\langle \cdot, \cdot \rangle \equiv$ inner product on $\mathscr{H}$). The same kind of uniqueness as occurs in the Kolmogorov theorem holds here.

Remark We will call $\{\phi(v)\}$ **the Gaussian process with covariance** $\langle \cdot, \cdot \rangle$.

Proof Pick $v_1, \ldots, v_n, \ldots$ an orthonormal basis for $\mathscr{H}$. Let $c(t)$ be defined for any sequence $t_1, \ldots, t_n, \ldots$ eventually zero, by

$$c(t) = \exp(-\tfrac{1}{2} \textstyle\sum t_i^2)$$

By the above remarks on when $\exp(-\frac{1}{2} \sum a_{ij} t_i t_j)$ is positive definite and on consistency, we can apply Kolmogorov's theorem to construct $(X, \mathscr{F}, \mu)$ and $\phi_1, \ldots, \phi_n, \ldots$, jointly Gaussian with covariance δ_{ij} (this is somewhat circumlocutory, one can just take $X = \mathbb{R}^\infty$, $\phi_i = x_i$, and

$$\mu = \bigotimes_{n=1}^{\infty} (2\pi)^{-1/2} \exp(-\tfrac{1}{2} x_n^2)\, dx_n$$

directly). Now given a finite sum $v = \sum_{i=1}^N \alpha_i v_i$, set $\phi(v) = \sum_{i=1}^N \alpha_i \phi_i$. Then $\int |\phi(v)|^2 \, d\mu = \| v \|^2$, so, by continuity, ϕ extends from finite sums to a map from $\mathscr{H}$ to $L^2(X, d\mu)$. It is easy to use continuity to see that

$$\int e^{i\phi(v)} \, d\mu = \exp(-\tfrac{1}{2} \|v\|^2)$$

so the ϕ's are jointly Gaussian with the proper covariance. Uniqueness of the process restricted to finite sums of the v_i follows from the uniqueness part of Kolmogorov's theorem and general uniqueness follows from the L^2 continuity deduced above. ∎

Corollary 2.4 Let $c(t, s)$ be a jointly continuous real-valued function on $K \times K$ where K is a separable topological space. Suppose that for any $t_1, \ldots, t_n \in K$, $c(t_i, t_j)$ is a positive semidefinite matrix. Then, there exists an essentially unique measure space $(X, \mathscr{F}, \mu)$ and a random variable $q(t)$ for each $t \in K$ so that the $q(t)$ are jointly Gaussian with covariance c.

Proof For each $t \in K$, introduce a formal symbol δ_t and consider the vector space of finite sums $\sum a_i \delta_{t_i}$. Define an inner product on this space by

$$(\textstyle\sum a_i \delta_{t_i}, \sum b_j \delta_{s_j}) = \sum \bar{a}_i b_j c(t_i, s_j)$$

By hypothesis, this is a positive semidefinite inner product, so by the usual procedure of quotienting by null vectors and completing we can form a Hilbert space, $\mathscr{H}$. The separability of K and continuity of c imply that $\mathscr{H}$ is separable. Let ϕ be the Gaussian process with covariance $(\cdot, \cdot)_{\mathscr{H}}$ and let $q(t) = \phi(\delta_t)$. ∎

Warning: $q(t)$ is *not*, in general, linear in t.

We want to give one final result in the abstract theory of Gaussian processes, a result due to Feldman [80a] and Hajek [121a]. It is essentially equivalent to a result of Shale [237] about the implementability of Bogoliubov ($\equiv$ symplectic) automorphisms in the theory of the free Bose field. Most field theorists are unaware of the work of Feldman–Hajek and most probabilists of the work of Shale!

Theorem 2.5 Let $\mathscr{H}$ be a real Hilbert space and let A be a bounded positive invertible operator on $\mathscr{H}$. A necessary and sufficient condition for there to exist a *single* measure space $(X, \mathscr{F})$ with functions $\{\phi(v)\}_{v \in \mathscr{H}}$ and two mutually absolutely continuous probability measures $d\mu$ and $d\nu$ so that the $\phi(v)$ are jointly Gaussian on $(X, \mathscr{F}, \mu)$ with covariance $(\cdot, \cdot)_{\mathscr{H}}$ and jointly Gaussian on $(X, \mathscr{F}, \nu)$ with covariance $(\cdot, A\cdot)_{\mathscr{H}}$ is that $A - 1$ be Hilbert–Schmidt.

A proof may be found, for example, in [258]. It is quite easy to prove; indeed, we will essentially prove it as Lemma 18.6. The useful direction is that $A - 1$ Hilbert–Schmidt implies mutual absolute continuity. The idea is to choose a basis for $\mathscr{H}$ with $A\phi_n = \alpha_n \phi_n$. Then formally $d\mu = \bigotimes (2\pi)^{-1/2} \exp(-\frac{1}{2}x_n^2)\, dx_n$ and $d\nu = \bigotimes (2\pi\alpha_n)^{-1/2} \exp(-\frac{1}{2}x_n^2/\alpha_n)\, dx_n$. The condition $\sum (\alpha_n - 1)^2 < \infty$ can be used to show that

$$\prod_{n=1}^{N} \alpha_n^{-1/4} \exp(-\tfrac{1}{4}(\alpha_n^{-1} - 1)x_n^2)$$

converges in $L^2(X, d\mu)$. Of course, the square of the limit turns out to be $d\nu/d\mu$.

There is an extensive probabilistic literature on properties of Gaussian processes; see, e.g., [61a, 82b, 82c, 85a, 180, 195a].

3. Some Fundamental Tools of Probability Theory

In this section, we present a number of related topics: Borel–Cantelli lemmas, the notions of independence and conditional expectation, Doob's martingale inequality, Lévy's maximal inequality, the individual ergodic theorem (without proof), and finally some calculations with Gaussian processes related to the above.

Definition Let A_n be a sequence of sets in a probability measure space. We define

$$\overline{\lim} A_n \equiv \bigcap_{m=1}^{\infty} \left(\bigcup_{n=m}^{\infty} A_n \right)$$

i.e., $x \in \overline{\lim} A_n$ if and only if x lies in infinitely many A_n's. One reason for the name is that if $A_n = \{x \mid f_n(x) > \lambda\}$ for some random variables f_n, then $\{x \mid (\overline{\lim} f_n)(x) > \lambda\} \subset \overline{\lim} A_n \subset \{x \mid (\overline{\lim} f_n)(x) \geq \lambda\}$.

Theorem 3.1 (first Borel–Cantelli lemma) Let A_n be a sequence of sets with $\sum_n E(A_n) < \infty$. Then $E(\overline{\lim} A_n) = 0$.

Proof Let χ_n be the characteristic function of A_n. Then $x \in \overline{\lim} A_n$ if and only if $\sum \chi_n(x) = \infty$, so we need only show that $\sum_n \chi_n(x) < \infty$ almost everywhere. But, by Fubini's theorem,

$$\int \left(\sum_n \chi_n \right) d\mu = \sum_n \int \chi_n \, d\mu = \sum_n E(A_n) < \infty$$

A fortiori, $\sum_n \chi_n < \infty$ a.e. ∎

The classical second Borel–Cantelli lemma is for independent A_n (see below for the definition). The following result is a special case of a theorem of Erdös–Rennyi [78]. Some related results can be found in [14, 223].

Theorem 3.2 (second Borel–Cantelli lemma) Let A_n be a sequence of sets with $\sum_n E(A_n) = \infty$. Suppose that there is a matrix α_{nm} defining a bounded operator on ℓ_2 so that

$$|E(A_n \cap A_m) - E(A_n)E(A_m)| \leq \alpha_{nm} E(A_n)^{1/2} E(A_m)^{1/2} \tag{3.1}$$

Then $E(\overline{\lim} A_n) = 1$.

Proof Let c be the norm of α and let $f_n = \chi_n - E(A_n)$ where χ_n is the characteristic function of A_n. Then

$$\begin{aligned}\int \left| \sum_{n=1}^{N} f_n(x) \right|^2 d\mu(x) &= \sum_{n,m=1}^{N} (E(A_n A_m) - E(A_n)E(A_m)) \\ &\leq \sum_{n,m=1}^{N} \alpha_{nm} E(A_n)^{1/2} E(A_m)^{1/2} \\ &\leq c \sum_{k=1}^{N} E(A_n)\end{aligned}$$

Thus, for any $a \geq 0$ and N so that $\sum_{n=1}^{N} E(A_n) \geq a$,

$$\mu\left\{x \,\middle|\, \sum_{n=1}^{\infty} \chi_n(x) \leq a\right\}\left[a - \sum_{n=1}^{N} E(A_n)\right]^2$$

$$\leq \mu\left\{x \,\middle|\, \sum_{n=1}^{N} \chi_n(x) \leq a\right\}\left[a - \sum_{n=1}^{N} E(A_n)\right]^2$$

$$\leq \int \left|\sum_{n=1}^{N} f_n(x)\right|^2 d\mu(x) \leq c \sum_{n=1}^{N} E(A_n)$$

since on the set where $\sum_{n=1}^{N} \chi_n \leq a$, we have that $-\sum_{n=1}^{N} f_n \geq -a + \sum_{n=1}^{N} E(A_n) \geq 0$. Fixing a and taking $N \to \infty$, we see that

$$\mu\left\{x \,\middle|\, \sum_{n=1}^{\infty} \chi_n(x) \leq a\right\} = 0$$

since $\sum_{n=1}^{\infty} E(A_n) = \infty$. Hence

$$\mu\left\{x \,\middle|\, \sum_{n=1}^{\infty} \chi_n(x) = \infty\right\} = 1 \quad \blacksquare$$

As an example of the power of these two results, we consider the following situation. Let f be a random variable and let $d\nu$ be its probability distribution. By **an independent sequence of copies of** f we mean $\otimes_{n=1}^{\infty} d\nu(x_n)$ on $\mathbb{R}^{\infty}$ (constructed à la Kolmogorov), which is to be thought of as a model of "sampling f." By f_n we mean the nth coordinate function. Suppose that f is unbounded with nonpathological behavior at ∞ and let a_n be defined by

$$\nu(a_n, \infty) = 1/n \tag{3.2}$$

One would expect to need about n trials to get a value of f in the interval (a_n, ∞); i.e., one would expect that $\overline{\lim}\, f_n/a_n = 1$ (which should be thought of as a fluctuation result since we will also have $\overline{\lim}\, f_n/(-a_n) = 1$ if $d\nu$ is an even measure).

Theorem 3.3 Let f_n be an independent sequence of copies of a random variable f. Let b_n be a sequence of numbers so that for each $\delta > 0$:

$$\sum_{n=1}^{\infty} E(f \geq (1 + \delta)b_n) < \infty \tag{3.3a}$$

$$\sum_{n=1}^{\infty} E(f \geq (1 - \delta)b_n) = \infty \tag{3.3b}$$

Then $\overline{\lim}\, f_n/b_n = 1$ with probability one.

Proof Let $A_n^{\pm\delta} = \{x \mid f_n(x) \geq (1 \pm \delta)b_n\}$. By (3.3a) and the first Borel–Cantelli lemma, almost every x is in only finitely many $A_n^{+\delta}$; i.e., $\overline{\lim}\ f_n/b_n \leq 1 + \delta$ with probability one. By the second Borel–Cantelli lemma, (3.3b), and the fact (take $\alpha_{nm} = \delta_{nm}$) that $E(A_n^{-\delta} \cap A_m^{-\delta}) = E(A_n^{-\delta})E(A_m^{-\delta})$ $(n \neq m)$ (we have a product measure), almost every x is in infinitely many $A_n^{-\delta}$; i.e., $\overline{\lim}\ f_n/b_n \geq 1 - \delta$ with probability one. Taking δ to zero through a countable set, we obtain the result. ∎

Remark (3.3b) is trivial for the a_n of (3.2) and "normally," (3.3a) will hold.

Example Let $f(x)$ have the probability distribution $(2\pi)^{-1/2}e^{-x^2/2}\,dx = d\nu$. Then

$$(2\pi)^{-1/2}a^{-1}(1 + a^{-2})^{-1}\exp(-a^2/2) \leq \nu(a, \infty) \leq (2\pi)^{-1/2}a^{-1}\exp(-a^2/2) \tag{3.4}$$

for $a > 0$. This may be seen as follows ([185, p. 4]):

$$\begin{aligned}\int_a^\infty \exp\left(\frac{-x^2}{2}\right)dx &\leq \int_a^\infty \frac{x}{a}\exp\left(\frac{-x^2}{2}\right)dx = a^{-1}\exp\left(\frac{-a^2}{2}\right)\\ &= -\int_a^\infty \frac{d}{dx}\left(x^{-1}\exp\left(\frac{-x^2}{2}\right)\right)dx\\ &= \int_a^\infty (1 + x^{-2})\exp\left(\frac{-x^2}{2}\right)dx\\ &\leq (1 + a^{-2})\int_a^\infty \exp\left(\frac{-x^2}{2}\right)dx\end{aligned}$$

Let $b_n = (2 \ln n)^{1/2}$. Then (3.4) shows that (3.3) holds. Thus for independent Gaussian trials, $\overline{\lim}\ f_n/(2 \ln n)^{1/2} = 1$ with probability one. The celebrated "law of the iterated logarithm" and some other limit theorems we prove in Section 7 are only one step beyond this simple example. An alternative to McKean's inequality (3.4) is the following inequality of Fernique [82c]:

$$(2\pi)^{-1/2}(a + 1)^{-1}\exp(-a^2/2) \leq \nu(a, \infty) \leq \tfrac{4}{3}(2\pi)^{-1/2}(a + 1)^{-1}\exp(-a^2/2) \tag{3.4$'$}$$

(3.4′) follows from $(t \geq 0)$

$$g'(t) \leq \exp(-\tfrac{1}{2}t^2) \leq \tfrac{4}{3}g'(t)$$

where $g(t) = -(t + 1)^{-1}\exp(-\tfrac{1}{2}t^2)$, for we can integrate this inequality from t to infinity. By elementary calculus this last inequality is equivalent to

$$\tfrac{3}{4} \leq (t^2 + t + 1)/(t + 1)^2 \equiv q(t) \leq 1$$

which is easy to check since $q'(t)$ has exactly one zero in $(0, \infty)$ at $t = 1$ and $q(0) = q(\infty) = 1$, $q(1) = \frac{3}{4}$. (3.4′) is clearly sharper than (3.4) for a small but not quite as sharp for a large.

* * *

Definition n random variables $f_1, \ldots, f_n$ are called **independent** if and only if their joint distribution $d\nu_{f_1, \ldots, f_n}$ is the product measure

$$d\nu_{f_1} \otimes \cdots \otimes d\nu_{f_n}$$

n events $A_1, \ldots, A_n$ are called **independent** if their characteristic functions are independent random variables.

For two events, A and B, their characteristic functions have a joint probability distribution supported at the four points (0, 0), (0, 1), (1, 0), and (1, 1), and it is easy to see that independence means $\mu(A \cap B) = \mu(A)\mu(B)$. Equation (3.1) can be interpreted as A_n and A_m are "asymptotically independent."

Suppose that one has the special situation where the underlying probability measure $d\mu$ can be realized as $d\mu_f \otimes d\nu$ where $d\mu_f$ is the distribution for f, and f becomes the first coordinate function in this realization. Clearly functions g of the second variable are independent of f, and given an arbitrary random variable $h(x, y)$, the variation of $\int h(x, y)\, d\nu(y)$ as x varies will be some kind of indication of how far h is from being independent of f. The conditional expectation we will now discuss is the analog of "integrating over y" in cases where $d\mu$ is not a product measure.

Definition Let $(X, \mathscr{F}, \mu)$ be a probability measure space and let Σ be a σ-field contained in $\mathscr{F}$. Let L_Σ^2 be the (closed) subspace of functions in $L^2(X, \mathscr{F}, d\mu)$ which are Σ-measurable and let P_Σ be the orthogonal projection from L^2 to L_Σ^2. For $f \in L^2(X, \mathscr{F}, d\mu)$, $P_\Sigma f$ is called the **conditional expectation** of f with respect to Σ written $E(f \mid \Sigma)$. If $g_1, \ldots, g_n$ are random variables, $E(f \mid g_1, \ldots, g_n)$ denotes the conditional expectation of f with respect to the σ-field generated by $g_1, \ldots, g_n$.

Remarks 1. Since $E(1 \mid \Sigma) = 1$ and $E(f \mid \Sigma) \geq 0$ if $f \geq 0$ (this follows from Remark 2 below), it follows that for $f \in L^1 \cap L^2$, $\|E(f \mid \Sigma)\|_1 \leq \|f\|_1$ so $E(\cdot \mid \Sigma)$ extends to L^1. The second remark also shows that $E(f \mid \Sigma) \geq E(g \mid \Sigma)$ if $f \geq g$.

2. $E(f \mid \Sigma)$ is determined by two conditions: That $E(f \mid \Sigma)$ is Σ-measurable and that

$$\int fg\, d\mu = \int E(f \mid \Sigma) g\, d\mu \tag{3.5}$$

for any Σ-measurable g. (3.5) implies, in particular, that

$$E(fh|\Sigma) = hE(f|\Sigma), \qquad \text{if} \quad h \text{ is } \Sigma\text{-measurable} \tag{3.6}$$

3. If g is the characteristic function of a set B, then

$$E(f|g) = (\mu(B)^{-1} \int_B f \, d\mu)g + (\mu(X \backslash B)^{-1} \int_{X \backslash B} f \, d\mu)(1 - g)$$

so that $E(f|g)(x)$ is the expectation of f with the additional information that we know $g(x)$ and nothing else. More generally $E(f|\Sigma)(x)$ can be interpreted as the expectation of f knowing the value of all Σ-measurable g's.

Proposition 3.4 f is independent of g if and only if $E(e^{i\alpha f}|g)$ is a constant for each real α.

Proof If f is independent of g, then $E(F(f)|g)$ is constant for any F. Conversely if $E(e^{i\alpha f}|g)$ is constant for all α, then the constant is necessarily $E(e^{i\alpha f})$ and

$$E(e^{i\alpha f + i\beta g}) = E[E(e^{i\alpha f}|g)e^{i\beta g}] \equiv E(e^{i\alpha f})E(e^{i\beta g})$$

for any α and β so, taking Fourier transforms, $\mu_{f,g} = \mu_f \otimes \mu_g$. ∎

* * *

Definition Let $f_1, \ldots, f_n$ be a sequence of random variables. We call them a **martingale** (respectively, **submartingale**) if and only if for each m, $E(|f_m|) < \infty$ and

$$E(f_m | f_1, \ldots, f_{m-1}) = f_{m-1}$$

(respectively, $E(f_m | f_1, \ldots, f_{m-1}) \geq f_{m-1}$ (pointwise)).

Remarks 1. It follows by induction that for $j < m$,

$$E(f_m | f_1, \ldots, f_j) = f_j \qquad (\text{resp.}, \geq f_j) \tag{3.7}$$

2. If $X_1, \ldots, X_n$ are random variables which have mean zero and are independent, then $f_j = \sum_{i=1}^{j} X_i$ is a martingale. Then Doob's inequality (below) can be interpreted as saying that a certain gambling strategy will not do any better than the strategy of just waiting n steps; see Feller [82].

Theorem 3.5 (Doob's inequality [61]) Let $\{f_n\}$ be a submartingale. Then for each $\lambda > 0$ and n:

$$E\left(\max_{0 \le j \le n} f_j \ge \lambda\right) \le \lambda^{-1} E(f_n^+) \tag{3.8}$$

where $f_n^+ = \max(f_n, 0)$.

Proof Let χ_j be the characteristic function of the set A_j of points where $f_1, \ldots, f_{j-1} < \lambda$ and $f_j \ge \lambda$. Let χ be the characteristic function of the set A with $\max_{0 \le j \le n} f_j \ge \lambda$. Since the A_j's are disjoint sets with union A, we have that

$$\begin{aligned}\lambda E(A) &= \sum_{j=1}^{n} \lambda E(A_j) \\ &\le \sum_{j=1}^{n} E(f_j \chi_j) \\ &\le \sum_{j=1}^{n} E(E(f_n | f_1, \ldots, f_j)\chi_j) \\ &= \sum_{j=1}^{n} E(f_n \chi_j) = E(f_n \chi) \le E(f_n^+)\end{aligned}$$

In the above, we used $f_j \ge \lambda$ in the first inequality, the submartingale relation (3.7) in the second inequality, the fact that χ_j is $f_1, \ldots, f_j$-measurable and (3.5) in the next equality, and the calculation $E(f_n \chi) \le E(f_n^+ \chi) \le E(f_n^+)$ in the final step. ∎

The exciting thing about the above inequality is that it extends to continuously indexed processes:

Definition Suppose that $\{q_t\}_{t \in [a,b]}$ is a family of random variables. We say it is a **martingale** (respectively, **submartingale**) if and only if

$$E(q_t | \{q_s | s \le u\}) = q_u$$

(respectively, $\ge q_u$) for all $a \le u \le t \le b$.

Notice that a fortiori, one has that

$$E(q_t | q_{s_1}, \ldots, q_{s_n}) = q_{s_n} \qquad (\text{resp.}, \ge q_{s_n})$$

if $s_1 \leq s_2 \cdots \leq s_n \leq t$ [check (3.5)!] so that for $s_1 \leq \cdots s_n \leq \cdots$ the sequence $f_n = q_{s_n}$ is a (sub)martingale. As a result,

$$E\left(\max_{0 \leq j \leq n} q_{s_n} \geq \lambda\right) \leq \lambda^{-1} E(q_{s_n}^+)$$

Taking the mesh of the points s_i to zero we have the following.

Corollary 3.6 Let $\{q_t\}_{t \in [a,b]}$ be a submartingale with continuous sample paths (i.e., $t \to q_t$ is almost everywhere continuous; this is "version dependent"). Then

$$E\left(\max_{0 \leq s \leq t} q_s \geq \lambda\right) \leq \lambda^{-1} E(q_t^+) \tag{3.8}$$

Remark Doob's inequality is just one of a number of interesting developments in the theory of martingales described in Doob's book [61]. For example using

$$E(F \geq \lambda) \leq \lambda^{-1} E(f_n; F \geq \lambda)$$

with $F = \max_{0 \leq j \leq n} f_j$, an inequality proven as an intermediate step above, one can show that

$$E(|F|^p)^{1/p} \leq p(p-1)^{-1} E(|f_n|^p)^{1/p}$$

as follows:

$$\begin{aligned}
E(|F|^p) &= p \int \lambda^{p-1} E(F \geq \lambda)\, d\lambda \\
&\leq p \int_0^\infty d\lambda\, \lambda^{p-2} \left[\int_{F(x) \geq \lambda} |f_n(x)|\, d\mu\right] \\
&= p \int d\mu(x) |f_n(x)| \left[\int_0^{F(x)} \lambda^{p-2}\, d\lambda\right] \\
&= \frac{p}{p-1} \int d\mu |f_n| |F|^{p-1} \\
&\leq \frac{p}{p-1} E(|f_n|^p)^{1/p} E(|F|^p)^{1/q}
\end{aligned}$$

where $q^{-1} = 1 - p^{-1}$, and we used Hölder's inequality in the last step.

* * *

There is another powerful inequality which has a proof similar to that of Doob's inequality:

Theorem 3.6.5 (Lévy's maximal inequality) Let $X_1, \ldots, X_n$ be n random variables with values in some vector space $\mathbb{R}^\nu$. Suppose that the joint probability distribution for $X_1, \ldots, X_n$ is invariant under any change of sign $X_i \to \varepsilon_i X_i$ (each $\varepsilon_i = \pm 1$). Let $S_j = X_1 + \cdots + X_j$. Then for any $\lambda > 0$,

$$E\left(\max_{1 \leq j \leq n} |S_j| \geq \lambda\right) \leq 2E(|S_n| \geq \lambda) \tag{3.8a}$$

If $\nu = 1$, then, in addition

$$E(\max S_j \geq \lambda) \leq 2E(S_n \geq \lambda) \tag{3.8b}$$

Proof We prove (3.8a); (3.8b) is similar. Let A_j be the set with $|S_1| < \lambda, \ldots, |S_{j-1}| < \lambda, |S_j| \geq \lambda$, and let $A = \bigcup A_j$. Let $T_j = X_1 + \cdots + X_j - X_{j+1} - \cdots - X_n$ and notice that

$$S_j = \tfrac{1}{2}S_n + \tfrac{1}{2}T_j$$

Thus

$$\{|S_j| \geq \lambda\} \subset \{|S_n| \geq \lambda\} \cup \{|T_j| \geq \lambda\}$$

so

$$\begin{aligned} E(A_j) &= E(A_j; |S_j| \geq \lambda) \\ &\leq E(A_j; |S_n| \geq \lambda) + E(A_j; |T_j| \geq \lambda) \\ &= 2E(A_j; |S_n| \geq \lambda) \end{aligned}$$

where we use the invariance under changing the sign of $X_{j+1}, \ldots, X_n$. Since the A_j are disjoint:

$$E(A) = \sum E(A_j) \leq 2 \sum E(A_j; |S_n| \geq \lambda) \leq 2E(|S_n| \geq \lambda)$$

which is (3.8a). ∎

For further discussion of the role of reflections in probability theory, see [85a, pp. 21–29] and [198a, Chapter 5].

* * *

Definition A map T on a probability measure space is called **ergodic** if and only if T is measure preserving and $T[A] = A$ for $A \in \mathscr{F}$ implies $\mu(A) = 0$ or 1.

Examples 1. ("Kolmogorov 01 law") Let $(Y, \mathscr{F}_0, \nu)$ be a probability measure space and let X be the two-sided infinite product of Y's with $d\mu = \bigotimes_{-\infty}^{\infty} d\nu$. Let $(Ty)_j = y_{j-1}$. Then T is measure preserving and ergodic. For given A with $T[A] = A$ and given ε, we can find n and B only depending on $y_{-n}, \ldots, y_n$ so that $\mu(A \triangle B) \le \varepsilon$ (such B's have characteristic functions L^2-dense in the characteristic functions of all measurable sets; see the proof of Kolmogorov's theorem). Let $B' = T^{2n+1}[B]$. Then B' is independent of B so $\mu(B \cap B') = \mu(B)^2$. Since

$$A \triangle (B \cap B') \subset (A \triangle B) \cup (A \triangle B')$$

we see that

$$\mu(A \triangle (B \cap B')) \le 2\varepsilon$$

so that

$$|\mu(B)^2 - \mu(B)| \le 3\varepsilon$$

thus

$$|\mu(A)^2 - \mu(A)| \le 4\varepsilon$$

Since ε is arbitrary, $\mu(A)^2 = \mu(A)$.

2. Suppose that T is measure preserving on $(X, \mathscr{F}, \mu)$ and that for each $[a, b] \subset \mathbb{R}$, we have a subalgebra $\mathscr{F}_{[a,b]}$ of $\mathscr{F}$ so that:

(i) the $\{\mathscr{F}_{[a,b]}\}$ generate $\mathscr{F}$, and $\mathscr{F}_I \subset \mathscr{F}_J$ if $I \subset J$,
(ii) if $a < b < c < d$, then

$$|\mu(B \cap C) - \mu(B)\mu(C)| \le f(c - b)$$

for all $B \in \mathscr{F}_{[a,b]}$ and $C \in \mathscr{F}_{[c,d]}$ with $f(t) \to 0$ as $t \to \infty$,
(iii) $T[\mathscr{F}_{[a,b]}] = \mathscr{F}_{[a+1,b+1]}$.

Then, as in Example 1, T is ergodic: For given A, we find $B \in \mathscr{F}_{[-n,n]}$ so that $\mu(A \triangle B) \le \varepsilon$. Then, if $B' = T^{2n+k}[B]$, we have that

$$|\mu(B \cap B') - \mu(B)| \le 3\varepsilon \qquad \text{so that} \qquad |\mu(B)^2 - \mu(B)| \le 3\varepsilon + f(k)$$

Taking $k \to \infty$, we find that $|\mu(B)^2 - \mu(B)| \le 3\varepsilon$, so $|\mu(A)^2 - \mu(A)| \le 4\varepsilon$.

In two places, we will need the following result which we state without proof; see Halmos [122] or Shields [238] for further discussion.

Theorem 3.7 (Birkhoff ergodic theorem) Let T be a map on a probability measure space which is measure preserving. Let $f \in L^1(X, d\mu)$. Then for almost every x, the limit

$$\lim_{n\to\infty} \frac{1}{n} \sum_{j=0}^{n-1} f(T^j x) \equiv g(x)$$

exists and $\int g(x)\, d\mu = \int f(x)\, d\mu$. If moreover T is ergodic, then g is the constant $\int f(x)\, d\mu$.

* * *

Next, we turn to the applicability of some of the above ideas to Gaussian random variables. First, we have the following trivial proposition.

Proposition 3.8 Two Gaussian random variables f and g are independent if and only if their covariance $(f, g) = \int fg\, d\mu$ is zero.

Proof The Fourier transform of their joint distribution is

$$\begin{aligned} M(t, s) \equiv E(e^{itf + isg}) &= \exp(-\tfrac{1}{2}\|tf + sg\|_2^2) \\ &= M(t, 0)M(0, s)\exp(-ts(f, g)) \end{aligned}$$

is a product if and only if $(f, g) = 0$. ∎

Slightly more subtle is the following.

Theorem 3.9 Let $\mathscr{H}$ be a real Hilbert space and let $\phi(\cdot)$ be the Gaussian process with covariance $(\cdot,\cdot)$ Let $\mathscr{M}$ be a closed subspace of $\mathscr{H}$, P the orthogonal projection onto $\mathscr{M}$, and $\Sigma_{\mathscr{M}}$ the σ-algebra generated by $\{\phi(v) | v \in \mathscr{M}\}$. Then for any $w \in \mathscr{H}$:

$$E(e^{i\phi(w)} | \Sigma_{\mathscr{M}}) = e^{i\phi(Pw)}e^{-1/2\|w\|^2}e^{1/2\|Pw\|^2} \tag{3.9}$$

Remark (3.9) is connected to Mehler's formula; see [258].

Proof Since the right-hand side (r.h.s.) of (3.9) is clearly $\Sigma_{\mathscr{M}}$-measurable, it is enough to check that

$$\int Fe^{i\phi(w)}\, d\mu = \int F(\text{r.h.s. of (3.9)})\, d\mu \tag{3.10}$$

for all $\Sigma_{\mathscr{M}}$-measurable F. By a limiting argument, we only need to check (3.10) for $F = G(\phi(v_1), \ldots, \phi(v_n))$ with $G \in \mathscr{S}(\mathbb{R}^n)$ and, therefore, using the Fourier transform, for $F = e^{i\sum t_i\phi(v_i)} = e^{i\phi(v)}$ with $v = \sum t_i v_i \in \mathscr{M}$. But by a direct calculation $\int e^{i\phi(v)}e^{i\phi(w)}\, d\mu = \exp(-\frac{1}{2}\|v + w\|^2)$. Since $(v, w) = (v, Pw)$ for $v \in \mathscr{M}$, (3.10) holds. ∎

(3.9) suggests we single out the object

$$:e^{i\phi(v)}: = e^{i\phi(v)}e^{+1/2\|v\|^2} \tag{3.11}$$

called the **Wick-ordered exponential**; more generally we define

$$:e^{\alpha\phi(v)}: = e^{\alpha\phi(v)}e^{-1/2\alpha^2\|v\|^2}$$

for any $\alpha \in \mathbb{C}$. Then (3.9) and its analog for general α read

$$E(:e^{\alpha\phi(w)}:|\Sigma_{\mathcal{M}}) = :e^{\alpha\phi(Pw)}: \tag{3.9'}$$

See [258] for further discussion of Wick ordering for Gaussian processes.

To apply Theorem 3.2 to Gaussian processes, we will need the following theorem.

Theorem 3.10 Let $\phi(\cdot)$ be the Gaussian process with covariance $(\cdot, \cdot)$ for some real Hilbert space $\mathcal{H}$. Let $\mathcal{M}$, $\mathcal{N}$ be two subspaces of $\mathcal{H}$ and let P, Q be the corresponding orthogonal projections and $\Sigma_{\mathcal{M}}$, $\Sigma_{\mathcal{N}}$ the corresponding σ-algebras. Then, for any $\Sigma_{\mathcal{N}}$-measurable f and $\Sigma_{\mathcal{M}}$-measurable g, both in L^2, we have that

$$|E(fg) - E(f)E(g)| \le \|PQ\| \, \|f\|_2 \, \|g\|_2 \tag{3.12}$$

with $\|\cdot\|_2$ the L^2-norm.

Proof Let $h = g - E(g)$. Then

$$\begin{aligned} |E(fg) - E(f)E(g)| &= |E(fh)| = |E(fE(h|\Sigma_{\mathcal{N}}))| \\ &\le \|f\|_2 \, \|E(h|\Sigma_{\mathcal{N}})\|_2 \end{aligned}$$

Since $\|h\|_2 \le \|g\|_2$, it suffices to show that for $\Sigma_{\mathcal{M}}$-measurable h with $E(h) = 0$, we have that

$$\|E(h|\Sigma_{\mathcal{N}})\|_2 \le \|PQ\| \, \|h\|_2 \tag{3.13}$$

By a limiting argument, we may suppose that $\mathcal{M}$ is finite dimensional.

Let $A = PQP$ and let $w_1, \ldots, w_n$ be an orthonormal basis for $\mathcal{M}$ of eigenvectors for A, $Aw_i = \alpha_i w_i$. Notice that $\sup_i \alpha_i = \|A\| = \|PQ\|^2$. By a limiting argument, we can suppose that

$$h = \int H(x):\exp[i(x_1\phi(w_1) + \cdots + x_n\phi(w_n)]: d^n x$$

for $H \in \mathcal{S}(R^n)$. Then using (3.9′) and the fact that $E(:e^{\alpha\phi(w)}: :e^{\beta\phi(v)}:) = e^{\alpha\beta(w, v)}$, we see that

$$\|h\|_2^2 = \int \overline{H}(x)H(y)e^{(x, y)} \, d^n x \, d^n y$$

$$\|E(h|\Sigma_{\mathcal{N}})\|_2^2 = \int \overline{H}(x)H(y)e^{(x, Ay)} \, d^n x \, d^n y$$

Now, expand the exponential in each of the above expressions and note that each term is positive, e.g.,

$$\int x^\beta y^\beta \overline{H}(x)H(y)\, d^n x\, d^n y = \left| \int x^\beta H(x)\, d^n x \right|^2$$

Moreover, in comparing the two series, all terms except for the $(\int H(x)\, d^n x)^2$ term for the second are smaller than for the first by a factor always bounded by $\sup_i \alpha_i$. But since $E(h) = 0$, $\int H(x)\, d^n x = 0$, so (3.13) is proven. ∎

Remarks 1. The above shows that to get equality in (3.12) and (3.13), we should take h so that $\int x^\beta H(x)$ is only nonzero for $\beta = (1, 0, 0, \ldots)$ (if $\sup \alpha_i = \alpha_1$); i.e., $H(x) = -\partial\delta(x)/\partial x_1$, or $h = \phi(w_1)$. With this choice one easily sees that the constants in (3.12) and (3.13) are best possible.

2. This result is more transparent if one first develops the $\Gamma(\cdot)$ functor, see [258].

* * *

Finally, we want to do some explicit calculation with Gaussian processes. We have already seen that

$$\int \exp(\alpha\phi(f))\, d\mu_0 = \exp(\tfrac{1}{2}\alpha^2(f, f)) \tag{3.14}$$

In general, one can explicitly do one-dimensional Gaussian integrals for functions which are exponentials of quadratic polynomials so one should be able to do the same for Gaussian processes. Let A be a trace class operator on $\mathscr{H}$. Then A has a canonical expansion [214]:

$$A = \sum_n \alpha_n (f_n, \cdot) g_n$$

where f_n, g_n are orthonormal and $\sum_n \alpha_n < \infty$, $\alpha_n \geq 0$. We define

$$(\phi, A\phi) \equiv \sum_n \alpha_n \phi(f_n)\phi(g_n) \tag{3.15}$$

since $\int |\phi(f_n)\phi(g_n)|\, d\mu_0 \leq 1$, the sum converges in L^1. If $A \geq 0$, then $g_n = f_n$ and $(\phi, A\phi) \geq 0$.

Theorem 3.11 Let $d\mu_0$ be the Gaussian process over a real Hilbert space, $\mathscr{H}$ with covariance $(\cdot, \cdot)$. Let A be a positive trace class operator. Then:

(a) $$\int \exp[-\tfrac{1}{2}(\phi, A\phi)]\, d\mu_0 = [\det(1 + A)]^{-1/2}$$

(b) $$d\nu = \frac{\exp[-\tfrac{1}{2}(\phi, A\phi)]\, d\mu_0}{\int \exp[-\tfrac{1}{2}(\phi, A\phi)]\, d\mu_0} \tag{3.16}$$

is the Gaussian process with covariance $(\cdot, (1 + A)^{-1}\cdot)$

(c) $$\int \exp[-\tfrac{1}{2}(\phi, A\phi) + \alpha\phi(f)]\, d\mu_0$$

$$= (\exp[\tfrac{1}{2}\alpha^2(f, (1 + A)^{-1}f)])[\det(1 + A)]^{-1/2} \tag{3.17}$$

Remark For A trace class

$$\det(1 + A) \equiv \prod_{n=1}^{\infty}(1 + \lambda_n(A))$$

where λ_n are the eigenvalues of A. (See [259] for further discussion.)

Proof (a) Choose coordinates x_n corresponding to $\phi(f_n)$ so that $d\mu_0 = \bigotimes (2\pi)^{-1/2}\exp(-\tfrac{1}{2}x_n^2)\, dx_n$. Let $F_N = \exp(-\tfrac{1}{2}\sum_{n=1}^{N}\alpha_n x_n^2)$ so that $F_N \downarrow \exp[-\tfrac{1}{2}(\phi, A\phi)]$. Then

$$\int F_N\, d\mu_0 = \prod_{n=1}^{N}\int (2\pi)^{-1/2}\exp[-\tfrac{1}{2}(1 - \alpha_n)x_n^2]\, dx_n = \prod_{n=1}^{N}(1 + \alpha_n)^{-1/2}$$

Taking $N \to \infty$ and using the monotone convergence theorem, (3.16) results.

(b) By the proof of (a), $d\nu = \lim[F_N\, d\mu_0 / \int F_N\, d\mu_0]$ is obviously Gaussian with $\int x_k x_l\, d\nu = (1 + \alpha_l)^{-1}\delta_{kl}$.

(c) Follows from (a), (b), and (3.14). ∎

Theorem 3.11 can be extended in two ways. First, $A \geq 0$ is not necessary. All that is needed is $1 + A > 0$. Moreover, with an elementary subtraction, one can extend to A Hilbert–Schmidt. Namely, define

$$:(\phi, A\phi): = \sum \alpha_n[\phi(f_n)^2 - 1]$$

Since $\int [\phi(f_n)^2 - 1][\phi(f_m)^2 - 1]\, d\mu_0 = 2\delta_{mn}$, the sum converges in $L^2(d\mu_0)$ so long as $\sum \alpha_n^2 < \infty$. Moreover, a simple extension of (a) shows that

$$\int \exp[-\tfrac{1}{2}:(\phi, A\phi):]\, d\mu_0 = \prod_{n=1}^{\infty}[(1 + \alpha_n)e^{-\alpha_n}]^{1/2}$$

$$\equiv [\det_2(1 + A)]^{-1/2}$$

(See [259] for discussion of $\det_2$.)

If one defines

$$((\phi - f), A(\phi - f)) \equiv (\phi, A\phi) - 2\phi(Af) + (f, Af)$$

then (3.17) implies that

$$\begin{aligned}\int \exp[-\tfrac{1}{2}((\phi - f), A(\phi - f))]\, d\mu_0 \\ = \det(1 + A)^{-1/2} \exp[-\tfrac{1}{2}(f, Af) + \tfrac{1}{2}(Af, (1 + A)^{-1}Af)] \\ = \det(1 + A)^{-1/2} \exp[-\tfrac{1}{2}(f, A(1 + A)^{-1}f)] \qquad (3.18)\end{aligned}$$

II

The Basic Processes

4. The Wiener Process, the Oscillator Process, and the Brownian Bridge

In this section, we define the three basic Gaussian processes whose perturbations will concern us in the remainder of this book. Intuitively (and even rigorously *in a strong sense*, see Section 17), the Wiener process, $b(t)$, is the limit of elementary random walks. An **elementary random walk** is defined as follows: Let ν be the measure $\{-1, 1\}$ with $\nu(-1) = \nu(1) = \frac{1}{2}$, and let $d\mu = \bigotimes_{n=1}^{\infty} d\nu_n$ on $\times_{n=1}^{\infty} \{-1, 1\}$. Let y_n be the nth coordinate function and $X_n = \sum_{m=1}^{n} y_m$. The family of random variables $\{X_n\}_{n=1}^{\infty}$ is called a random walk. Since $E(y_n y_m) = \delta_{nm}$ we have $E(X_n^2) = n$ so that the variables $\tilde{X}_n = n^{-1/2} X_n$ at least have a chance of having a limit. In fact the well-known De Moivre–Laplace limit theorem (special case of the central limit theorem) says that $\tilde{X}_n$ approaches a Gaussian:

Theorem 4.1 (central limit theorem) Let $\{y_n\}_{n=1}^{\infty}$ be a family of independent, identically distributed random variables with $E(y_n) = 0$, $E(y_n^2) = 1$. Let

$$\tilde{X}_n = n^{-1/2} \sum_{m=1}^{n} y_m$$

Then $\tilde{X}_n$ approaches a Gaussian random variable X_∞ with variance one in the sense, that for any continuous, bounded function on $\mathbb{R}$,

$$E(f(\tilde{X}_n)) \to (2\pi)^{-1/2} \int e^{-y^2/2} f(y)\, dy$$

Proof By a simple limiting argument (going through $f \in \mathscr{S}$ as an intermediate step), it suffices to prove that $E(e^{i\alpha\tilde{X}_n}) \to e^{-\alpha^2/2}$ for all real α. But

$$E(e^{i\alpha\tilde{X}_n}) = [F(\alpha/\sqrt{n})]^n$$

where $F(\beta) = E(e^{i\beta y})$. Since $E(y_n) = 0$, $E(y_n^2) = 1$, we have that

$$(F(\beta) - 1)\beta^{-2} \to -\tfrac{1}{2}$$

as $\beta \to 0$ (by using the dominated convergence theorem) so that

$$E(e^{i\alpha\tilde{X}_n}) = \left[1 - \frac{1}{2}\frac{\alpha^2}{n} + o\left(\frac{1}{n}\right)\right]^n = \exp\left(-\frac{1}{2}\alpha^2\right) + o(1)$$

by the compound interest formula. ∎

Intuitively, $b(t)$ is $\lim_{n\to\infty} n^{-1/2}X_{[nt]}$ (where $[a]$ = integral part of a). Thus, each $b(t)$ should be Gaussian of variance t. To find the covariance, we note that $X_n - X_m$ and X_m are independent if $m < n$. Thus, we expect that $b(t) - b(s)$ should be independent of $b(s)$ for $s < t$; i.e.,

$$E(b(s)(b(t) - b(s))) = 0$$

so $E(b(t)b(s)) = s$ for $s < t$. Therefore, we define the following.

Definition The **Wiener process** (or **Brownian motion**) is the family $\{b(t)\}_{0 \le t}$ of Gaussian random variables with covariance $E(b(t)b(s)) = \min(s, t)$.

Of course, to be able to make this definition, we need the following.

Lemma 4.2 Let $0 \le s_1 \le s_2 \le \cdots \le s_n$ and let $z_1, \ldots, z_n \in \mathbb{C}$. Then $\sum_{j,i=1}^n \bar{z}_i z_j \min(s_i, s_j) \ge 0$.

Proof Note that (with $s_0 = 0$)

$$\sum_{j,i=1}^{n} \bar{z}_i z_j \min(s_i, s_j) = \sum_{i=1}^{n} (s_i - s_{i-1}) \left| \sum_{j=i}^{n} z_j \right|^2$$

is obviously positive. ∎

Remark This lemma can also be proven by noting that $b(t)$ really is the limit of $n^{-1/2}X_{[nt]}$ as far as joint probability distributions of finitely many b's are concerned.

Notice that

$$E((b(t) - b(s))^2) = t - s \qquad (t > s) \tag{4.1}$$

Moreover, by writing $b(t) = b(s) + (b(t) - b(s))$ and using the independence of $b(t) - b(s)$ from $\{b(u)\}_{0 \le u \le s}$, we see that

$$E(b(t) \mid \{b(u)\}_{0 \le u \le s}) = b(s) \qquad (t > s) \tag{4.2}$$

so that $b(t)$ is a martingale.

An important property of Brownian motion is that it continually starts afresh; i.e., for each fixed s, the process $\tilde{b}(t) = b(t + s) - b(s)$ for $t > 0$ has the same joint distributions as b and is independent of $b(s)$; i.e., at any given time a particle following the paths $b(t)$ stops and except for the addition of $b(s)$ follows the exact same paths as a particle beginning at $s = 0$. Later (Theorem 7.9) we will prove an even stronger version of this property of starting afresh.

There are a large number of results known characterizing Brownian motion in terms of fairly weak conditions. Typical is the following result, which we will not use and which we state without proof. (It is Theorem 5.1 of [86].)

Theorem 4.3 If $x(t)$ is a process for $0 \le t$ with a continuous version so that for $0 \le s \le t$

$$E(x(t) - x(s) \mid \{x(u)\}_{0 \le u \le s}) = 0$$

$$E((x(t) - x(s))^2 \mid \{x(u)\}_{0 \le u \le s}) = t - s$$

Then $x(t)$ is Brownian motion.

Remarks 1. The proof begins by noticing the above conditions, say, that $x(t)$ and $x^2(t) - t$ are martingales. We will see later (Section 7) that more generally certain special polynomials $(:x(t)^n:)$ are martingales.

2. An interesting realization of Brownian motion comes from the fact that if χ_s is the characteristic function of $[0, s]$, then $\langle \chi_s, \chi_t \rangle_{L^2} = \min(s, t)$. Thus if Φ is the Gaussian process associated to $L^2(0, \infty)$, then $b(s) = \Phi(\chi_s)$ is Brownian motion. Φ is often called *white noise* and the formal relation $db/ds = \Phi(s)$ is often expressed by saying that "the derivative of Brownian motion is white noise."

A second main Gaussian process we will consider is the following.

Definition The **oscillator process** is the family $\{q(t)\}_{-\infty < t < \infty}$ of Gaussian random variables with covariance $E(q(t)q(s)) = \frac{1}{2}\exp(-|t - s|)$.

To be sure this is a legitimate definition, we note the following.

Lemma 4.4 Let $s_1, \ldots, s_n$ be real and $z_1, \ldots, z_n \in \mathbb{C}$. Then

$$\sum_{j,i=1}^{n} \bar{z}_i z_j(\tfrac{1}{2}\exp(-|s_i - s_j|)) \geq 0$$

Proof By direct calculation of the Fourier transform of $e^{-|t|}$ and the Fourier inversion formula,

$$\tfrac{1}{2}e^{-|t-s|} = (2\pi)^{-1} \int (k^2 + 1)^{-1} e^{ik(t-s)}\, dk$$

so that

$$\sum_{j,i=1}^{n} \bar{z}_i z_j(\tfrac{1}{2}\exp(-|s_i - s_j|)) = (2\pi)^{-1} \int (k^2 + 1)^{-1} \left| \sum_{j=1}^{n} z_j e^{iks_j} \right|^2 dk$$

which is clearly positive. ∎

From a probabilistic point of view, the oscillator process is "natural" as the only "invariant Gaussian Markov process" (up to changes of scale); see Corollary 4.11. Our interest comes from the fact that it is a "path integral for the harmonic oscillator" as we shall see. Often, the process we have called the oscillator process is called the "Ornstein–Uhlenbeck velocity process" since Uhlenbeck and Ornstein [281] introduced a process $x(t)$ with differentiable paths so that $dx/dt = q$.

We defer the definition of the third major Gaussian process, the Brownian bridge.

Our interest in the Wiener process comes from the fact (responsible for its invention by Wiener [286]) that it is intimately connected with the semigroup e^{-tH_0} where $H_0 = -\frac{1}{2}d^2/dx^2$. Given the connections of $b(t)$ with random walks, of random walks with diffusion, of diffusion with H_0 via the diffusion equation, this should not be surprising. e^{-tH_0} has the integral kernel:

$$P_t(x, y) = (2\pi t)^{-1/2} \exp\left(-\frac{1}{2t}|x - y|^2\right) \tag{4.3}$$

Since $P_t(x, \cdot)$ is in L^2 and varies continuously in L^2 as x is varied, $\int P_t(x, y) f(y)\, dy$ is a continuous function for any $f \in L^2$. Using this continuity, we define $(e^{-tH_0}f)(x)$ for *every* x (a priori it is only defined almost everywhere).

Theorem 4.5 Let $f_1, \ldots, f_{n-1} \in L^\infty(\mathbb{R})$, $f_n \in L^2 \cap L^\infty(\mathbb{R})$, and let $0 < s_1 < \cdots < s_n$. Then

$$E(f_1(b(s_1)) \cdots f_n(b(s_n))) = (e^{-t_1 H_0} f_1 e^{-t_2 H_0} f_2 \cdots e^{-t_n H_0} f_n)(0) \tag{4.4}$$

where $t_1 = s_1$, $t_2 = s_2 - s_1$, $\ldots$, $t_n = s_n - s_{n-1}$, where f_n is viewed as an element of the Hilbert space L^2 and where $f_1, \ldots, f_{n-1}$ are viewed as bounded multiplication operators on L^2.

Proof It clearly suffices to show that the joint probability distribution of $(b(s_1), \ldots, b(s_n))$ is

$$P_{t_1}(0, x_1) P_{t_2}(x_1, x_2) \cdots P_{t_n}(x_{n-1}, x_n)\, d^n x \tag{4.5}$$

But since $b(s_1), b(s_2) - b(s_1), \ldots, b(s_n) - b(s_{n-1})$ are mutually independent Gaussian random variables of variance $t_1, t_2, \ldots, t_n$, their joint distribution is

$$P_{t_1}(0, y_1) P_{t_2}(0, y_2) \cdots P_{t_n}(0, y_n)\, d^n y$$

so that (4.5) holds since the Jacobian of the change of variables $y_1 = x_1$, $y_2 = x_2 - x_1$, $y_3 = x_3 - x_2, \ldots$ is 1. ∎

It is easy to generalize Theorem 4.5 to several dimensions:

Definition **ν-dimensional Brownian motion, $\mathbf{b}(t)$** ($t \geq 0$), is the family of $\mathbb{R}^\nu$-valued "random variables" whose ν-components are ν independent Brownian motions; i.e., $(b_j(t))_{0<t;\, 1 \leq j \leq \nu}$ are Gaussian random variables with covariance

$$E(b_j(t) b_k(s)) = \delta_{jk} \min(t, s)$$

An identical calculation to that in Theorem 4.5 proves the following theorem.

Theorem 4.6 Let $H_0 = -\frac{1}{2}\Delta$ on $L^2(\mathbb{R}^\nu)$. Let $f_1, \ldots, f_{n-1} \in L^\infty(\mathbb{R}^\nu)$, $f_n \in L^2 \cap L^\infty$ and let $0 < s_1 < \cdots < s_n$. Then

$$E(f_1(\mathbf{b}(s_1)) \cdots f_n(\mathbf{b}(s_n))) = (e^{-t_1 H_0} f_1 \cdots e^{-t_n H_0} f_n)(\mathbf{0}) \tag{4.4$'$}$$

where $t_1 = s_1$, $t_2 = s_2 - s_1, \ldots, t_n = s_n - s_{n-1}$.

To find the analog of (4.4) for the oscillator process, we let

$$L_0 = -\tfrac{1}{2} d^2/dx^2 + \tfrac{1}{2}x^2 - \tfrac{1}{2}$$

and $\Omega_0(x) = \pi^{-1/4} e^{-(1/2)x^2}$ so that $L_0 \Omega_0 = 0$ and $\int |\Omega_0|^2\, dx = 1$. Notice that $x\Omega_0$ is also an eigenvalue of L_0: $L_0(x\Omega_0) = (x\Omega_0)$ and that $\int x^2 |\Omega_0|^2\, dx = \frac{1}{2}$.

Theorem 4.7 Let $f_0, \ldots, f_n \in L^\infty(\mathbb{R})$ and let $-\infty < s_0 < \cdots s_n < \infty$. Then

$$E(f_0(q(s_0)) \cdots f_n(q(s_n))) = (\Omega_0, f_0 e^{-t_1 L_0} f_1 \cdots e^{-t_n L_0} f_n \Omega_0) \tag{4.6}$$

where $t_i = s_i - s_{i-1}$.

Proof Fix $t_1, \ldots, t_n$. We first claim that there is a Gaussian probability measure, G, on $\mathbb{R}^{n+1}$ so that

$$\text{r.h.s. of (4.6)} = \int f_0(x_0) \cdots f_n(x_n)\, dG(x) \tag{4.7}$$

For, using the Trotter product formula,

$$\text{r.h.s. of (4.6)} = \lim_{m \to \infty} (\Omega_0, f_0(e^{-t_1 H_0/m_1} e^{-t_1 W/m_1})^{m_1} f_1 \cdots f_n \Omega_0)$$

with $W(x) = \frac{1}{2}(x^2 - 1)$. For each fixed $\mathbf{m} = (m_1, \ldots, m_n)$, this is of the form

$$\int f_0(x_0) \cdots f_n(x_n)\, dG_{\mathbf{m}}(x, y)$$

where $G_{\mathbf{m}}$ is a Gaussian measure in x and auxiliary variables y obtained by putting together the explicit Gaussian kernel of e^{-tH_0}, the Gaussian Ω_0, and the Gaussian in e^{-tW}. The partial integral $\int_y dG_{\mathbf{m}}(x, y)$ of Gaussians is again a Gaussian and a limit of Gaussians is a Gaussian, so (4.7) holds. To find the covariance of the dG in (4.7), we note that for $i < j$,

$$\int x_i x_j\, dG \equiv (x\Omega_0, e^{-(s_j - s_i)L_0} x\Omega_0) = \tfrac{1}{2} e^{-(s_j - s_i)}$$

since $L_0(x\Omega_0) = (x\Omega_0)$ and $(x\Omega_0, x\Omega_0) = \frac{1}{2}$. This shows that dG is just the joint probability distribution of $q(s_0), \ldots, q(s_n)$. ∎

The above theorem yields an explicit kernel $Q_t(x, y)$ for e^{-tL_0}; i.e.,

$$(e^{-tL_0} f)(x) = \int Q_t(x, y) f(y)\, dy$$

For

$$(g, e^{-tL_0} f) = \int f(y) \Omega_0^{-1}(y) g(x) \Omega_0^{-1}(x)\, dG(x, y)$$

where dG is the Gaussian measure on $\mathbb{R}^2$ with covariance matrix

$$\frac{1}{2}\begin{pmatrix} 1 & e^{-t} \\ e^{-t} & 1 \end{pmatrix}$$

Thus, inverting this matrix and using (2.6′)

$$dG(x, y) = \pi^{-1}(1 - e^{-2t})^{-1/2} \exp[-(1 - e^{-2t})^{-1}(x^2 + y^2 - 2e^{-t}xy)]\, dx\, dy$$

or

$$Q_t(x, y) = \pi^{-1/2}(1 - e^{-2t})^{-1/2} \times \exp\{-[1 - e^{-2t}]^{-1}[(x^2 + y^2)\tfrac{1}{2}(1 + e^{-2t}) - 2e^{-t}xy]\}$$

This is known as Mehler's formula and this proof is essentially due to Doob [61].

* * *

The formulas (4.4) and (4.4′) clearly distinguish the point zero; to restore the translation invariance of H_0, we introduce some measures which are not normalized (although they are still positive).

Definition Let dx be Lebesgue measure on $\mathbb{R}^\nu$ and let $(B, \mathscr{B}, Db)$ be the measure space for ν-dimensional Brownian motion. By **Wiener measure**, we mean the measure $d\mu_0$ on $\mathbb{R}^\nu \times B$ given by $dx \otimes Db$. We let $\boldsymbol{\omega}(t) \equiv \mathbf{x} + \mathbf{b}(t)$.

Remark We use Db, Dq for the measures associated with b and q to avoid confusion with the stochastic differentials db of Chapter V.

Theorem 4.8 Let $d\mu_0$ be Wiener measure. Let $f_0, \ldots, f_n \in L^\infty(\mathbb{R}^\nu)$ with $f_0, f_n \in L^2$. Let $0 \le s_0 < s_1 \cdots s_{n-1} < s_n$. Then

$$\int f_0(\boldsymbol{\omega}(s_0)) \cdots f_n(\boldsymbol{\omega}(s_n))\, d\mu_0(\boldsymbol{\omega}) = (\bar{f}_0, e^{-t_1 H_0} f_1 \cdots e^{-t_n H_0} f_n) \tag{4.8}$$

where $t_i = s_i - s_{i-1}$.

Proof Suppose first that $s_0 = 0$. Then writing $\boldsymbol{\omega} = \mathbf{x} + \mathbf{b}$ and using (4.4) to do the b integration, we see that

$$\text{l.h.s. of (4.8)} = \int f_0(x)(e^{-t_1 H_0} f_1 \cdots e^{-t_n H_0} f_n)(x)\, dx$$

which is the right-hand side of (4.8). If $s_0 > 0$, we let g_R be the characteristic function of the ball of radius R in $\mathbb{R}^\nu$. Then, by the dominated convergence theorem

$$\text{l.h.s. of (4.8)} = \lim_{R \to \infty} \int g_R(\boldsymbol{\omega}(0)) f_0(\boldsymbol{\omega}(s_0)) \cdots f_n(\boldsymbol{\omega}(s_n))\, d\mu_0(\boldsymbol{\omega})$$

while the right-hand side is

$$\lim_{R\to\infty} (\bar{f}_0 e^{-s_0 H_0} g_R, e^{-t_1 H_0} f_1 \cdots e^{-t_n H_0} f_n)$$

since $\bar{f}_0 e^{-s_0 H_0} g_R$ converges to $\bar{f}_0$ in L^2 by a simple application of the monotone convergence theorem. Thus, the general case of (4.8) follows from the special case with $s_0 = 0$. ∎

Remarks 1. For reasons of symmetry and also for emphasizing the analogy with the oscillator process, one can consider $\mathbb{R}^\nu \times B \times B$ with measure $dx \otimes Db \otimes Db$, with b_1 (respectively, b_2) the first (respectively, second) Brownian coordinate function. One then defines, $\boldsymbol{\omega}(t) = \mathbf{x} + \mathbf{b}_1(t)$ for $t \geq 0$ and $\mathbf{x} + \mathbf{b}_2(-t)$ for $t \leq 0$. (4.8) then holds for all t's and the joint (nonprobability) distribution of $(\boldsymbol{\omega}(s_1), \ldots, \boldsymbol{\omega}(s_n))$ is time translation invariant.

2. If we prove some "translation invariant" statement like continuity for the Brownian paths with probability one, then, by Fubini's theorem, it automatically holds for $\boldsymbol{\omega}$ on a set whose complement has μ_0-measure zero.

The next special measure which concerns us is formally just

$$\delta(\boldsymbol{\omega}(0) - \mathbf{a})\delta(\boldsymbol{\omega}(t) - \mathbf{b})\, d\mu_0$$

which is not a probability measure, but rather one of mass $P_t(\mathbf{a}, \mathbf{b}) \equiv \prod_{i=1}^{\nu} P_t(a_i, b_i)$.

Definition Fix $\mathbf{a}, \mathbf{c} \in \mathbb{R}^\nu$ and $t > 0$. For $0 \leq s \leq t$ define random variables $\boldsymbol{\omega}(t)$ with the joint distribution for $(\boldsymbol{\omega}(s_1), \ldots, \boldsymbol{\omega}(s_n))$ $(0 \leq s_1 \leq s_2 \leq \cdots \leq s_n \leq t)$

$$P_t(\mathbf{a}, \mathbf{c})^{-1}[P_{t_1}(\mathbf{a}, \mathbf{x}_1)P_{t_2}(\mathbf{x}_1, \mathbf{x}_2) \cdots P_{t_n}(\mathbf{x}_{n-1}, \mathbf{x}_n)P_{t_{n+1}}(\mathbf{x}_n, \mathbf{c})]\, d^{n\nu}x$$

where $t_1 = s_1$, $t_2 = s_2 - s_1, \ldots, t_{n+1} = t - s_n$, and $P_0(\mathbf{x}, \mathbf{y}) = \delta(\mathbf{x} - \mathbf{y})$. The consistency and normalization conditions follow from

$$\int P_t(\mathbf{x}, \mathbf{y})P_s(\mathbf{y}, \mathbf{z})\, d^\nu y = P_{t+s}(\mathbf{x}, \mathbf{z})$$

Let $d\nu_{\mathbf{a}, \mathbf{c}; t}$ be the corresponding probability measure and let us define **conditional Wiener measure** by

$$d\mu_{0, \mathbf{a}, \mathbf{c}; t}(\boldsymbol{\omega}) = P_t(\mathbf{a}, \mathbf{c})\, d\nu_{\mathbf{a}, \mathbf{c}; t}$$

Using the explicit form of the distribution of finitely many $\boldsymbol{\omega}(s)$'s one sees that

$$\int f(\boldsymbol{\omega})\, d\mu_0 = \iint\left[\int f(\boldsymbol{\omega})\, d\mu_{0,\mathbf{a},\mathbf{c};t}\right] d^\nu a\, d^\nu c \tag{4.9}$$

so long as f is measurable with respect to the σ-algebra generated by $\{\boldsymbol{\omega}(s) \mid 0 \leq s \leq t\}$. Similarly,

$$E(f(\mathbf{b})) = \iint\left[\int f(\boldsymbol{\omega})\, d\mu_{0,\mathbf{0},\mathbf{a};t}\right] d^\nu a \tag{4.10}$$

for the Wiener process. (4.9) and (4.10) have an interpretation in terms of conditional expectations. For example, (4.10) says that

$$E(f(\mathbf{b}) \mid \mathbf{b}(t))(\mathbf{a}) = \int f(\boldsymbol{\omega})\, d\mu_{0,\mathbf{0},\mathbf{a};t}$$

where $E(\cdot \mid \mathbf{b}(t))(\mathbf{a})$ means the value of $E(\ldots)$ at those points with $\mathbf{b}(t) = \mathbf{a}$. [More precisely, it is defined for almost every $\mathbf{a}$ with respect to $d\nu(\mathbf{a})$, the joint distribution of $\mathbf{b}(t)$, by

$$E(\cdot\, g(\mathbf{b}(t))) = \int g(\mathbf{a}) E(\cdot \mid \mathbf{b}(t))(\mathbf{a})\, d\nu(\mathbf{a})$$

for all measurable g.]

We will occasionally use the symbol

$$E(\cdot \mid \boldsymbol{\omega}(0) = \mathbf{a}, \boldsymbol{\omega}(t) = \mathbf{c})$$

for expectations with respect to $d\nu_{\mathbf{a},\mathbf{c};t}$. There is a useful way of representing all the $d\nu_{\mathbf{a},\mathbf{c};t}$ processes at once.

Definition The **Brownian bridge**, $\{\alpha(s)\}_{0 \leq s \leq 1}$ is the Gaussian process with covariance

$$E(\alpha(s)\alpha(t)) = s(1 - t) \qquad (0 \leq s \leq t \leq 1)$$

Rather than directly showing that the required covariance is positive definite, we note that

$$\tilde{\alpha}(s) = b(s) - sb(1)$$

is a family of Gaussian random variables with $(s \leq t)$

$$E(\tilde{\alpha}(s)\tilde{\alpha}(t)) = s + st - 2st = s(1 - t)$$

Moreover, $E(\tilde{\alpha}(s)b(1)) = 0$ so that, by Proposition 3.8, one can write $(0 \leq s \leq 1)$

$$b(s) \doteq \alpha(s) + sb(1) \qquad \text{(independent sum)} \tag{4.11}$$

where $\doteq$ means that the distributions on both sides are the same.

Now let $\boldsymbol{\alpha}(s)$ be ν-independent copies of $\alpha(s)$. Using (4.11) and the definition of $\boldsymbol{\omega}$, one immediately sees that

$$\boldsymbol{\omega}(s) \doteq \left(1 - \frac{s}{t}\right)\mathbf{a} + \frac{s}{t}\mathbf{c} + \sqrt{t}\,\boldsymbol{\alpha}\left(\frac{s}{t}\right) \tag{4.12}$$

for the $d\nu_{\mathbf{a},\mathbf{b};t}$ process. Thus, (4.9) becomes

$$\int f(\boldsymbol{\omega}(s))\,d\mu_0 = \int f\left(\left(1 - \frac{s}{t}\right)\mathbf{a} + \frac{s}{t}\mathbf{c} + \sqrt{t}\,\boldsymbol{\alpha}\left(\frac{s}{t}\right)\right) P_t(\mathbf{a}, \mathbf{c})\, d^\nu a\, d^\nu c\, D\alpha \tag{4.13}$$

* * *

We want to close this section with a brief discussion of the Markov property as it applies to the processes and results just discussed. We emphasize that these notions will not be used in the remainder of this book. See Dynkin [67] for further discussions of Markov processes.

Definition A stochastic process $\{x(t)\}_{a \leq t \leq b}$ is called **Markovian** if and only if for $a \leq u \leq t$, the conditional expectation $E(f(x(t)) | \{x(s)\}_{a \leq s \leq u})$ is measurable with respect to $x(u)$. Colloquially, this says that the future depends on the past only through the present.

Theorem 4.9 Let $\{x(t)\}_{a \leq t \leq b}$ be a Gaussian stochastic process with (pointwise) strictly positive covariance $C(t, s)$. Then $\{x(t)\}$ is Markovian if and only if $C(t, s) = f(t)g(s)$ $(t > s)$ for suitable functions f and g.

Proof Let $t \geq u \geq s$. We first show that the process is Markovian if and only if for such triples:

$$C(t, s)C(u, u) = C(t, u)C(u, s) \tag{4.14}$$

For the Markov property is equivalent to the condition that

$$E(:e^{i\alpha x(t)}: | \{x(s)\}_{a \leq s \leq u})$$

is $x(u)$-measurable. By (3.9), this is equivalent to the fact that, in the notation of Corollary 2.4, the projection of δ_t onto the span of $\{\delta_s\}_{a \leq s \leq u}$ in the $C(\cdot, \cdot)$-inner product is just a multiple of δ_u, and this is equivalent to the fact that

$\delta_t - E_u(\delta)$ is orthogonal to δ_s in the C-inner product, where $E_u(\delta_t) = C(u, t)C(u, u)^{-1}\delta_u$. Thus, the Markov property is equivalent to (4.14).

Now, clearly if $C(t, s) = f(t)g(s)$ $(t \geq s)$, then (4.14) holds. Conversely, let (4.14) hold and pick some u and let

$$f(t) = \begin{cases} C(t, u) & (t \geq u) \\ C(t, t)C(u, u)/C(u, t) & (t \leq u) \end{cases}$$

$$g(s) = \begin{cases} C(s, s)/C(s, u) & (s \geq u) \\ C(s, u)/C(u, u) & (s \leq u) \end{cases}$$

Using (4.14) and considering the three cases $t \geq s \geq u$, $t \geq u \geq s$, and $u \geq t \geq s$, one easily sees that $C(t, s) = f(t)g(s)$ for $t \geq s$. ▮

Remarks 1. The formulas for $f(t)$ and $g(s)$ can be guessed by supposing $g(u) = 1$ and that $C(t, s) = f(t)g(s)$.

2. Some kind of restriction on C is needed to deduce $C(t, s) = f(t)g(s)$ from (4.14). For example, take $C(t, u) = \delta_{tu}$.

3. A sufficient condition for C to be strictly positive, is that it be jointly continuous and $C(t, t) > 0$ for all t. For C is then uniformly continuous so $C(t, s) > \varepsilon$ for $|t - s| < \delta$. Given any $t < s$ find $t < s_1 < \cdots < s_k < s$ with $|s_1 - t| < \delta, |s_2 - s_1| < \delta, \ldots, |s - s_k| < \delta$. By (4.14)

$$C(t, s) = \frac{C(t, s_1)C(s_1, s_2)\cdots C(s_k, s)}{C(s_1, s_1)\cdots C(s_k, s_k)}$$

is strictly positive.

Corollary 4.10 $\{b(t)\}_{t \geq 0}$ and $\{q(t)\}_{-\infty < t < \infty}$ are Markov processes.

Corollary 4.11 Let $\{x(t)\}$ be a Gaussian stochastic process which is time invariant [i.e., the joint distribution of $x(t_1), \ldots, x(t_n)$ agrees with that of $x(t + t_1), \ldots, x(t + t_n)$], continuous in quadratic mean [$E(x(t)x(s))$ continuous], and Markovian. Then for suitable $\alpha, \beta > 0$, $x(t) = \alpha q(\beta t)$ where q is the oscillator process.

Proof By the time invariance and symmetry of $C(t, u)$, we have that $C(t, u) = h(|t - u|)$ for a suitable continuous function h. The Markov property implies (4.14) and thus

$$h(t)h(0) = h(t - u)h(u)$$

for $t > u > 0$. This and continuity imply that

$$h(t) = h(0)e^{-\gamma|t|}$$

The Schwarz inequality implies $\gamma \geq 0$. Take $\beta = \gamma$ and $\alpha = (2h(0))^{1/2}$. ▮

There is a close connection between formulas (4.4) and (4.6) and the Markov property. Given any stochastic process, $\{x(t)\}$ we can define the spaces $L_t^2 = L^2(\mathbb{R}, d\nu_t)$ where ν_t is the probability distribution of $x(t)$ and for $t \leq s$ define

$$U(t, s): L_s^2 \to L_t^2 \quad \text{by the formula} \quad U(t, s)[f(x(s))] = E(f(x(s)) \mid x(t)).$$

The Markov property says that for $v \leq t \leq s$,

$$E(f(x(s)) \mid x(t)) = E(f(x(s)) \mid x(t), x(v))$$

so, since $E(E(\cdot \mid x(t), x(v)) \mid x(v)) = E(\cdot \mid x(v))$ for any process, we see that for Markov processes

$$U(v, t)U(t, s) = U(v, s) \qquad (v \leq t \leq s)$$

In the case of the oscillator process where the L_t^2 are isomorphic and $U(t, s) = U(s - t)$, we see that $U(t, s) = e^{-(s-t)B}$ for some generator B of a contraction semigroup; this is clearly why (4.6) holds. (4.4) is more complicated but similar. Moreover, one can go quite easily from (4.4), (4.6) to the Markov property. For example, (4.6) implies that for $t \geq u$;

$$E(f(q(t)) \mid \{q(s)\}_{s \leq u}) = \Omega_0(q(u))^{-1}(e^{-(t-u)L_0} f \Omega_0)(q(u))$$

which is clearly $q(u)$-measurable. In this way, one easily sees that the $P(\phi)_1$-processes to be constructed in Section 6 are Markovian. Notice that the above expectation has the form $(e^{-tB}f)$ on $L^2(\mathbb{R}, d\nu)$ where $d\nu = \Omega_0^2\, dx$ and $B = \Omega_0^{-1} L_0 \Omega_0$.

5. Regularity Properties—1

We will deduce the existence of continuous versions for the Wiener and oscillator processes from the following beautiful result ("Kolmogorov's lemma"):

Theorem 5.1 Let $\{x(t)\}_{a \leq t \leq b}$ be a stochastic process obeying

$$E(|x(t + h) - x(t)|^p) \leq K|h|^{1+r}$$

for some K, some $r < p$, and all t, h with $a \leq t < t + h \leq b$. Fix $0 < \alpha < r/p$. Then

$$|x(t) - x(s)| \leq C(x)|t - s|^\alpha \tag{5.1}$$

for all dyadic rational t and s where $C(x)$ is finite almost everywhere. In particular, x has a Hölder continuous version [i.e., one can find $x(t)$ defined for all points with the correct joint distributions and for which (5.1) holds for *all* t and s].

Proof We prove (5.1); the version question is then solved via method (a) of Section 2 (p. 14). Without loss, suppose that $a = 0, b = 1$. Let $\varepsilon = r - \alpha p$. Then

$$E(|x(t+h) - x(t)| \geq |h|^{\alpha}) \leq |h|^{-\alpha p}[E(|x(t+h) - x(t)|^{p})]$$
$$\leq K|h|^{1+\varepsilon}$$

so that

$$E\left(\left|x\left(\frac{k+1}{2^n}\right) - x\left(\frac{k}{2^n}\right)\right| \geq 2^{-n\alpha}\right) \leq K2^{-n}2^{-n\varepsilon}$$

Thus

$$\sum_{n=1}^{\infty} \sum_{k=0}^{2^n-1} E\left(\left|x\left(\frac{k+1}{2^n}\right) - x\left(\frac{k}{2^n}\right)\right| \geq 2^{-n\alpha}\right) < \infty$$

so, by the first Borel–Cantelli lemma, there exists a random integer $\nu(x)$, almost everywhere finite so that

$$\left|x\left(\frac{k+1}{2^n}\right) - x\left(\frac{k}{2^n}\right)\right| \leq 2^{-n\alpha}; \qquad n \geq \nu(x), \quad k = 0, \ldots, 2^n - 1$$

Now let $n \geq \nu(x)$ and let t be a dyadic rational in $[k/2^n, (k+1)/2^n]$. Write $t = k/2^n + \sum_{i=1}^{n} \gamma_i/2^{n+i}$; each $\gamma_i = 0$ or 1. Then

$$\left|x(t) - x\left(\frac{k}{2^n}\right)\right| \leq \sum_{i=1}^{m} \gamma_i 2^{-\alpha(n+i)} \leq d2^{-n\alpha} \tag{5.2a}$$

where $d = (1 - 2^{-\alpha})^{-1}$. Similarly

$$\left|x(t) - x\left(\frac{k+1}{2^n}\right)\right| \leq d2^{-n\alpha} \tag{5.2b}$$

Now let t and h be dyadic rationals with $h \leq 2^{-\nu(x)}$. Take n with $2^{-n-1} \leq h < 2^{-n}$ and k so that $k/2^{n+1} \leq t < k + 1/2^{n+1}$. Then $k + 1/2^{n+1} \leq t + h < k + 3/2^{n+1}$. It follows that

$$|x(t+h) - x(t)| \leq (2d+1)2^{-(n+1)\alpha} \leq (2d+1)h^{\alpha}$$

Since, for fixed x this holds for all sufficiently small h, we have that (5.1) holds. ∎

Theorem 5.2 The Wiener process, ν-dimensional Brownian motion, and the oscillator process all have Hölder continuous paths of any order $\alpha < \frac{1}{2}$.

Proof We consider the Wiener case. The ν-dimensional result is then immediate. The oscillator case is similar. For any Gaussian variable X:

$$E(|X|^p) = C_p E(|X|^2)^{p/2} \tag{5.3}$$

(5.3) follows by scaling once we know that the expectation is finite which is trivial. From (5.3) and the fact that $b(t+h) - b(t)$ has variance h, we see that

$$E(|b(t+h) - b(t)|^p) = C_p h^{p/2}$$

We can, for fixed $p > 2$, choose α arbitrarily close to $\frac{1}{2} - p^{-1}$ so by taking p large, we can obtain α's arbitrarily close to $\frac{1}{2}$. ∎

By pushing the ideas of the proof of Theorem 5.1 to their extreme (see [183, Section 1.6]) one can obtain Lévy's precise law for the local smoothness [167]:

Theorem 5.3 With probability one

$$\varlimsup_{\substack{0 \le t_1 < t_2 \le 1 \\ t = t_2 - t_1 \downarrow 0}} \frac{|b(t_2) - b(t_1)|}{[2t \ln(t^{-1})]^{1/2}} = 1$$

for the Wiener process. In particular, the Wiener process is (everywhere) Hölder continuous of order $\frac{1}{2}$ on a set of measure zero.

Remark We will see below that $b(t)$ and $q(t)$ have the same local behavior.

Lévy's law only assures us that $b(t)$ is not Hölder continuous of order $\frac{1}{2}$ at some point and not that it fails to be Hölder continuous of order $\frac{1}{2}$ at all points [indeed, we shall see in Section 7 that the behavior at a fixed point s differs from the behavior over all s in (0, 1); explicitly

$$\varlimsup_{t \downarrow 0} \frac{|b(t+s) - b(s)|}{[2t \log_2 (t^{-1})]^{1/2}} = 1$$

with probability one]. The following is a slight generalization of the proof that Dvoretsky *et al.* [63] gave of the celebrated result of Payley, Wiener and Zygmund [199] that $b(t)$ is everywhere nondifferentiable; the result below is also in [199].

Theorem 5.4 Fix $\alpha > \frac{1}{2}$. Then with probability one, $b(t)$ is nowhere Hölder continuous of order α; i.e., for almost every b,

$$\inf_{0 \le t \le 1} \left[\overline{\lim_{h \to 0}} |b(t+h) - b(h)| \, |h|^{-\alpha} \right] = \infty \tag{5.4}$$

In particular, $b(t)$ is nowhere differentiable.

Proof Fix an integer k with $k(\alpha - \frac{1}{2}) > 1$. Suppose that $b(t)$ is a path with the left-hand side of (5.4) finite. Then, there is a t, a C, and an h_0 with

$$|b(t+h) - b(t)| \le C|h|^\alpha$$

for all h with $|h| \le h_0$. Pick m so that $(k-1)/m < h_0$. Given n let $i = [tn] + 1$ so that $i/n, (i+1)/n, \ldots, (i+k)/n \in [t, t+h_0]$ if $n > m$. Then

$$\left| b\left(\frac{i+j}{n}\right) - b\left(\frac{i+j-1}{n}\right) \right| \le Cn^{-\alpha}[|k|^\alpha + |k+1|^\alpha]$$

for $j = 1, \ldots, k$ (since $|b(s) - b(u)| \le C[|s-t|^\alpha + |u-t|^\alpha]$ by the triangle inequality). Thus

$$\{b \mid b \text{ somewhere Hölder continuous of order } \alpha\}$$
$$\subset \bigcup_{D>1} \bigcup_{m>1} \bigcap_{n \ge m} \bigcup_{0 \le i \le n-k+1} \left\{ b \,\middle|\, \left| b\left(\frac{j}{n}\right) - b\left(\frac{j-1}{n}\right) \right| \le Dn^{-\alpha} \text{ for } j = i+1, \ldots, i+k \right\}$$

It thus suffices to show that

$$E\left(\bigcap_{n \ge m} \bigcup_{0 \le i \le n-k+1} \left\{ b \,\middle|\, \left| b\left(\frac{j}{n}\right) - b\left(\frac{j-1}{n}\right) \right| \le Dn^{-\alpha};\ j = i+1, \ldots, i+k \right\} \right) = 0$$

This follows from

$$\lim_{n \to \infty} n \left[E\left(b \,\middle|\, \left| b\left(\frac{1}{n}\right) \right| \le Dn^{-\alpha} \right) \right]^k = 0 \tag{5.5}$$

since the $b(j/n) - b((j-1)/n)$ are independent and distributed identically to $b(1/n)$. Now by scaling

$$\begin{aligned} E(b \,||b(1/n)| \le Dn^{-\alpha}) &= E(b \,||b(1)| \le Dn^{-\alpha+1/2}) \\ &= O(n^{-(\alpha-1/2)}) \end{aligned}$$

since the distribution of $b(1)$ is continuous. Since $k(\alpha - \frac{1}{2}) > 1$, (5.5) holds. ∎

Finally, we turn to showing that $q(t)$ and $b(t)$ have the same local behavior. We give two distinct proofs of this fact. The first relies on an elementary observation that

$$q(t) \doteq e^{-t}b(e^{2t})/\sqrt{2} \tag{5.6}$$

where $\doteq$ means that both sides have the same joint distributions. To prove (5.6), we need only check covariances since both sides are Gaussian. For $t > s$, we have that

$$\tfrac{1}{2}e^{-t}e^{-s}\min(e^{2t}, e^{2s}) = \tfrac{1}{2}e^{s-t}$$

From (5.6) we immediately conclude that Theorems 5.3 and 5.4 extend to $q(t)$. Expressions like (5.6) occur with functions of b on both sides. Since $q(-t) \doteq q(t)$ trivially, (5.6) suggests that $b(t)$ and $b(1/t)$ are related. In fact

$$b(1/t) \doteq t^{-1}b(t) \tag{5.7}$$

since $\min(1/t, 1/s) = (ts)^{-1}\min(t, s)$. More trivially

$$b(at) \doteq a^{1/2}b(t) \tag{5.8}$$

Notice that (5.7) and (5.8) extend to ν-dimensional Brownian motion since they hold for each component.

The second way of seeing that $b(t)$ and $q(t)$ have the same local behavior is to apply Theorem 2.5:

Theorem 5.5 Let Db (respectively, Dq) denote the measure on paths $\omega(t)$ associated to Brownian motion (respectively, the oscillator process). Let $\Sigma_{(a,b)}$ denote the σ-algebra generated by $\{\omega(t) \mid a < t < b\}$. Then for $0 < a < b < \infty$, $Db \restriction \Sigma_{(a,b)}$ and $Dq \restriction \Sigma_{(a,b)}$ are mutually absolutely continuous.

Proof Pick $c > b$ and on $[a, b]$ consider the four processes with covariances:

$$\begin{aligned}
K_\infty(t, s) &= \tfrac{1}{2}e^{-|t-s|}\\
K_c(t, s) &= \tfrac{1}{2}e^{-|t-s|} - \tfrac{1}{2}(e^{2c} - 1)^{-1}\{e^{s+t} - e^{s-t} - e^{t-s} + e^{2c}e^{-t-s}\}\\
A_\infty(t, s) &= \min(t, s)\\
A_c(t, s) &= \min(t, s) - c^{-1}ts
\end{aligned}$$

The positive definiteness of K_c and A_c follows if one notices K_c (respectively, A_c) is the kernel of $(B + 1)^{-1}$ (respectively, B^{-1}) where B is the operator $-d^2/dx^2$ on $L^2(0, c)$ with $u(0) = u(c) = 0$ boundary conditions. We will prove the mutually absolute continuity of the pairs (K_c, A_c), (A_c, A_∞), (K_c, K_∞) by application of Theorem 2.5 from which the result then follows.

Let $(f, g)_{A,c} = (f, B^{-1}g)_{L^2}$, $(f, g)_{K,c} = (f, (B+1)^{-1}g)_{L^2}$, and let $(\cdot, \cdot)_{K,c}$ also denote the corresponding Hilbert space, etc. Then

$$(f, g)_{A,c} - (f, g)_{K,c} = (f, Cg)_{A,c}$$

where $C = B^{1/2}(B^{-1} - (B+1)^{-1})B^{1/2} = (B+1)^{-1}$. For C to be Hilbert–Schmidt on $(\cdot, \cdot)_{A,c}$, it is necessary and sufficient that C be Hilbert–Schmidt on L^2 and this is obvious since its eigenvalues go as n^{-2}. It is clear that $1 - C = B(B+1)^{-1}$ is invertible as a map on $(\cdot, \cdot)_{K,c}$ since B is invertible. This shows that K_c and A_c are equivalent.

Next we note that for $\eta \in (\cdot, \cdot)_{A,\infty}$

$$(\eta, \eta)_{A,c} = (\eta, \eta)_{A,\infty} - c^{-1}(\eta, \delta_b)_{A,\infty}(\delta_b, \eta)_{A,\infty}$$

where δ_b has norm $(\delta_b, \delta_b)^{1/2} = b^{1/2}$. Thus

$$(\eta, \eta)_{A,c} = (\eta, (1 - Q)\eta)_{A,\infty}$$

where Q is rank 1 (and so Hilbert–Schmidt) and $\|Q\| = bc^{-1} < 1$. It follows that A_∞ and A_c are equivalent.

The proof of the equivalence of K_∞ and K_c is similar. The difference is rank 2 with δ_a and δ_b both involved. ∎

Remarks 1. The above proof was suggested by work of Guerra *et al.* [120, 121] on the $P(\phi)_2$ field theory. In the language of those papers, K_∞ is the free field of mass $m = 1$, K_c has added Dirichlet boundary conditions at 0 and c, A_∞ is the free field mass 0 with a Dirichlet condition at 0, and A_c has an extra Dirichlet condition at c.

The three equivalences above have analogs in the two-dimensional case but, e.g., A_c and K_c are *not* equivalent in four or more dimensions since $(-\Delta_D + 1)^{-1}$ is not Hilbert–Schmidt if $-\Delta_D$ is the Laplacian with Dirichlet conditions on the boundary of some bounded set. We emphasize that in this paragraph "dimension" refers to the dimension of the parameter t (see Section 24 for discussion of multidimensional t's) and not to the number of components of **b**.

2. For $c = 1$, A_c is the covariance of the Brownian bridge. Thus we have also shown that Db and $D\alpha$ restricted to times in $[0, d]$ with $d < 1$ are mutually absolutely continuous.

6. The Feynman–Kac Formula

Our main goal in this section is to give a number of variants of a basic formula relating e^{-tH} to path integrals when H is of the form $H_0 + V(x)$ or $L_0 + V(x)$. The basic formula is illustrated by the following result for which we now give two independent proofs. Third and fourth proofs appear in Sections 14 and 16.

Theorem 6.1 Let $V \in C_\infty(\mathbb{R}^\nu)$, the continuous functions vanishing at ∞, and let $d\mu_0$ be the ν-dimensional Wiener measure. Let $H = H_0 + V$ as a self-adjoint operator on $D(H_0)$. Then

$$(f, e^{-tH}g) = \int \overline{f(\omega(0))}g(\omega(t)) \exp\left(- \int_0^t V(\omega(s))\, ds\right) d\mu_0(\omega) \qquad (6.1)$$

Remark Since $V(\omega(s))$ is almost everywhere (in ω) continuous, $\int_0^t V(\omega(s))\, ds$ can be taken as a Riemann integral and so is ω-measurable as a limit of Riemann sums.

First proof By the Trotter product formula:

$$(f, e^{-tH}g) = \lim_{n\to\infty} (f, (e^{-tH_0/n}e^{-tV/n})^n g)$$

so by Theorem 4.8:

$$(f, e^{-tH}g) = \lim_{n\to\infty} \int \overline{f(\omega(0))}g(\omega(t)) \exp\left[-\frac{t}{n}\sum_{j=0}^{n-1} V\left(\omega\left(\frac{tj}{n}\right)\right)\right] d\mu_0(\omega) \quad (6.2)$$

Since ω is almost everywhere continuous,

$$\frac{t}{n}\sum_{j=0}^{n-1} V\left(\omega\left(\frac{jt}{n}\right)\right) \to \int_0^t V(\omega(s))\, ds$$

as $n \to \infty$ for almost every ω. Moreover, the integrand in (6.2) is dominated by $|f(\omega(0))||g(\omega(t))|\exp(t\|V\|_\infty)$ which is L^1 since

$$\int |f(\omega(0))||g(\omega(t))|\, d\mu_0 = (|f|, e^{-tH_0}|g|) < \infty$$

Thus, by the dominated convergence theorem, (6.1) holds. ∎

Second proof For fixed V and t, the right-hand side of (6.1) clearly obeys (by Hölder's inequality)

$$|(\text{r.h.s. of } (6.1))| \le e^{t\|V\|_\infty}\|f\|_2\,\|g\|_2$$

since

$$\int |f(\omega(0))|^2\, d\mu_0 = \|f\|_2^2, \qquad \int |g(\omega(t))|^2\, d\mu_0 = \|g\|_2^2$$

Thus we can find an operator A_t so that

$$\text{r.h.s. of (6.1)} = (f, A_t g)$$

We want to show that $A = e^{-tH}$.

Next notice that

$$\exp\left[-\int_0^t V(\omega(s))\,ds\right] = 1 - \int_0^t V(\omega(u)) \exp\left[-\int_u^t V(\omega(s))\,ds\right] du \quad (6.3)$$

(6.3) can be proven by integrating the perfect differential in the integral on the right-hand side. Multiply both sides of (6.3) by $\overline{f(\omega(0))}g(\omega(t))$ and integrate with respect to $d\mu_0$ to find

$$(f, A_t g) = (f, e^{-tH_0} g) - \int_0^t \alpha_u\,du$$

with

$$\begin{aligned}\alpha_u &= \int \overline{f(\omega(0))} V(\omega(u)) \exp\left[-\int_u^t V(\omega(s))\,ds\right] g(\omega(t))\,d\mu_0 \\ &= \int \overline{E(f(\omega(0))\,|\,\omega(s), u \leq s)} V(\omega(u)) \exp\left[-\int_u^t V(\omega(s))\,ds\right] g(\omega(t))\,d\mu_0 \\ &= (f, e^{-uH_0} V A_{t-u} g)\end{aligned}$$

where $E(\cdot\,|\,\omega(s), u \leq s)$ stands for the L^2-projection onto those functions measurable with respect to $\{\omega(s)\,|\,u \leq s\}$ (this remark is needed since $d\mu_0$ is *not* a probability measure). In the above, we use

$$E(f(\omega(0))\,|\,\omega(s), u \leq s) = (e^{-uH_0} f)(\omega(u))$$

(which follows from Theorem 4.8), and we use the translation covariance of $d\mu_0$. We have thus proven that

$$A_t = e^{-tH_0} - \int_0^t e^{-uH_0} V A_{t-u}\,du$$

and from this one can conclude that $A_t = e^{-tH}$; e.g., one can iterate the above equation and obtain an expansion for A_t which is identical to the Dyson–Phillips expansion for e^{-tH}. ▮

Remark Formula (6.1) is a variant of a result of Kac [138] (see also [139]) who was trying to understand Feynman [83]. Both proofs appear to be well known to probabilists. In the mathematical physics literature, the first proof appeared in [189] and the second in [120].

Once we have (6.1) for smooth V's it is easy to extend it to a wide class of V's (note that $f \in L^1_{\text{loc}}(\mathbb{R}^\nu \setminus G)$ means $\int_A |f|\, d^\nu x < \infty$, for any compact subset of $\mathbb{R}^\nu \setminus G$):

Theorem 6.2 Let V be a potential so that $V_+ \in L^1_{\text{loc}}(\mathbb{R}^\nu \setminus G)$ where G is a closed set of measure zero and V_- obeys $Q(H_0) \supset Q(V_-)$ and

$$|(\phi, V_- \phi)| \leq \alpha(\phi, H_0 \phi) + \beta(\phi, \phi)$$

for some $\alpha < 1$ and all $\phi \in Q(H_0)$. Let $H = H_0 + V$ as a sum of forms on $Q(H_0) \cap Q(V)$. Then (6.1) holds.

Remark Since V and ω are measurable, $V(\omega(s))$ is a measurable function of s, so modulo the question of convergence of the integral $\int_0^t V(\omega(s))\, ds \equiv g(\omega)$ can be defined for each ω as a Lebesgue integral. That the integral converges or diverges to $+\infty$ almost everywhere and that g is measurable as a function of ω follows from the proof below which establishes that it is a pointwise limit of a sequence of measurable functions.

Proof Suppose first that $V \in L^\infty$ and let $V_n = h_n(V * j_n)$ where $j_n(x) = n^\nu \phi(xn)$ and $h_n = \phi(x/n)$ with $\phi \in C_0^\infty$, $0 \leq \phi \leq 1$, $\int \phi(x)\, dx = 1$, and $\phi(0) = 1$. As $n \to \infty$, $V_n(x) \to V(x)$ for $x \notin K$, some set of measure zero. For each fixed t, $\{\omega \mid \omega(t) \in K\}$ has measure zero since the distribution of $\omega(t)$ is just Lebesgue measure. Thus by Fubini's theorem with respect to $d\mu_0 \otimes dt$, $\{(\omega, t) \mid \omega(t) \in K\}$ has measure zero. It follows, again by Fubini's theorem, that for almost every ω, $\{t \mid \omega(t) \in K\}$ has Lebesgue measure zero. Thus, by the dominated convergence theorem, $\int_0^t V_n(\omega(s))\, ds \to \int_0^t V(\omega(s))\, ds$ almost everywhere in ω. Thus as $n \to \infty$, the right-hand side of (6.1) converges. Since $H_0 + V_n \to H_0 + V$ in strong resolvent sense (since they converge on a common core; see [152, 214] for discussion of strong resolvent convergence), the left-hand side converges. This establishes (6.1) for $V \in L^\infty$.

Now, let V be an arbitrary function of the type allowed in the theorem and let

$$V_{n,m}(x) = \begin{cases} V(x), & -n < V(x) < m \\ 0, & \text{otherwise} \end{cases}$$

Then, first taking $n \to \infty$ and then $m \to \infty$, both sides of (6.1) converge, the left-hand side converges by application of monotone convergence theorems for forms [152, 221, 253, 254] and the right-hand side by monotone convergence theorems for integrals. ∎

Remark If one interprets e^{-tH} in the proper way for nondensely defined closed forms [154, 245] then the condition $V_+ \in L^1_{\text{loc}}(R \setminus G)$ can be dropped—it was only needed to assure that H was densely defined.

By mimicking the above proofs, one easily shows that the following is true.

Theorem 6.3 Let V be a potential so that $V_+ \in L^1_{\text{loc}}(\mathbb{R})$ and so that V_- obeys $Q(L_0) \supset Q(V_-)$

$$|(\phi, V_- \phi)| \leq \alpha(\phi, L_0 \phi) + \beta(\phi, \phi)$$

for some $\alpha < 1$ and all $\phi \in Q(L_0)$. Let $L = L_0 + V$ as a sum of forms on $Q(L_0) \cap Q(V)$. Then

$$(f\Omega_0, e^{-tL} g\Omega_0) = \int \overline{f(q(0))} g(q(t)) \exp\left[-\int_0^t V(q(s))\, ds\right] Dq \tag{6.4}$$

for all $f, g \in L^2(\mathbb{R}, \Omega_0^2\, dx)$.

Corollary Under the above hypotheses:

$$\inf \operatorname{spec}(L) = -\lim_{t\to\infty} t^{-1} \ln\left[\int \exp\left(-\int_0^t V(q(s))\, ds\right) Dq\right] \tag{6.5}$$

Proof In case L has an eigenvalue E (with $L\Omega = E\Omega$) at the bottom of its spectrum, the proof is immediate from (1.9) since $(\Omega, \Omega_0) > 0$ (for Ω_0 is pointwise strictly positive and Ω is pointwise positive, see [217]). In general, we proceed as follows: Choosing $f = g = 1$ in (6.4) we see that

$$\overline{\lim}\, t^{-1} \ln[\cdot] \leq -\inf \operatorname{spec}(L)$$

On the other hand, pick f and g with $f^2\Omega_0, g^2\Omega_0 \in L^2(\mathbb{R}, dx)$. Then, by the Schwarz inequality and (6.4)

$$(f\Omega_0, e^{-tL} g\Omega_0) \leq \left(\int |f(q(0))|^2 |g(q(t))|^2 \exp\left[-\int_0^t V(q(s))\, ds\right] Dq\right)^{1/2} \times \left(\int \exp\left[-\int_0^t V(q(s))\, ds\right] Dq\right)^{1/2}$$

so that

$$\underline{\lim}\{t^{-1} \ln(f\Omega_0, e^{-tL} g\Omega_0)\} \leq -\tfrac{1}{2} \inf \operatorname{spec}(L) + \tfrac{1}{2} \underline{\lim}\, t^{-1} \ln[\cdot]$$

Varying over all f and g we see that

$$-\tfrac{1}{2} \inf \operatorname{spec}(L) \leq \tfrac{1}{2} \underline{\lim}\, t^{-1} \ln[\cdot]$$

completing the proof. ∎

Remark The above "general case" proof is patterned after a field theoretic argument in Seiler–Simon [235].

This completes the general discussion of Feynman–Kac formulas; in the remainder of this section, we will present some complementary material: Feynman–Kac formula for Db and $d\mu_{0,\mathbf{a},\mathbf{b},t}$; a second proof of Mehler's formula; definition of $P(\phi)_1$-processes, and the use of Feynman–Kac formulas to do calculations with Wiener measure.

* * *

Formulas (4.9) and (6.1) imply that e^{-tH} is an integral operator with kernel

$$e^{-tH}(\mathbf{a}, \mathbf{b}) = \int \exp\left(- \int_0^t V(\omega(s))\, ds \right) d\mu_{0,\mathbf{a},\mathbf{b};t} \tag{6.6}$$

(6.6) can only be interpreted a priori as holding almost everywhere in **a**, **b**. The right-hand side is defined for each **a** and **b** but the left-hand side is only defined almost everywhere. We want to investigate when $e^{-tH}(\mathbf{a}, \mathbf{b})$ and the right-hand side of (6.6) are continuous so one can interpret (6.6) pointwise. We begin by studying when the related formula:

$$(e^{-tH}f)(0) = \int \exp\left(- \int_0^t V(b(s))\, ds \right) f(b(t))\, Db \tag{6.7}$$

is valid.

We first study all operators on $C(\mathbb{R}^\nu)$, the bounded continuous functions on $\mathbb{R}^\nu$. The operator

$$(e^{-t\tilde{H}_0}f)(x) = \int P_t(x, y) f(y)\, dy$$

is easily seen to define a continuous semigroup on $C(\mathbb{R}^\nu)$. Moreover (4.4) holds if $f_1, \ldots, f_n$ lie in $C(\mathbb{R}^\nu)$ and H_0 is replaced by $\tilde{H}_0$, for the proof of Theorem 4.5 goes through without change. If $V \in C(\mathbb{R}^\nu)$, then $\tilde{H}_0 + V$ defined on $D(\tilde{H}_0)$ is the generator of an exponentially bounded semigroup on $C(\mathbb{R}^\nu)$, so using the Trotter product formula for semigroups on $C(\mathbb{R}^\nu)$, we conclude that (6.7) holds (the convergence is now in $\|\cdot\|_\infty$ and thus pointwise). Moreover if $f \in L^2 \cap C(\mathbb{R}^\nu)$, then the two definitions of $e^{-tH}f$ agree (this follows, e.g., from the equality of the Trotter approximations). Thus the following theorem is proven.

Theorem 6.4 If $V \in C(\mathbb{R}^\nu)$, and $\tilde{H} = \tilde{H}_0 + V$ as an operator sum on $C(\mathbb{R}^\nu)$, then (6.7) is valid for $f \in C(\mathbb{R}^\nu)$. In particular, if $f \in L^2 \cap C(\mathbb{R}^\nu)$, then $e^{-tH}f$ (in L^2-sense) is continuous and (6.7) holds for the continuous representative.

This result ignores the fact that e^{-tH} has a tendency to be smoothing; indeed $\text{Ran}(e^{-tH}) \subset C^\infty(H)$ and under fairly general hypotheses ([150, 243] and Section 25), $C^\infty(H)$ consists of continuous functions. When $\nu \leq 3$, $D(H_0)$ is already contained in $C(\mathbb{R}^\nu)$ and this makes proofs easier; for this reason we state the following for $\nu \leq 3$. Using the machinery of [243], an extension exists for $\nu \geq 4$.

Theorem 6.5 Let $\nu \leq 3$. Let $V \in L^2(\mathbb{R}^\nu) + L^\infty(\mathbb{R}^\nu)$ and let $H = H_0 + V$ as an operator sum on $D(H_0)$. Then $e^{-tH}f$ is continuous for each $t > 0$ and $f \in L^2$, and (6.7) holds for the continuous representative of $e^{-tH}f$.

Proof Let j_n be an approximate identity as in the proof of Theorem 6.2. Let $j_n * V = V_n$, so that (6.7) holds for $f \in C(\mathbb{R}^\nu) \cap L^2$, and $H_n \equiv H_0 + V_n$ by Theorem 6.4. Now $\{e^{-tH_n}\}$ and e^{-tH} are uniformly bounded from L^2 to $D(H_n)$ and so to $D(H_0)$. Since $D(H_0)$ is continuously embedded in $C(\mathbb{R}^\nu)$, the maps e^{-tH_n} and e^{-tH} are equi-continuous from L^2 to $C(\mathbb{R}^\nu)$. It follows that (6.7) for $f \in C(\mathbb{R}^\nu)$, and H_n implies the result for all $f \in L^2$ and H_n since the left-hand side is L^2-continuous in f by the above and the right-hand side is continuous by Hölder's inequality.

If we take $n \to \infty$, the right-hand side of (6.7) converges by Hölder's inequality and the dominated convergence theorem. Thus, we need only prove that for fixed $f \in L^2$, $e^{-tH_n}f$ converges to $e^{-tH}f$ *in* $C(\mathbb{R}^\nu)$. This follows if we show that $(H_0 + 1)[e^{-tH_n}f]$ converges to $(H_0 + 1)[e^{-tH}f]$. But since H_n converges to H in a strong resolvent sense, we know that $(H_n + 1)(e^{-tH_n}f) \to (H + 1)e^{-tH}f$. By a simple argument $V_n(e^{-tH_n}f) \to V(e^{-tH}f)$. Now use $H_0 = H_n - V_n$. ∎

We remark, that in any event, (6.1) implies that under the hypotheses of Theorem 6.2

$$(e^{-tH}f)(x) = \int \exp\left(-\int_0^t V(x + b(s))\, ds\right) f(x + b(t))\, Db \tag{6.8}$$

almost everywhere in x.

Theorem 6.6 Let $V \in C(\mathbb{R}^\nu)$. Then e^{-tH} has an integral kernel $e^{-tH}(\mathbf{a}, \mathbf{b})$ jointly continuous in $\mathbf{a}$, $\mathbf{b}$ and $t > 0$, and (6.6) holds for *all* $\mathbf{a}$, $\mathbf{b}$, $t > 0$.

Proof This is a simple exercise in the use of (4.12)–(4.13). Namely, let $\boldsymbol{\alpha}$ be the Brownian bridge and note that since $V \in C(\mathbb{R}^\nu)$,

$$Q(\mathbf{a}, \mathbf{b}, t) = E\left(\exp\left(-\int_0^t V\left(\left(1 - \frac{s}{t}\right)\mathbf{a} + \frac{s}{t}\mathbf{b} + \sqrt{t}\,\boldsymbol{\alpha}\left(\frac{s}{t}\right)\right) ds\right)\right)$$

is jointly continuous in **a**, **b**, t by the almost everywhere continuity of $\boldsymbol{\alpha}(s)$ and the dominated convergence theorem. But by (4.12)

$$\text{r.h.s. of (6.6)} = Q(\mathbf{a}, \mathbf{b}, t)(2\pi t)^{-\nu/2} \exp(-|\mathbf{a} - \mathbf{b}|^2/2t) \tag{6.9}$$

This proves the result. ▮

Remark In Section 25, we shall show that $e^{-tH}(\mathbf{a}, \mathbf{b}) \in L^\infty(\mathbb{R}^{2\nu})$ for a very general class of potentials V.

* * *

The Feynman–Kac formula also provides another proof of Mehler's formula for the integral kernel of e^{-tL_0}. By the Feynman–Kac formulas (and a slight extension of Theorem 6.6)

$$(e^{-tL_0})(a, b) = \int \exp\left(-\tfrac{1}{2}\int_0^t (\omega(\tau)^2 - 1)\, d\tau\right) d\mu_{0, a, b; t}$$

Using (4.12) and (4.13) we see that

$$(e^{-tL_0})(a, b) = (2\pi t)^{-1/2} e^{-(a-b)^2/2t} e^{-t/2}$$
$$\times \int \exp\left(-\frac{1}{2}\int_0^1 (g(s) + \sqrt{t}\,\alpha(s))^2 t\, ds\right) D\alpha$$

where $g(s) = a(1 - s) + bs$ and α is the Brownian bridge. Now, as noted in the proof of Theorem 5.5, the covariance, $s(1 - u)$ $(s \le u)$, of the Brownian bridge is the integral kernel of the inverse of $-d^2/ds^2$ with vanishing boundary conditions at zero and one. This operator has eigenvalues $(n\pi)^2$ with eigenfunctions

$$\phi_n(s) = \sqrt{2}\sin(n\pi s) \qquad (n = 1, 2, \ldots)$$

If $\lambda_n \equiv (n\pi)^{-1}$, we can write

$$\alpha(s) \doteq \sum_{n=1}^{\infty} \lambda_n x_n \phi_n(s)$$

with the x_n's Gaussian with covariance δ_{nm} (for we need only check that $E([\sum \lambda_n x_n \phi_n(s)][\sum \lambda_m x_m \phi_m(u)]) = \sum \lambda_n^2 \phi_n(s)\phi_n(u) = s(1 - u)$ for $s \le u$). A direct calculation shows that $g(s) = \sum g_n \phi_n(s)$ with

$$g_n = \begin{cases} \sqrt{2}\lambda_n(a + b), & n \text{ odd} \\ \sqrt{2}\lambda_n(a - b), & n \text{ even} \end{cases}$$

Thus

$$\int \exp\left(-\frac{1}{2}\int_0^1 (g(s) + \sqrt{t}\,\alpha(s))^2 t\,ds\right) D\alpha$$

$$= \int \exp\left(-\frac{1}{2}\sum_{n=1}^{\infty}(t^2\lambda_n^2)[x_n + g_n\lambda_n^{-1}t^{-1/2}]^2\right) Dx$$

with $Dx = \prod_{n=1}^{\infty}(2\pi)^{-1/2}\exp(-\frac{1}{2}x_n^2)\,dx_n$. Using (3.18), we find that

$$(e^{-tL_0})(a, b) = (2\pi t)^{-1/2}e^{-t/2}f(t)\exp[-\tfrac{1}{2}A(t)(a^2 + b^2) - B(t)(ab)]$$

where

$$f(t) = \left[\prod_{n=1}^{\infty}(1 + \lambda_n^2 t^2)\right]^{1/2}$$

$$A(t) = t^{-1} + \sum_{n=1}^{\infty} 2\lambda_n^2 t(1 + t^2\lambda_n^2)^{-1}$$

$$B(t) = t^{-1} + \sum_{n=1}^{\infty}(-1)^{n+1}2\lambda_n^2 t(1 + t^2\lambda_n^2)^{-1}$$

The identification with Mehler's formula now follows from the formulas:

$$\sinh x = x\prod_{n=1}^{\infty}(1 + \lambda_n^2 x^2)$$

$$(\sinh x)^{-1} = x^{-1} + \sum_{n=1}^{\infty}(-1)^{n+1}2\lambda_n^2 x(1 + x^2\lambda_n^2)^{-1}$$

$$\frac{\cosh x}{\sinh x} = x^{-1} + \sum_{n=1}^{\infty} 2\lambda_n^2 x(1 + x^2\lambda_n^2)^{-1}$$

(These formulas are the Weierstress–Hadamard factorization of $\sinh x$ and the Mittag–Leffler expansions of $(\sinh x)^{-1}$ and $\coth x$.) If one deals directly with Green's functions for $-d^2/dx^2$ and does not discretize, one can avoid these arcane formulas at the cost of doing slightly more involved integrals.

* * *

We will use P as a generic symbol for a potential obeying the hypothesis of Theorem 6.3 with the additional property that

$$E(P) \equiv \inf\operatorname{spec}(L_0 + P)$$

is a simple eigenvalue with an associated strictly positive eigenvector Ω_P. Most reasonable V's will obey this extra condition (see, e.g., [217]) and, in particular, P can be any polynomial which is bounded from below. Set $\hat{L} = L_0 + P - E(P)$.

Definition **The $P(\phi)_1$-process** is the stochastic process with joint distribution of $q(t_1), \ldots, q(t_n)$ $(t_1 < \cdots < t_n)$:

$$\Omega_P(x_1)\Omega_P(x_n)(e^{-s_1\hat{L}})(x_1, x_2) \cdots (e^{-s_{n-1}\hat{L}})(x_{n-1}, x_n)$$

where $(e^{-s\hat{L}})(a, b)$ is the integral kernel of $e^{-s\hat{L}}$ (which is certainly defined as a measure) and $s_i = t_{i+1} - t_i$. $d\nu_P$ denotes the corresponding measure.

Consistency of the measure follows from $e^{-a\hat{L}}e^{-b\hat{L}} = e^{-(a+b)\hat{L}}$ and $e^{-s\hat{L}}\Omega_P = \Omega_P$. Thus, the oscillator process is just the $P(\phi)_1$-process for $P = 0$ and the definition is just made so that a suitable version of (4.6) holds for $L_0 + P - E(P)$. There are two versions of Feynman–Kac relevant to the $P(\phi)_1$-process. The first is the following.

Theorem 6.7 Let $g_P = \Omega_P(\Omega_0)^{-1}$. Then the measure $d\nu_P$ restricted to $\Sigma_{[a,b]}$, the σ-algebra generated by $\{q(t)\}_{a \le t \le b}$ is absolutely continuous with respect to the oscillator measure, Dq, and

$$d\nu_P(q) \restriction \Sigma_{[a,b]} = F_{[a,b]}(q; P) Dq \restriction \Sigma_{[a,b]} \tag{6.10a}$$

$$F_{[a,b]}(q; P) = g_P(q(a))g_P(q(b))e^{E(P)(b-a)} \exp\left[-\int_a^b P(q(s))\, ds\right] \tag{6.10b}$$

Proof One must show that

$$\int G\, d\nu_P = \int G F_{[a,b]}\, Dq$$

for any function G, of $\{q(t) | a \le t \le b\}$. When $G = h_1(q(a))h_2(q(b))$ for suitable h_1, h_2, this is just Theorem 6.3, and for $G = \prod_{i=1}^n h_i(g(t_i))$, it follows from a simple extension of Theorem 6.3. Since every $\Sigma_{[a,b]}$-measurable G is a limit of sums of such products, the result is proven. ∎

Since we have just seen that all $P(\phi)_1$-processes are locally mutually absolutely continuous, one can ask naturally about global absolute continuity; for convenience we think of all processes on $(\dot{\mathbb{R}})^{(-\infty,\infty)}$, although one can realize most of them on $\mathscr{S}'(\mathbb{R})$ or some other similar space.

Theorem 6.8 ([223]) If $P_1 - P_2$ is nonconstant, then $d\nu_{P_1}$ and $d\nu_{P\delta}$ are mutually singular as measures on $(\dot{\mathbb{R}})^{(-\infty,\infty)}$.

Proof We first claim that $d\nu_{P_1}$ and $d\nu_{P_2}$ are distinct on functions of $q(0)$. For the distribution of $q(0)$ is $\Omega_P^2\,dx$ which therefore determines Ω_P and so $P - E(P) = \frac{1}{2}\Omega_P^{-1}(\Omega_P'') - \frac{1}{2}x^2 + \frac{1}{2}$.

Next we claim that every $P(\phi)_1$-process restricted to $\{q(n)\,|\,n$ an integer$\}$ is ergodic with respect to the map $U: q(i) \to q(i+1)$; see Example 2 preceding Theorem 3.7.

Now pick some set $A \subset \mathbb{R}$ with $d\nu_{P_1}(q(0) \in A) \equiv a_1 \neq a_2 \equiv d\nu_{P_2}(q(0) \in A)$. Then, by Theorem 3.7,

$$\frac{1}{n}\sum_{i=1}^{n} q(i) \to a_j$$

almost everywhere with respect to $d\nu_{P_j}$. Thus, we have disjoint sets which support the $d\nu_{P_j}$. ▮

We will see later how to distinguish the $d\nu_{P_j}$ more explicitly. As a final result on $P(\phi)_1$-processes note:

Theorem 6.9 For any G

$$\int G\,d\nu_P = \lim_{T\to\infty} Z_T^{-1} \int G \exp\left(-\int_{-T}^{T} P(q(s))\,ds\right) Dq \tag{6.11a}$$

where

$$Z_T = \int \exp\left(-\int_{-T}^{T} P(q(s))\,ds\right) Dq \tag{6.11b}$$

Proof It suffices to consider G of the form $\prod_{i=1}^{n} f_i(q(t_i))$ with $t_1 < \cdots < t_n$. In that case for T with $-T < t_1 < t_n < T$:

$$\int G \exp\left(-\int_{-T}^{T} P(q(s))\,ds\right) Dq = (\Omega_0, e^{-s_0 L} f_1(q) e^{-s_1 L} \cdots f_n(q) e^{-s_n L}\Omega_0)$$

where $L = L_0 + P$ and $s_0 = T + t_1$, $s_1 = t_2 - t_1, \ldots, s_{n-1} = t_n - t_{n+1}$, $s_n = T - t_n$. (1.10) completes the proof. ▮

Equations (6.11) give the study of the $P(\phi)_1$-process something of a statistical mechanical flavor; see Sections 12 and 19.

* * *

In these notes, we will be primarily interested in the direction of the Feynman–Kac formula that says one can study e^{-tH} using Wiener integrals. However, one can turn this analysis around and use e^{-tH} to study Brownian motion. The following example (suggested to me by Donsker) illustrates this nicely.

For each $t > 0$, consider the function of the Brownian path b, given by

$$U_t(b) = t^{-1}|\{s\,|\,0 \le s \le t; b(s) \ge 0\}| \tag{6.12}$$

where $|\cdot|$ is Lebesgue measure; i.e., $U_t(b)$ is the fraction of time that the path is to the right of zero. Think of U_t as random variables. Since $\{t^{-1/2}b(st)\}_{0\le s\le 1}$ has the same joint distributions as $\{b(s)\}_{0\le s\le 1}$ the U_t's all have the same probability distribution $d\mu(x)$ which we want to compute. Let V be the function which is zero (respectively, one) for $x < 0$ (respectively, $x \ge 0$). Clearly,

$$\exp\left[-\int_0^t V(b(s))\,ds\right] = \exp[-tU_t(b)]$$

so by the Feynman–Kac formula in the form (6.7) and Theorem 6.5:

$$\int e^{-tU_t(b)}f(b(t))\,Db = \int e^{-tH}(x, 0)f(x)\,dx$$

where $H = -\frac{1}{2}(d^2/dx^2) + V$. On both sides of the last expression, we can choose $f_n \in L^2$ converging monotonically upwards to one. Since $e^{-tH}(x, 0)$ and $e^{-tU_t(b)}$ are positive, we see that

$$\int e^{-ty}\,d\mu(y) = \int e^{-tH}(x, 0)\,dx$$

Multiplying both sides by e^{-ta} and integrating dt (using Fubini's theorem to interchange orders), we find that

$$\int (y + a)^{-1}\,d\mu(y) = \int (H + a)^{-1}(x, 0)\,dx \tag{6.13}$$

Now, one can easily compute $(H + a)^{-1}(x, 0) = g(x; a)$ explicitly since it satisfies $-\frac{1}{2}g'' + Vg + ag = \delta$ with the boundary conditions $g \to 0$ at $\pm\infty$. The net result is

$$(H + a)^{-1}(x, 0) = \begin{cases} \alpha(a)e^{x\sqrt{2a}} & (x \le 0) \\ \alpha(a)e^{-x\sqrt{2a+2}} & (x \ge 0) \end{cases}$$

with $\alpha(a) = \sqrt{2}/(\sqrt{a} + \sqrt{a + 1})$. Thus, integrating on the right-hand side of (6.13), we see that

$$\int (y + a)^{-1}\,d\mu(y) = \frac{1}{\sqrt{a(a + 1)}} \equiv F(a)$$

Now $F(a)$ is analytic in the plane cut from -1 to 0 with $F(a) \to 0$ at infinity. The Cauchy integral formula thus implies that

$$\begin{aligned} F(a) &= \pi^{-1} \int_{-1}^{0} (x - a)^{-1} \operatorname{Im}(F(x + i0))\, dx \\ &= \pi^{-1} \int_{0}^{1} (y + a)^{-1} [\sqrt{y(1 - y)}]^{-1}\, dy \end{aligned}$$

since F is pure imaginary for $-1 < x < 0$. Thus, we have proven the following.

Theorem 6.10 The random variable U_t of (6.12) has probability distribution

$$d\mu(x) = \begin{cases} \pi^{-1}(x(1 - x))^{-1/2}\, dx, & 0 < x < 1 \\ 0, & \text{otherwise} \end{cases}$$

Since $\int_0^\alpha d\mu(x) = 2\pi^{-1} \text{Arc sin} \sqrt{\alpha}$, this result is often called the **Arcsin law**.

7. Regularity and Recurrence Properties—2

In this section, we continue our study of the regularity of Brownian paths and study certain global properties. The first result we will prove is Khintchine's [157] famous law of the iterated logarithm:

Theorem 7.1 For one-dimensional Brownian motion

$$\overline{\lim_{t \downarrow 0}} \frac{b(t)}{[2t \log_2(t^{-1})]^{1/2}} = 1 \tag{7.1}$$

with probability one where $\log_2(y) = \ln(\ln y)$.

Before proving this result, we state a number of results which follow from (7.1). Since $b(t + s) - b(s) \doteq b(t)$ and since b and q and any $P(\phi)_1$-process are locally absolutely continuous (Theorems 5.5 and 6.7):

Theorem 7.2 For one-dimensional Brownian motion, for any $P(\phi)_1$-process and any fixed s:

$$\overline{\lim_{t \downarrow 0}} \frac{b(s \pm t) - b(s)}{[2t \log_2(t^{-1})]^{1/2}} = 1$$

$$\overline{\lim_{t \downarrow 0}} \frac{q(s \pm t) - q(s)}{[2t \log_2(t^{-1})]^{1/2}} = 1$$

with probability one.

The result for $b(s-t)-b(s)$ follows from that for $b(s+t)-b(s)$ if we notice that $b(2s-t)-b(2s) \equiv \bar{b}(t)$ has $\bar{b}(t) \doteq b(t)$ for $0 \le t \le 2s$.

The equivalences (5.6) and (5.7) immediately give the following.

Theorem 7.3 For one-dimensional Brownian motion

$$\overline{\lim_{t\to\infty}} \frac{b(t)}{[2t \log_2(t)]^{1/2}} = 1 \tag{7.2}$$

with probability one.

Theorem 7.4 For the oscillator process:

$$\overline{\lim_{t\to\infty}} \frac{q(t)}{(\ln t)^{1/2}} = 1 \tag{7.3}$$

with probability one.

To get (7.3), we use the fact that $\ln(2t)/\ln(t) \to 1$ as $t \to \infty$. (7.2) and (7.3) show explicitly that $Db \restriction \Sigma_{(a,\infty)}$ and $Dq \restriction \Sigma_{(a,\infty)}$ are mutually singular measures for each $a > 0$. One can also distinguish various kinds of $P(\phi)_1$-processes by similar means; e.g., if $P(x) = x^{2m}$, then

$$\overline{\lim_{n\to\infty}} \frac{q(n)}{(\ln n)^{1/(m+1)}} = \alpha_m$$

(with α_m a nonzero constant) for almost every $P(\phi)_1$-path. See [223] where "higher order oscillations" are discussed; e.g., for the oscillator process an explicit $f(t)$ is exhibited with

$$\overline{\lim}\, f(t)[q(t) - (\ln t)^{1/2}] = 1$$

(earlier references to this subject may also be found in [223]). Now $b(t) \doteq -b(t)$ and $q(t) \doteq -q(t)$. Thus the following theorem holds.

Theorem 7.5 In (7.1), (7.2), and (7.3) b (respectively, q) may be replaced by $-b$ (respectively, $-q$). Equivalently, $\overline{\lim}$ and "$=1$" may be replaced by $\underline{\lim}$ and "$=-1$."

These results are another indication of the roughness of Brownian motion. They imply that $b(t) = 0$ for infinitely many t's near zero, since $b(t)$ is positive and negative infinitely often. In fact, one can show that, with probability one, $\{t \mid b(t) = 0\}$ is a nonempty perfect set with Hausdorff dimension $\frac{1}{2}$ (see [136]), and, in particular, it is uncountable. In multidimensions, one has the following.

Theorem 7.6 For ν-dimensional Brownian motion

$$\overline{\lim_{t\downarrow 0}} \frac{|\mathbf{b}(t)|}{[2t \log_2(t^{-1})]^{1/2}} = 1 \tag{7.4}$$

$$\overline{\lim_{t\to\infty}} \frac{|\mathbf{b}(t)|}{[2t \log_2(t)]^{1/2}} = 1 \tag{7.5}$$

with probability one.

Proof Let $\{\mathbf{e}_n\}_{n=1}^{\infty}$ be a dense set of vectors on the unit sphere. By (7.1)

$$\overline{\lim_{t\downarrow 0}} \frac{\mathbf{b}(t)\cdot \mathbf{e}_n}{[2t \log_2(t^{-1})]^{1/2}} = 1$$

for each n with probability one. Since $|\mathbf{b}(t)| \geq |\mathbf{b}(t)\cdot \mathbf{e}_n|$, the $\overline{\lim}$ in (7.4) is greater than or equal to one. On the other hand, if the $\overline{\lim}$ were some $a > 1$, then by a compactness argument $\overline{\lim}\, \mathbf{b}(t)\cdot \mathbf{e}_n \geq \frac{1}{2}(1 + a)$ for some $\mathbf{e}_n$. This proves (7.4). The proof of (7.5) is similar. ∎

Remarks 1. One curious feature of the ν-dimensional case is the following: In one dimension, the set of limit points of $b(t)/[2t \log_2(t^{-1})]^{1/2}$ as $t \downarrow 0$ is clearly $[-1, 1]$ since b is continuous and we know that the $\overline{\lim}$ ($\underline{\lim}$) is 1 (-1). One can ask about the limit points in ν-dimensions. Since the components are independent, one would naively expect that the set of limit points is just the ν-fold product of $[-1, 1]$. But this cannot be since (7.4) implies that the limit points lie in the unit ball! In fact the entire unit ball occurs.

2. The law of the iterated logarithm has been proven for many Gaussian processes; see, e.g., [196a].

We now turn to the proof of (7.1). The key to understanding why (7.1) holds comes from (7.3) and its comparison to the example following Theorem 3.3: If the $q(t)$ were independent, then $\lim q(n)/\sqrt{\ln n}$ would be one. The point is that the $q(n)$ are "almost independent" so that this is still true. But $q(t)$ for t near n is not significantly different from $q(n)$. The "easy" part of the proof of Theorem 7.1 just follows this intuition; note that by (5.6) and (5.9), $q(n)$ and $b(e^{-2n})$ are related:

Lemma 7.7 For one-dimensional Brownian motion

$$\overline{\lim_{n\to\infty}}\, b(e^{-n})/[2e^{-n} \log n]^{1/2} = 1$$

with probability one and, in particular, the $\overline{\lim}$ in (7.1) is greater than or equal to one with probability one.

Proof Let $x_n = e^{+n/2}b(e^{-n})$. Then, the x_n are Gaussian with covariance $e^{-|n-m|/2}$. Let A_n be an event depending on x_n. Then, by (3.12):

$$|E(A_n \cap A_m) - E(A_n)E(A_m)| \le e^{-|n-m|/2}E(A_n)^{1/2}E(A_m)^{1/2}$$

Using the strong Borel–Cantelli lemma, Theorem 3.2, one can now mimic the proof of Theorem 3.3 and the example following to see that $\overline{\lim}\, x_n/(2 \ln n)^{1/2} = 1$ which completes the proof. ∎

The hard part of proving (7.1) involves showing that sampling the points $t_n = e^{-n}$ does not result in a smaller $\overline{\lim}$ than sampling all the times. We will give two different proofs. The first gets control with martingale inequalities:

First proof of $\overline{\lim} \le 1$ (following [183]) In the inner product $(\delta_t, \delta_s) = \min(t, s)$, the projection of δ_t onto $[\delta_s | s \le u]$ with $u < t$ is δ_u. It follows from (3.9) that

$$E(:e^{\alpha b(t)}: | b(s);\, s \le u) = :e^{\alpha b(u)}:$$

i.e., $:e^{\alpha b(t)}: = e^{\alpha b(t) - \alpha^2 t/2}$ is a positive martingale. Thus by Doob's inequality (3.8)

$$\begin{aligned} E\left(\max_{s \le t}\left[b(s) - \frac{\alpha s}{2}\right] > \beta\right) &= E\left(\max_{s \le t} :e^{\alpha b(s)}: > e^{\alpha\beta}\right) \\ &\le e^{-\alpha\beta}E(:e^{\alpha b(t)}:) \\ &= e^{-\alpha\beta} \end{aligned}$$

Let $h(t) = [2t \log_2(t^{-1})]^{1/2}$ and choose $0 < \theta < 1$ and $0 < \delta < 1$. Let $t_n = \theta^{n-1}$ and choose $\alpha_n = (1 + \delta)\theta^{-n}h(\theta^n)$, $\beta_n = h(\theta^n)/2$ so $\alpha_n\beta_n = (1 + \delta)\log_2 \theta^{-n}$ and $e^{-\alpha_n\beta_n} = \exp(-(1 + \delta)[\ln n + \log_2 \theta^{-1}]) = cn^{-1-\delta}$. Thus by the first Borel–Cantelli lemma, for $n \ge N(b)$ (with $N(b) < \infty$ almost everywhere),

$$\max_{s \le t_n}\left[b(s) - \frac{\alpha_n s}{2}\right] \le \beta_n$$

Let $t \le \theta^{N(b)-1}$ and suppose that $t_{n+1} < t \le t_n \le e^{-e}$. Then

$$\begin{aligned} b(t) &\le \max_{s \le t_n} b(s) \le \beta_n + \frac{1}{2}\alpha_n\theta^{n-1} \\ &= \left[\frac{1}{2} + \frac{1 + \delta}{2\theta}\right]h(t_{n+1}) \le \left[\frac{1 + \delta}{2\theta} + \frac{1}{2}\right]h(t) \end{aligned}$$

since $h(t)$ is monotone increasing for $t \leq e^{-e}$. Thus

$$\overline{\lim} \left(\frac{b(t)}{h(t)}\right) \leq \frac{1}{2} + \frac{1 + \delta}{2\theta}$$

Letting $\delta \downarrow 0$ and $\theta \uparrow 1$, the result follows. ∎

Remark The above implies that $:b(t)^n:$ is a martingale where $:x^n: = (d^n:e^{\alpha x}:/d\alpha^n)|_{\alpha=0}$ for Gaussian x; see [258] for discussion of $:x^n:$.

The second proof depends on the following remarkable fact whose proof we defer.

$$E\left(\max_{0 \leq s \leq t} b(s) \geq \lambda\right) = 2E(b(t) \geq \lambda) \qquad \lambda \geq 0 \tag{7.6}$$

Second proof of $\overline{\lim} \leq 1$ (following [86]) By (7.6) and (3.4)

$$E\left(\sup_{0 \leq t \leq t_n} b(t) \geq x_n\sqrt{t_n}\right) = \sqrt{2/\pi} \int_{x_n}^{\infty} e^{-(1/2)y^2}\, dy \leq \sqrt{2/\pi}\, \frac{e^{-x_n^2/2}}{x_n}$$

Pick $\delta > 0$ and then $\theta \in (0, 1)$ with $(1 + \delta)^2\theta > 1$, $t_n = \theta^n$, and $x_n = (1 + \delta)h(t_{n+1})/\sqrt{t_n}$ where $h(t) = [2t \log_2(t^{-1})]^{1/2}$ so that

$$x_n = (1 + \delta)[2\theta \ln n + c]^{1/2}$$

and thus since $(1 + \delta)^2\theta > 1$,

$$\frac{e^{-x_n^2/2}}{x_n} \leq cn^{-\alpha}, \qquad \alpha > 1$$

Let $A_n = \{b(t) > (1 + \delta)h(t) | \text{some } t \in [t_{n+1}, t_n]\}$. Then $b \in A_n$ and n so large that h is monotone on $[0, t_n]$ implies $\sup_{0 \leq t \leq t_n} b(t) \geq (1 + \delta)h(t_{n+1}) \geq x_n\sqrt{t_n}$, so by the first Borel–Cantelli lemma, $b \notin A_n$ for all n sufficiently large; i.e., with probability one

$$\overline{\lim_{t \downarrow 0}} \frac{b(t)}{h(t)} \leq 1 + \delta$$

Taking $\delta \downarrow 0$, the result is proven. ∎

Notice that the full power of (7.6) was not used; all that was needed was

$$E\left(\max_{0 \leq s \leq t} b(s) \geq \lambda\right) \leq 2E(b(t) \geq \lambda)$$

This is just Lévy's inequality, Theorem 3.6.5, in this situation: the hypotheses of that theorem are applicable once one discretizes time since a product of Gaussians is clearly invariant under sign changes. More significantly, Lévy's inequality is applicable to n-dimensional Brownian motion yielding

$$E\left(\max_{0 \le s \le t} |\mathbf{b}(s)| \ge \lambda\right) \le 2E(|\mathbf{b}(t)| \ge \lambda) \tag{7.6$'$}$$

for n-dimensional Brownian motion. Of course the fact that equality holds in (7.6) is interesting.

We turn now to proving (7.6). A second proof will be given below following Lemma 7.10. To understand why it should be true, consider an elementary random walk X_n. The analog is (N, k positive integers)

$$E\left(\max_{0 \le n \le N} X_n \ge k\right) = 2E(X_N > k) + E(X_N = k) \tag{7.7}$$

(7.7) comes from the fact that if $\max_{0 \le n \le N} X_n \ge k$, it is equally likely that also $X_N > k$ or also $X_N < k$, since we can reflect the path about the first n where $X_n = k$. This idea is easy to make rigorous. One lets

$$A_j = \{X | X_j = k, X_0, \dots, X_{j-1} < k\}$$

and notes that

$$E((X_N - X_j) > 0, A_j) = E((X_N - X_j) < 0, A_j)$$

since A_j is independent of $X_N - X_j$, and $X_N - X_j$ is even. Thus, summing over j

$$E\left(X_N > k, \max_{0 \le n \le N} X_n \ge k\right) = E\left(X_N < k, \max_{0 \le n \le N} X_N \ge k\right)$$

which easily yields (7.7). The intuition is the same for Brownian motion but the fact that t is continuous makes it difficult to find analogs of A_j. The key to finding a suitable way is to discretize. It is useful to introduce the following.

Definition $\mathscr{B}_t$ is the σ-field generated by $\{b(s) | s \le t\}$; $\mathscr{B}_\infty = \bigcup_{t<\infty} \mathscr{B}_t$.

Definition A **stopping time** is a function τ with values in $[0, \infty]$ of the Brownian path with the property that $\{b | \tau(b) < t\} \in \mathscr{B}_t$ and so that $E(\tau < \infty) > 0$.

Roughly speaking, a stopping time is a function given by "$\tau(b)$ is the smallest time, t, so that ... has occurred before t."

Examples 1. Fix $\lambda > 0$. Let $\tau(b) = \inf\{t \leq 1 \mid b(t) \geq \lambda\}$. If $\max_{0 \leq t \leq 1} b(t) < \lambda$, set $\tau(b) = 1$. Then, for $t < 1$:

$$\{b \mid \tau(b) \leq t\} = \bigcap_{m=1}^{\infty} \left[\bigcup_{n=1}^{\infty} \bigcup_{\{k \mid k/2^n \leq t\}} \left\{ b \,\middle|\, b\left(\frac{k}{2^n}\right) \geq \lambda - \frac{1}{m} \right\} \right]$$

is obviously in $\mathscr{B}_t$. As a result $\{b \mid \tau(b) < t\} = \bigcup_n \{b \mid \tau(b) \leq t - 1/n\}$ is in $\mathscr{B}_t$. Thus τ is a stopping time.

2. $\tau(b) = \inf\{t \mid b(t) = 0, \ |b(s)| \geq 1 \text{ for some } s \in [0, t]\}$. As above, this "first return time" is a stopping time.

3. Given any stopping time τ, define its **discretization** $\tau^{(n)}$ by

$$\tau^{(n)}(b) = k/2^n \qquad \text{if} \quad (k-1)/2^n \leq \tau(b) < k/2^n$$

Then, for $t \in (k/2^n, (k+1)/2^n]$:

$$\{b \mid \tau^{(n)}(b) < t\} = \{b \mid \tau(b) < k/2^n\} \in \mathscr{B}_t$$

Thus, $\tau^{(n)}$ is a stopping time. Notice also, that

$$\{b \mid \tau^{(n)}(b) = k/2^n\} = \{b \mid \tau(b) < k/2^n\} \setminus \{b \mid \tau(b) < (k-1)/2^n\}$$

is $\mathscr{B}_{k/2^n}$-measurable.

We can now prove (7.6).

Theorem 7.8 For Brownian motion and $\lambda \geq 0$,

$$E\left(\max_{0 \leq s \leq t} b(s) \geq \lambda\right) = 2E(b(t) \geq \lambda) \tag{7.6}$$

Proof Since $b(s) \doteq t^{1/2} b(st^{-1})$, we can suppose that $t = 1$. Let τ be the stopping time of Example 1 above and let $\tau^{(n)}$ be its discretization modified so that $\tau^{(n)}(b) = 1$ if $\tau(b) \geq (2^n - 1)/2^n$. Let $f \in C(\mathbb{R})$. We claim that

$$E(f(b(1) - b(\tau^{(n)}))) = E(f(b(\tau^{(n)}) - b(1))) \tag{7.8}$$

For introducing the symbol

$$E(f; A) = \int_A f \, Db$$

we have that

$$\begin{aligned}
E(f(b(1)-b(\tau^{(n)}))) &= \sum_{k=0}^{2^n} E\left(f(b(1)-b(\tau^{(n)}));\ \tau^{(n)}=\frac{k}{2^n}\right)\\
&= \sum_{k=0}^{2^n} E\left(f\left(b(1)-b\left(\frac{k}{2^n}\right)\right);\ \tau^{(n)}=\frac{k}{2^n}\right)\\
&= \sum_{k=0}^{2^n} E\left(f\left(b(1)-b\left(\frac{k}{2^n}\right)\right)\right)E\left(\tau^{(n)}=\frac{k}{2^n}\right)\\
&= \sum_{k=0}^{2^n} E\left(f\left(b\left(\frac{k}{2^n}\right)-b(1)\right)\right)E\left(\tau^{(n)}=\frac{k}{2^n}\right)\\
&= E(f(b(\tau^{(n)})-b(1)))
\end{aligned}$$

where we have used the facts that $b(1)-b(k/2^n)$ is independent of $\mathscr{B}_{k/2^n}$, that $(\tau^{(n)}=k/2^n)\in\mathscr{B}_{k/2^n}$ and that $b(1)-b(k/2^n)$ is even. As $n\to\infty$, $\tau^{(n)}\downarrow\tau$ so $b(\tau^{(n)})\to b(\tau)$ by the continuity of paths. Thus, by the dominated convergence theorem:

$$E(f(b(1)-b(\tau)))=E(f(b(\tau)-b(1)))$$

or

$$\begin{aligned}
&E(f(b(1)-\lambda);\ \tau<1)+f(0)E(\tau=1)\\
&\quad=E(f(\lambda-b(1));\ \tau<1)+f(0)E(\tau=1)
\end{aligned}$$

The continuity of the distribution of $b(1)$ then allows us to let f be the characteristic function of $(0,\infty)$. Thus,

$$E(b(1)>\lambda;\ \tau<1)=E(b(1)<\lambda;\ \tau<1)$$

Since

$$E(b(1)=\lambda;\ \tau<1)+E(\tau=1,\ b(1)=\lambda)=E(b(1)=\lambda)=0$$

we have

$$E(\tau<1)=2E(b(1)>\lambda;\ \tau<1)=2E(b(1)>\lambda)\quad\blacksquare$$

Stopping times will be useful in our study of recurrence; especially useful is Theorem 7.9 due to Dynkin [67] and Hunt [130] which says that "Brownian motion starts afresh at stopping times." As a preliminary, we need the following.

Definition Let τ be a stopping time. Then

$$\mathscr{B}_{\tau^+}\equiv\{B\in\mathscr{B}_\infty\,|\,B\cap(\tau<t)\in\mathscr{B}_t\text{ for all }t\}$$

Examples 1. τ is $\mathscr{B}_{\tau^+}$-measurable as is $b(\tau)$ if we set $b(\tau) = \infty$ when $\tau = \infty$. Also $\int_0^\tau f(s)F(b(s))\,ds$ is $\mathscr{B}_{\tau^+}$-measurable.

2. If t_0 is a fixed positive number, then $\tau(b) \equiv t_0$ is a stopping time and $\mathscr{B}_{\tau^+} = \bigcap_{s \geq t_0} \mathscr{B}_s$ explaining the notation.

Definition Let B be an event with $E(B) > 0$. Then the measure μ_B given by

$$\mu_B(A) = \mu(B \cap A)/\mu(B)$$

is called the ***B* conditional measure**. Given n random variables $f_1, \ldots, f_n$, their joint distribution with respect to μ_B is called their **joint distribution conditional on *B***.

Theorem 7.9 (Dynkin–Hunt) Let τ be a stopping time for Brownian motion. Let $\tilde{b}(t) \equiv b(t + \tau) - b(\tau)$. Then, conditional on $\tau < \infty$, $\tilde{b}(t)$ has identical joint distribution to $b(t)$. Moreover, the $\tilde{b}(t)$ are independent, conditional on $\tau < \infty$, of $\mathscr{B}_{\tau^+}$.

Proof The theorem asserts that

$$E(F(\tilde{b}(t)); \tau < \infty) = E(F(b(t)))E(\tau < \infty)$$

and for $B \in \mathscr{B}_{\tau^+}$:

$$E(F(\tilde{b}(t)); B \cap (\tau < \infty))E(\tau < \infty) = E(B \cap (\tau < \infty))E(F(\tilde{b}(t)); \tau < \infty)$$

This pair of statements is clearly implied by

$$E(F(\tilde{b}(t)); (\tau < \infty) \cap B) = E(F(b(t)))E((\tau < \infty) \cap B) \tag{7.9}$$

It suffices to prove (7.9) when

$$F(b(t)) = G(b(t_1), \ldots, b(t_n))$$

with $G \in C(\mathbb{R}^n)$. Define

$$g(t) = G(b(t_1 + t) - b(t), \ldots, b(t_n + t) - b(t))$$

Then

$$\begin{aligned} E(F(\tilde{b}(t)); (\tau < \infty) \cap B) &= E(g(\tau); (\tau < \infty) \cap B) \\ &= \lim_{n \to \infty} E(g(\tau^{(n)}); (\tau^{(n)} < \infty) \cap B) \\ &= \lim_{n \to \infty} \left(\sum_{k=0}^{\infty} E\left(g\left(\frac{k}{2^n}\right); \left(\tau^{(n)} = \frac{k}{2^n}\right) \cap B\right)\right) \\ &= \lim_{n \to \infty} \left(\sum_{k=0}^{\infty} E(g(0))E\left(\left(\tau^{(n)} = \frac{k}{2^n}\right) \cap B\right)\right) \\ &= E(g(0))E((\tau < \infty) \cap B) \end{aligned} \tag{7.10}$$

where (7.10) follows from the facts that

$$(\tau^{(n)} = k/2^n) \cap B = ((\tau < k/2^n) \cap B) \backslash (\tau < (k-1)/2^n) \in \mathscr{B}_{k/2^n}$$

since $B \in \mathscr{B}_{\tau^+}$ and τ is a stopping time and that $b(t + k/2^n) - b(k/2^n)$ is independent of $\mathscr{B}_{k/2^n}$ and identically distributed to $b(t)$. ▮

* * *

We next want to study the question of recurrence for Brownian paths. In $\nu = 1$ dimension, we know that $\overline{\lim}\, b(t) = \infty$ while $\underline{\lim}\, b(t) = -\infty$. It follows that, with probability one, $b(t)$ returns infinitely often to any given set, since it must sweep through the whole real line infinitely often. We want to examine the analogous question in detail when $\nu \geq 2$. We will give two presentations of the basic facts, one after Lemma 7.21 using stopping times and harmonic functions; the major tool in our first approach will be the use of Dirichlet operators:

Definition Let S be a closed subset of $\mathbb{R}^\nu$. Consider the quadratic form with form domain $C_0^\infty(\mathbb{R}^\nu \backslash S)$ and form

$$(f, H_{\mathrm{D};S} f) = \tfrac{1}{2}(f, (-\Delta) f)$$

The form closure of this form defines a self-adjoint operator $H_{\mathrm{D};S}$ on $L^2(\mathbb{R}^\nu \backslash S)$. We let $P_{\mathrm{D};S}(\mathbf{x}, \mathbf{y}; t)$ and $G_{\mathrm{D};S}(\mathbf{x}, \mathbf{y}; \alpha)$ denote the kernels of $e^{-tH_{\mathrm{D};S}}$ and $(H_{\mathrm{D};S} + \alpha)^{-1}$. (A priori these are only distributions, but using elliptic regularity and the form of their equation, one shows they are C^∞ for $\mathbf{x} \neq \mathbf{y}$ in $\mathbb{R}^\nu \backslash S$, and of course, they are positive; see, e.g., [121, 217].) We let $G_{\mathrm{D};S}(\mathbf{x}, \mathbf{y}; 1) \equiv G_{\mathrm{D};S}(\mathbf{x}, \mathbf{y})$.

For S's with sufficiently large interiors, we expect that (χ_S is the characteristic function of S)

$$(H_0 + \mu\chi_S + z)^{-1} \to (H_{\mathrm{D};S} + z)^{-1} \tag{7.11}$$

strongly as $\mu \to \infty$, for all $z \notin [0, \infty)$. We call S **regular** if $P_{\mathrm{D};S}(\mathbf{x}, \mathbf{y}; t)$ is continuous on all of $\mathbb{R}^\nu \times \mathbb{R}^\nu$ (set equal to zero if either $\mathbf{x}$ or $\mathbf{y} \in S$) and if (7.11) converges *pointwise* for the integral kernels. When S is a nice set like the exterior of a ball, it is easy to see that $H_{\mathrm{D};S}$ is the classical Dirichlet operator and that S is regular.

Lemma 7.10 Let S be a regular closed set so that the Lebesgue measure of ∂S is zero and $\mathbf{0} \notin S$. Then

$$E(\mathbf{b}(s) \notin S^{\mathrm{int}} \,|\, 0 \leq s \leq t) = \int P_{\mathrm{D};S}(\mathbf{x}, \mathbf{0}; t)\, d\mathbf{x} \tag{7.12}$$

Proof Let χ be the characteristic function of S^{int}. Then $\int_0^t \chi(\mathbf{b}(s))\,ds > 0$ if and only if $\mathbf{b}(s) \in S^{\text{int}}$ for some $s \in (0, t]$ since paths are continuous. Let g_n be a sequence in C_0^∞ with $g_n(\mathbf{x}) \nearrow 1$ for each $\mathbf{x}$. Then

$$\begin{aligned}
\text{l.h.s. of (7.12)} &= \lim_{n\to\infty} E(g_n(\mathbf{b}(t));\, \mathbf{b}(s) \notin S^{\text{int}}, 0 \le s \le t) \\
&= \lim_{n\to\infty} \lim_{\lambda\to\infty} E\left(\exp\left[-\lambda \int_0^t \chi(\mathbf{b}(s))\,ds\right] g_n(\mathbf{b}(t))\right) \\
&= \lim_{n\to\infty} \lim_{\lambda\to\infty} (\exp[-t(H_0 + \lambda\chi)]g_n)(\mathbf{0}) \\
&= \lim_{n\to\infty} \int P_{\mathrm{D};S}(\mathbf{0}, \mathbf{y}; t) g_n(\mathbf{y})\,dy \\
&= \text{r.h.s. of (7.12)}
\end{aligned}$$

In the first two and last steps above, we use the monotone convergence theorem. In the third step, we use Theorem 6.5 and we use regularity in the fourth step. ∎

Example We can now give an alternate proof of (7.6). For let $S = [\lambda, \infty) \subset R$. Then, by the method of images:

$$P_{\mathrm{D};S}(x, y; t) = \begin{cases} P(x, y; t) - P(2\lambda - x, y; t), & x, y < \lambda \\ 0, & x < \lambda \le y \end{cases}$$

Thus, by (7.12):

$$\begin{aligned}
E\left(\max_{0\le s\le t} b(s) \le \lambda\right) &= \int P_{\mathrm{D};S}(x, 0; t)\,dx \\
&= \int_{-\infty}^{\lambda} [P(x, 0; t) - P(2\lambda - x, 0; t)]\,dx \\
&= E(b(t) \le \lambda) - E(2\lambda - b(t) \le \lambda) \\
&= E(b(t) \le \lambda) - E(b(t) \le -\lambda) \\
&= E(-\lambda \le b(t) \le \lambda) = 1 - 2E(b(t) \ge \lambda)
\end{aligned}$$

which is (7.6).

Theorem 7.11 Let S be a ball of radius r_0 about a point $\mathbf{y}$ with $|y| = R > r_0$. Then, for ν-dimensional Brownian motion:

$$E(\mathbf{b}(t) \in S \mid \text{some } t) = \begin{cases} 1, & \nu = 1, 2 \\ (r_0/R)^{\nu-2}, & \nu \ge 3 \end{cases}$$

Remarks 1. The fact that for R fixed, and r_0 small, this "hitting probability" is $O(r_0^{\nu-2})$ not $O(r_0^{\nu-1})$ (= cross-section) is another expression of the "wigglyness" of Brownian paths; see Section 22.

2. An alternative proof of this theorem is given in the Aside following Lemma 7.21. That proof uses hitting probability ideas in place of Green's functions.

Proof Let $\rho(s) = E(\mathbf{b}(t) \notin S \,|\, 0 \le t \le s)$. Since $\rho(s) \to E(\mathbf{b}(t) \notin S \,|\, \text{all } t)$, as $s \to \infty$, we have that

$$\begin{aligned} E(\mathbf{b}(t) \notin S \,|\, \text{all } t) &= \lim_{\alpha \downarrow 0} \int_0^\infty \alpha e^{-\alpha s} \rho(s)\, ds \\ &= \lim_{\alpha \downarrow 0} \int_{\mathbb{R}^\nu} \int_0^\infty \alpha e^{-\alpha s} P_{\mathrm{D};S}(\mathbf{0}, \mathbf{x}; s)\, dx\, ds \\ &= \lim_{\alpha \downarrow 0} \int_{\mathbb{R}^\nu} \alpha G_{\mathrm{D};S}(\mathbf{0}, \mathbf{x}; \alpha)\, dx \end{aligned}$$

Now, by scaling

$$G_{\mathrm{D};S}(\mathbf{x}, \mathbf{y}; \alpha) = \alpha^{(\nu-2)/2} G_{\mathrm{D};\alpha^{1/2}S}(\alpha^{1/2}\mathbf{x}, \alpha^{1/2}\mathbf{y}; 1)$$

So using translation covariance, we see that

$$E(\mathbf{b}(t) \notin S \,|\, \text{all } t) = \lim_{\alpha \downarrow 0} \int G_{\mathrm{D};S(\alpha)}(\mathbf{x}, \mathbf{y}_\alpha)\, dx \tag{7.13}$$

where $\mathbf{y}_\alpha = \alpha^{1/2}\mathbf{y}$ and $S(\alpha)$ is the sphere of radius $\alpha^{1/2} r_0$ *about* 0. If we replace S with the empty set, we see for comparison that

$$1 = \lim_{\alpha \downarrow 0} \int G_0(\mathbf{x}, \mathbf{y}_\alpha)\, dx \tag{7.14}$$

For $\nu = 1$, one easily sees that the right-hand side of (7.13) goes pointwise to zero (use the explicit formula for G_{D}) and so the integral goes to zero since $G_{\mathrm{D}} \le G_0$. For $\nu \ge 2$, G_{D} and G_0 have spherical harmonic expansions, and

$$\int d\Omega_x\, G \ldots (\mathbf{x}, \mathbf{y}) = g \ldots (|x|, |y|)$$

where $d\Omega_x$ is the measure on the sphere and

$$g_0(r, r') = r^{-1/2(\nu-1)} (r')^{-1/2(\nu-1)} A_\nu(r_<) B_\nu(r_>)$$

where $r_< = \min(r, r')$; $r_> = \max(r, r')$ and A_ν (respectively, B_ν) is a suitably normalized solution of

$$-u'' + \tfrac{1}{4} r^{-2}(\nu - 1)(\nu - 3)u + 2u = 0$$

which is regular at $r = 0$ (respectively, $O(e^{-r\sqrt{2}})$ at ∞). If $g_{D;a}$ comes from the Green's function with a Dirichlet condition on the sphere of radius a, then

$$g_{D;a}(r, r') = r^{-1/2(\nu-1)}(r')^{-1/2(\nu-1)}B_\nu(r_>)\left[A_\nu(r_<) - \frac{A_\nu(a)B_\nu(r_<)}{B_\nu(a)}\right]$$

Thus,

$$\lim_{\alpha\downarrow 0} g_0(r, \alpha^{1/2}R) = r^{-1/2(\nu-1)}B_\nu(r)\lim_{x\downarrow 0}[x^{-1/2(\nu-1)}A_\nu(x)]$$

$$\lim_{\alpha\downarrow 0} g_{D;\alpha^{1/2}r_0}(r, \alpha^{1/2}R) = r^{-1/2(\nu-1)}B_\nu(r)$$

$$\times \lim_{x\downarrow 0}\left[x^{-1/2(\nu-1)}\left(A_\nu(x) - \frac{A_\nu(\gamma x)B_\nu(x)}{B_\nu(\gamma x)}\right)\right]$$

with $\gamma = r_0/R$. Thus, dividing (7.13) by (7.14)

$$E(b(t) \notin S \mid \text{all } t) = 1 - \lim_{x\downarrow 0}\left[\frac{A_\nu(\gamma x)B_\nu(x)}{B_\nu(\gamma x)A_\nu(x)}\right] \tag{7.15}$$

If $\nu \geq 3$, then $A_\nu \sim x^{1/2(\nu-1)}$ and $B_\nu \sim x^{1/2(3-\nu)}$ for x small, so the limit in (7.15) is $\gamma^{\nu-2}$. If $\nu = 2$, then $A_\nu \sim x^{1/2}$ and $B_\nu \sim x^{1/2}\ln x$, so the limit in (7.15) is $\lim_{x\downarrow 0}(\ln x)/\ln(\gamma x) = 1$. ∎

Example (Friedman [87]) Let f_n be a sequence of bounded functions on $\mathbb{R}^\nu$ with $\text{supp } f_n \subset \{x \mid |x| \leq 1/n\}$. Friedman noted that for $\nu \geq 4$, $-\Delta + f_n \to -\Delta$ in strong resolvent sense irrespective of how fast $\|f\|_\infty$ grows! For $C_{00}^\infty = \{g$ which are in C_0^∞, $g \equiv 0$ near $0\}$ is a core for $-\Delta$ if $\nu \geq 4$ and for $g \in C_{00}^\infty$:

$$\|[(-\Delta + f_n + i)^{-1} - (-\Delta + i)^{-1}](-\Delta + i)g\|$$
$$= \|(-\Delta + f_n + i)^{-1}f_n g\| = 0$$

for n large. Since $\{(-\Delta + i)g\}$ is dense by the core statement, the strong resolvent result is true. This assertion is false for $\nu \leq 3$ by explicit example, but Friedman did prove it for $\nu = 2, 3$ if the f_n's are positive. His proof is expressed quite nicely in terms of path integrals: the key fact is that when $\nu \geq 2$ for fixed T, $Q \equiv \{\omega \mid \omega(t) = 0, \text{ some } t \in [0, T]\}$ has Wiener measure zero. For $\nu \geq 3$, this follows easily from Theorem 7.11; for $\nu = 2$ a different argument is available: see the Aside following Lemma 7.21. If $\omega \notin Q$, then by the continuity of paths, $\inf_{0\leq s\leq T}|\omega(s)| > 0$, so for n sufficiently large $\int_0^T f_n(\omega(s))\, ds = 0$. It follows that for any $g, h \in L^2$,

$$\overline{g(\omega(T))}h(\omega(0))\exp\left(-\int_0^T f_n(\omega(s))\, ds\right) \to \overline{g(\omega(T))}h(\omega(0)) \quad \text{as} \quad n \to \infty$$

for almost every ω. Since $f_n \geq 0$, we can apply the dominated convergence theorem to conclude that

$$(g, \exp(-T(-\Delta + f_n))h) \to (g, \exp(T\Delta)h) \qquad \text{as} \quad n \to \infty$$

this implies the strong resolvent convergence.

Theorem 7.11 and stopping time arguments are what is needed to complete the study of recurrence. As a preliminary, we note that if $\tilde{\mathbf{b}}(t)$ is a copy of ν-dimensional Brownian motion on a space with some additional *independent* coordinates, x, and if $R(x)$ is a random rotation on $\mathbb{R}^\nu$ depending only on x, then $R(x)\tilde{\mathbf{b}}(t)$ is also a copy of Brownian motion. This follows since for each fixed x, $E_{\tilde{b}}(f(R(x)\tilde{\mathbf{b}}(t))) = E(f(\mathbf{b}(t)))$ so by independence

$$E_{\tilde{b};\, x}(f(R(x)\tilde{\mathbf{b}}(t))) = E(f(\mathbf{b}(t)))$$

As an example, let τ be a stopping time which is everywhere finite and let $R_\tau(\mathbf{b})$ be a rotation with $R_\tau(\mathbf{b})\mathbf{b}(\tau) = (|\mathbf{b}(\tau)|, 0, \ldots, 0)$. Then $R_\tau(\mathbf{b})(\mathbf{b}(t + \tau) - \mathbf{b}(\tau))$ is a copy of Brownian motion.

Theorem 7.12 Let $\mathbf{b}$ be ν-dimensional Brownian motion and suppose that $\nu \geq 3$. Then

$$\lim_{t \to \infty} |\mathbf{b}(t)| = \infty$$

with probability one. In particular, for any bounded S, $\{t \mid \mathbf{b}(t) \in S\}$ is a bounded set.

Proof Define stopping times τ_n by

$$\tau_n = \inf\{s \mid |\mathbf{b}(s)| = 2^n\}$$

Then $\tau_n < \infty$ everywhere by (7.5). Since $R_{\tau_n}(\mathbf{b}(t + \tau_n) - \mathbf{b}(\tau_n)) \equiv \tilde{\mathbf{b}}_n(t)$ is a copy of Brownian motion, $|\mathbf{b}(\tau_n)| = 2^n$, and $\mathbf{b}(s) \in \{x \mid |x| < r\}$ for some $s > \tau_n$ if and only if $\tilde{\mathbf{b}}_n(t) \in \{x \mid |x + (2^n, 0, \ldots, 0)| < r\}$ for some $t > 0$, we see that for $n > \ln r/\ln 2$

$$E(|\mathbf{b}(s)| < r \mid \text{some } s > \tau_n) = (r2^{-n})^{\nu - 2}$$

Thus, by the first Borel–Cantelli lemma, there is, for almost every $\mathbf{b}$, an $n(\mathbf{b})$ so that $|\mathbf{b}(s)| \geq r$ for all $s \geq \tau_{n(\mathbf{b})}(\mathbf{b})$. Thus, since each $\tau_n < \infty$, $\underline{\lim} |\mathbf{b}(s)| \geq r$. Since r is arbitrary, the theorem is proven. ∎

Theorem 7.13 Let $\mathbf{b}$ be ν-dimensional Brownian motion and suppose that $\nu \leq 2$. Then for any nonempty open set S

$$|\{t \mid b(t) \in S\}| = \infty$$

for almost every b where $|\cdot|$ indicates Lebesgue measure. In particular, the above set is unbounded and for any $\mathbf{a} \in \mathbb{R}^\nu$

$$\varliminf_{t\to\infty} |\mathbf{b}(t) - \mathbf{a}| = 0$$

Proof It suffices to take S a ball, say of radius r about $\mathbf{x}_0$. Let $\tau = \inf\{t \,|\, |\mathbf{b}(t) - \mathbf{x}_0| \le r/2\}$. Then τ is a stopping time, so it suffices to show the result for all r and $\mathbf{x}_0 = 0$, since then $|\{t \,|\, |\mathbf{b}(\tau + t) - \mathbf{b}(\tau)| \le r/2\}| = \infty$. Finally, by scaling we can suppose that $r = 1$ also.

Define stopping times $\tau_1, \sigma_1, \tau_2, \sigma_2, \ldots$ inductively by

$$\begin{aligned} \tau_1 &= \inf\{t \,|\, |\mathbf{b}(t)| = 2\} \\ \sigma_n &= \inf\{t > \tau_n \,|\, |\mathbf{b}(t)| = \tfrac{1}{2}\} \\ \tau_n &= \inf\{t > \sigma_{n-1} \,|\, |\mathbf{b}(t)| = 2\} \end{aligned}$$

It is not hard to see that each stopping time is almost everywhere finite: For $\tau_n < \infty$ given $\sigma_{n-1} < \infty$ on account of (7.5) and $\sigma_n < \infty$ since $\tilde{\mathbf{b}}(t) = R_{\tau_n}(\mathbf{b}(t + \tau_n) - \mathbf{b}(\tau_n))$ is a Brownian motion and $\sigma_n < \infty$ is equivalent to

$$E(\tilde{\mathbf{b}}(t) \in \{x \,|\, |x + (2, 0)| \le 1\}; \text{some } t) = 1$$

which follows from Theorem 7.11. Using the Dynkin–Hunt theorem, $\{\sigma_n - \tau_n\}$ and $\{\tau_n - \sigma_{n-1}\}$ are two mutually independent families of identically distributed independent random variables. Therefore,

$$T_n = |\{t \,|\, |\mathbf{b}(t)| \le 1; \tau_n < t < \tau_{n+1}\}|$$

are strictly positive random variables which are independent and identically distributed. Let $S_n = \min(T_n, 1)$. Since $T_n > 0$, $E(S_n) > 0$. Clearly $E(|S_n|) < \infty$, $E(|S_n|^2) < \infty$. Thus, by the strong law of large numbers (see Lemma 7.14 below),

$$\frac{1}{n} \sum_{j=1}^{n} S_j \to E(S_1)$$

with probability one and, in particular:

$$\sum_{j=1}^{\infty} S_j = \infty$$

with probability one. Since

$$|\{t \,|\, |b(t)| \le 1\}| \ge \sum_{j=1}^{\infty} T_j \ge \sum_{j=1}^{\infty} S_j$$

the result is proven. ∎

In the above, we needed the following lemma.

Lemma 7.14 (strong law of large numbers) Let $\{X_n\}_{n=1}^{\infty}$ be a family of independent, identically distributed random variables and suppose that $E(|X_1|) < \infty$, $E(|X_1|^2) < \infty$. Then

$$\frac{1}{n}\sum_{j=1}^{n} X_j \to E(X_1)$$

with probability one.

Proof By replacing X_j by $X_j - E(X_j)$, we can suppose $E(X_j) = 0$. Notice that in this case, if Y is the sum of any kX's, then

$$E(Y^2) = kE(X_1^2) \tag{7.16}$$

In particular, if $Y_n = \sum_{j=1}^{n} X_j$, then

$$E(|2^{-n}Y_{2^n}|^2) = 2^{-n}E(X_1^2)$$

so that

$$E(|2^{-n}Y_{2^n}| \geq \varepsilon) \leq \varepsilon^{-2}2^{-n}E(X_1^2)$$

Thus, by the first Borel–Cantelli lemma, $|2^{-n}Y_{2^n}| < \varepsilon$, eventually; i.e., $\lim_{n\to\infty} 2^{-n}Y_{2^n} = 0$.

Now let $Z_j^{(n)} = 2^{-n+1}\sum_{i=1}^{j} X_{i+2^{n-1}}$. Then

$$E((Z_j^{(n)})^2 \,|\, (Z_{j-1}^{(n)})^2) = (Z_{j-1}^{(n)})^2 + E((2^{-n+1}X_1)^2) > (Z_{j-1}^{(n)})^2$$

so $(Z_j^{(n)})^2$ is a submartingale. Doob's inequality (3.8), thus, implies that

$$\begin{aligned} E\left(\max_{0\leq j\leq 2^{n-1}} |Z_j^{(n)}|^2 \geq \varepsilon^2\right) &\leq \varepsilon^{-2}E((Z_{2^{n-1}}^{(n)})^2) \\ &= 4\varepsilon^{-2}2^{-n}E(X_1^2) \end{aligned}$$

Thus, by the first Borel–Cantelli lemma (again), $|Z_j^{(n)}| < \varepsilon$ for $n \geq N_0$, $j = 1, \ldots, 2^{n-1}$. Let $2^{n-1} \leq m < 2^n$. Then

$$m^{-1}|Y_m| \leq 2^{-n+1}|Y_{2^{n-1}}| + |Z_{m-2^{n-1}}^{(n)}| \leq 2\varepsilon$$

for n sufficiently large. ▌

Remarks 1. The law of large numbers also follows from the Birkhoff ergodic theorem (Theorem 3.7) and the Kolmogorov 01 law (Example 1 before Theorem 3.7).

2. One can dispense with the condition $E(|X_1|^2) < \infty$; see [158].

As a final set of results involving recurrence, we want to note that while one-dimensional Brownian paths spend an infinite time in any given bounded set, there is a significant probability that they will take a long time to reach the set. Explicitly, we want to show that for the stopping times used in the proof of Theorem 7.13, $E(\tau_n - \tau_{n-1}) = \infty$. This fact prevents an amusing paradox from taking place. While $|\{t \,||b(t)| \le 1\}|$ is infinite if $v = 1$ or 2, it is fairly clear that $t^{-1}|\{s\,||b(s)| \le 1;\ s \le t\}|$ should go to zero as $t \to \infty$ for the Brownian path should forget where it began; i.e.,

$$\lim t^{-1}|\{s\,||b(s)| \le 1; s \le t\}| = \lim t^{-1}|\{s\,|b(s) \in [n, n+2]; s \le t\}|$$

for any n. If $E(\tau_n - \tau_{n-1})$ were finite, then we would have $\tau_n/n \to E(\tau_2 - \tau_1)$ by the strong law of large numbers. Since $T_n \le \tau_{n+1} - \tau_n$, we would also have $(\sum_{j=1}^n T_j)/n \to E(T_1)$ so that $\tau_{n+1}^{-1} \sum_{j=1}^n T_j \to E(T_1)/E(\tau_2 - \tau_1)$ in contradiction to intuition, which says that $t^{-1}|\cdot| \to 0$. Fortunately, $E(\tau_n - \tau_{n-1}) = \infty$, so there is no problem.

Proposition 7.15 In one dimension,

$$t^{1/2}E(b(s) \le 1 | 0 \le s \le t) \to (2/\pi)^{1/2}$$

as $t \to \infty$.

Proof As already noted in the Example following Lemma 7.10, Equation (7.6) is equivalent to

$$E\left(\max_{0 \le s \le t} b(s) \le 1\right) = E(-1 \le b(t) \le 1)$$

$$= (2\pi t)^{-1/2} \int_{-1}^{1} \exp\left(\frac{-x^2}{2t}\right) dx$$

which proves the result since $\exp(-x^2/2t) \to 1$ pointwise as $t \to \infty$. ∎

Actually, one can explicitly compute the probability distribution for the stopping time which gives the first time that b hits one:

Proposition 7.15′ The stopping time

$$\tau = \inf\{s\,|b(s) = 1\}$$

has the distribution

$$(2\pi t^3)^{-1/2} \exp(-(2t)^{-1})\, dt$$

In particular, as $t \to \infty$,

$$t^{1/2}E(b(s) \le 1 | 0 \le s \le t) = t^{1/2}E(\tau > t) \to (2/\pi)^{1/2}$$

First proof ([272]) Clearly

$$f(t) \equiv E(\tau > t) = E\left(\max_{0 \le s \le t} b(s) < 1\right)$$

so by (7.6):

$$\begin{aligned} f(t) &= E(|b(t)| \le 1) \\ &= 2\int_0^1 (2\pi t)^{-1/2} \exp\left(\frac{-x^2}{2t}\right) dx \\ &= 2\int_0^{t^{-1/2}} (2\pi)^{-1/2} \exp\left(\frac{-y^2}{2}\right) dy \end{aligned}$$

Thus f is differentiable and

$$-\frac{df}{dt} = \left(\frac{2}{\pi}\right)^{1/2} \exp(-(2t)^{-1}) \frac{d}{dt}(-t^{-1/2}) = (2\pi t^3)^{-1/2} \exp(-(2t)^{-1})$$

which establishes the probability distribution for τ. ∎

Second proof ([183]) Introduce stopping times $\tau_{N,m}$ by

$$\tau_{N,m} = \begin{cases} \dfrac{k}{2^m} & \text{if} \quad (k-1)2^{-m} \le \tau < k2^{-m}; \quad k \le N \\ N2^{-m} & \text{if} \quad N2^{-m} \le \tau \end{cases}$$

We will first prove that for $\gamma > 0$,

$$E(:\exp \gamma b(\tau_{N,m}):) = 1 \tag{7.17}$$

For

$$\begin{aligned} E(:\exp \gamma b(\tau_{N,m}):) = {} & E(:\exp \gamma b(\tau_{N-1,m}):; \tau < (N-1)/2^m) \\ & + E(:\exp \gamma b(N2^{-m}):; \tau \ge (N-1)/2^m) \end{aligned}$$

But $b(N^{2-m}) - b((N-1)2^{-m})$ is independent of $\tau \ge (N-1)/2^m$ since $\tau \ge (N-1)/2^m$ is measurable with respect to $\mathscr{B}_{(N-1)/2^m}$. Thus

$$\begin{aligned} & E(:\exp \gamma b(N2^{-m}):; \tau \ge (N-1)/2^m) \\ & \quad = E(:\exp \gamma b((N-1)2^{-m}):; \tau \ge (N-1)/2^m)) \end{aligned}$$

i.e., we have shown that

$$E(:\exp \gamma b(\tau_{N,m}):) = E(:\exp \gamma b(\tau_{N-1,m}):)$$

so (7.17) holds by taking N to one where $b(\tau) = b(1/2^n)$, and the result is evident. Now, set $N = 2^m M$ with M an integer and take $m \to \infty$. Pointwise $b(\tau_{N,m}) \to b(\tilde{\tau}_M)$ where $\tilde{\tau}_M \equiv \min(\tau, M)$. Moreover

$$:\exp(\gamma b(\tau_{N,m})): \le \max_{0 \le s \le M} \exp(\gamma b(s))$$

$$= \exp\left(\gamma \max_{0 \le s \le M} b(s)\right) \equiv F(b)$$

But, by (7.6)

$$E(F(b)) = -\int_0^\infty e^{\gamma\lambda}\, d\left[E\left(\max_{0 \le s \le M} b(s) > \lambda\right)\right]$$

$$= 1 + 2\gamma \int_0^\infty e^{\gamma\lambda} E(b(M) > \lambda)\, d\lambda < \infty$$

Thus, by the dominated convergence theorem, we can conclude that

$$E(:\exp(\gamma b(\tilde{\tau}_M)):) = 1$$

Now $\tau < \infty$, by the law of the iterated logarithm so $b(\tilde{\tau}_M) \to b(\tau) = 1$ as $M \to \infty$. Moreover, *by the definition of* τ, $b(\tilde{\tau}_M) \le 1$, so $:\exp(\gamma b(\tilde{\tau}_M)): \le e^\gamma$ and thus we can take $M \to \infty$ and conclude that

$$E(:\exp(\gamma b(\tau)):) = 1$$

But $b(\tau) = 1$, so

$$E(\exp(-\tfrac{1}{2}\gamma^2\tau)) = \exp(-\gamma) \tag{7.18}$$

As in the inversion procedure used to reach Theorem 6.10, one can go from this to $E((a + \tau)^{-1})$ and from there to the claimed distribution. ∎

Remarks 1. Similarly, one finds that the distribution for first hitting $a > 0$ is $(2\pi t^3)^{-1/2} a \exp(-a^2/2t)\, dt$, and that for σ, the first time that $|b(s)| = 1$, one has that

$$E(\exp(-\tfrac{1}{2}\gamma^2\sigma)) = [\cosh(\gamma)]^{-1}$$

2. We were especially careful about the use of the dominated convergence theorem, because one can consider the stopping time

$$\eta = \min\{s \mid b(s) = 1;\ b(t) \ge 2, \text{ some } t < s\}$$

Clearly $\eta > \tau$ so since $b(\eta) = b(\tau) = 1$,

$$E(:\exp(\gamma b(\eta)):) < E(:\exp(\gamma b(\tau)):) = 1$$

But the proof above goes through most of the way! In fact $E(:\exp(\gamma b(\tilde{\eta}_M)):) = 1$. But as $M \to \infty$, $:\exp(\gamma b(\tilde{\eta}_M)):$ is no longer bounded by an L^1-function invalidating the use of the dominated convergence theorem. One can explicitly compute $E(:\exp(\gamma b(\eta)):)$ since η has the same distribution as the stopping time $\mu = \inf\{s \mid b(s) = 3\}$, since we can reflect the path about $b = 2$ for times past the first time the path hits two. Thus,

$$\begin{aligned} E(:\exp(\gamma b(\eta)):) &= e^{\gamma} E(\exp(-\tfrac{1}{2}\gamma^2 \eta)) \\ &= e^{\gamma} E(\exp(-\tfrac{1}{2}\gamma^2 \mu)) \\ &= e^{-2\gamma} E(:\exp(\gamma b(\mu)):) = e^{-2\gamma} \end{aligned}$$

as above.

3. The argument in the second proof can be abstracted to the following.

Theorem 7.15A Let $f(t)$ be a (sub-respectively, super-) martingale for Brownian motion with $\sup_{t,b} |f(t, b)| < \infty$. Then, for any stopping time, τ:

$$E(f(\tau)) = E(f(0)) \quad (\geq; \text{respectively}, \leq)$$

As in the above proof, one can relax the supremum requirement by additional argument and conditions on τ.

4. One can directly obtain that $E(\tau; \tau < t)$ diverges like $t^{1/2}$ from (7.18) which implies that

$$\begin{aligned} E(\tau \exp(-\alpha\tau)) &= -\frac{d}{d\alpha} \exp(-(2\alpha)^{1/2}) \\ &= (2\alpha)^{-1/2} \exp(-(2\alpha)^{1/2}) \end{aligned}$$

Now use a Tauberian theorem (see Section 10) to obtain $\lim_{t\to\infty} t^{1/2} E(\tau > t)$.

Corollary 7.16 The stopping times σ, τ of Theorem 7.13 obey

$$E(\sigma_n - \tau_n) = \infty$$

Proof Let $\tilde{\mathbf{b}}(t) = R_{\tau_n}(\mathbf{b}(t + \tau_n) - \mathbf{b}(\tau_n))$
Then $\sigma_n - \tau_n \geq \inf(s \mid \tilde{b}_1(s) \leq -\tfrac{3}{2})$ so

$$g(t) \equiv E(\sigma_n - \tau_n \geq t) \geq E\left(\max_{0 \leq s \leq t} b(s) \leq \tfrac{3}{2}\right) = \tfrac{3}{2}(2/\pi)^{1/2} t^{-1/2} + o(t^{-1/2})$$

As a result

$$E(\sigma_n - \tau_n) = -\int_0^{\infty} t\, dg = \int_0^{\infty} g(t)\, dt = \infty \quad \blacksquare$$

The reason that it may take very long for $b(s)$ to reach one is not that the path gets "trapped" in $(-1, 1)$, but rather that it makes excursions far to the left:

Proposition 7.17

(a)
$$E\left(\max_{0 \le s \le t} |b(s)| \le 1\right) = E(b(t) \in A) - E(b(t) \in B) \tag{7.19}$$

where

$$A = (-1, 1) \cup (3, 5) \cup (-3, -5) \cup \cdots$$

and

$$B = (1, 3) \cup (-1, -3) \cup (5, 7) \cup \cdots$$

(b)
$$E\left(\max_{0 \le s \le t} |b(s)| \le 1\right) = \frac{4}{\pi} \sum_{k=1}^{\infty} (2k + 1)^{-1}(-1)^k e^{-\pi^2(2k+1)^2 t/8} \tag{7.20}$$

(c)
$$E(-a \le b(s) \le 1 | 0 \le s \le t) = O(e^{-t\alpha(a)})$$

where $\alpha(a) = \pi^2/(2(a + 1)^2)$.

Remarks 1. (7.19) is good for t small and (7.20) for t large.

2. (c) verifies our remark, that for large times, $b(s)$ can only stay below one by going far to the left, indeed, it must go a distance $O(t^{1/2})$ to the left.

Proof By Lemma 7.10,

$$E(-a \le b(s) \le 1 | 0 \le s \le t) = \int_{-a}^{1} P_{\mathrm{D};a}(x, 0; t)\, dx$$

where $P_{\mathrm{D};a}$ is the propagator for $\frac{1}{2}$ times Laplacian with Dirichlet conditions at one and $-a$. (7.19) comes from writing $P_{\mathrm{D};a}$ by the method of images and (7.20) by expanding $P_{\mathrm{D};a}$ in its explicit eigenfunction expansion and noting that this expansion is uniformly convergent so that it can be used for $P(x, 0; t)$ and the sum and integral can be interchanged. (c) comes from the same kind of eigenfunction expansion noting that the lowest eigenvalue of $-\frac{1}{2}\, d^2/dx^2$ on $L^2(-a, 1)$ with Dirichlet conditions is $\pi^2/(2(a + 1)^2)$. ∎

One can ask how $|\{t \,||b(t)| \le 1; t \le T\}|$ diverges as $T \to \infty$ when $\nu = 1, 2$ (we have already remarked it cannot be as fast as T). We will answer the weaker question concerning how $E(|\{s\,||b(s)| \le 1;\ s \le t\}|)$ diverges as $t \to \infty$.

Proposition 7.18 For $\nu = 1$:

$$\lim_{t \to \infty} t^{-1/2} E(|\{s\,||b(s)| \le 1; s \le t\}|) = c$$

and for $\nu = 2$:

$$\lim_{t\to\infty}(\ln t)^{-1}E(|\{s\,||b(s)| \leq 1; s \leq t\}|) = d$$

for suitable nonzero c, d.

Proof Let χ be the characteristic function of $[-1, 1]$. Then, for each b

$$|\{s\,||b(s)| \leq 1; s \leq t\}| = \int_0^t \chi(b(s))\,ds$$

Thus, by a Feynman–Kac formula:

$$\begin{aligned} q(t) &\equiv E(|\{s\,||b(s)| \leq 1; s \leq t\}|) \\ &= \int_0^t (e^{-sH_0}\chi)(0)\,ds \end{aligned}$$

Thus $q'(t) \geq 0$ and as $\alpha \downarrow 0$:

$$\int_0^\infty e^{-\alpha t} q'(t)\,dt = ((H_0 + \alpha)^{-1}\chi)(0) \sim \begin{cases} c_1 \alpha^{-1/2} & (\nu = 1) \\ d_1 \ln(\alpha^{-1}) & (\nu = 2) \end{cases}$$

for explicit c_1, d_1 depending on the explicit small α behavior of the integral kernel for $(H_0 + \alpha)^{-1}$. The $\nu = 1$ result follows from the Tauberian theorem (Theorem 10.3) and the $\nu = 2$ result from an extension of that theorem. ∎

Remark The same method shows that in $\nu \geq 3$ dimensions,

$$E(|\{s\,|\,b(s) \in S\}|) < \infty$$

for any bounded S. This implies that almost everywhere $|\{s\,|\,b(s) \in S\}| < \infty$, which is not quite as strong as the fact that $|b(s)| \to \infty$ almost everywhere.

* * *

Finally we should like to discuss self-intersection of Brownian paths. In 1940, Lévy [169] proved that almost every two-dimensional Brownian path had a self-intersection, i.e., a pair t, s so that $\mathbf{b}(t) = \mathbf{b}(s)$; and Kakutani [145] proved that this could not happen if $\nu \geq 5$. The cases $\nu = 3, 4$ were settled in [63]; the question of n-fold points with $n \geq 3$ was settled in [64, 66]. These results are summarized in the following.

Definition A continuous curve $\{\mathbf{b}(t)|0 \leq t < \infty\}$ is said to have an **n-fold point** if and only if there exist $t_1, \ldots, t_n$ distinct with $\mathbf{b}(t_1) = \cdots = \mathbf{b}(t_n)$.

Theorem 7.19 Let $\mathbf{b}$ be ν-dimensional Brownian motion. Then

(a) If $\nu \leq 3$, almost every $\mathbf{b}(t)$ has infinitely many double points,
(b) If $\nu \geq 4$, almost every $\mathbf{b}(t)$ has no double points,
(c) If $\nu \leq 2$, almost every $\mathbf{b}(t)$ has infinitely many n-fold points for each n,
(d) If $\nu = 3$, almost every $\mathbf{b}(t)$ has no triple points.

We will give only the proof of (a) when $\nu = 3$ (this automatically proves the result for $\nu = 2$ since a double point for $(b_1(t), b_2(t), b_3(t))$ is automatically a double point for $(b_1(t), b_2(t))$). See [63, 64, 66] for (*b*), (*c*), (*d*), respectively.

Definition Let K be a compact subset of $\mathbb{R}^\nu$. We define the **hitting probability** by:

$$h(\mathbf{y}, K) = E(\mathbf{y} + \mathbf{b}(s) \in K \,|\, \text{some } s \in [0, \infty))$$

Theorem 7.20 $h(\cdot, K)$ is a harmonic function on $\mathbb{R}^\nu \backslash K$.

Proof Given $\mathbf{y} \notin K$, pick R so that $\{\mathbf{x} \,|\, |\mathbf{x} - \mathbf{y}| \leq R\} \cap K = \varnothing$. Let $\delta < R$ and define a stopping time τ by

$$\tau(b) = \inf\{s \,|\, |\mathbf{b}(s)| = \delta\}$$

$\tau < \infty$ almost everywhere by the law of the iterated logarithm. Moreover, $\mathbf{y} + \mathbf{b}(s) \notin K$ for $s \leq \tau(\mathbf{b})$ so that if $\tilde{\mathbf{b}}(s) = \mathbf{b}(s + \tau) - \mathbf{b}(\tau)$,

$$\begin{aligned} h(\mathbf{y}, K) &= E(\mathbf{y} + \mathbf{b}(s) \in K \,|\, \tau(\mathbf{b}) \leq s < \infty) \\ &= E(\mathbf{y} + \mathbf{b}(\tau(\mathbf{b})) - \tilde{\mathbf{b}}(s) \in K \,|\, 0 < s < \infty) \\ &= \int h(\mathbf{y} + \mathbf{x}; K) \, dS_\delta(\mathbf{x}) \end{aligned}$$

where dS_δ is the normalized invariant measure on the sphere of radius δ. The last equality comes from the fact that $\tilde{\mathbf{b}}$ is a Brownian motion independent of $\mathbf{b}(\tau)$ (Theorem 7.9), that $|\mathbf{b}(\tau)| = \delta$, and that $\mathbf{b}(\tau)$ has a rotationally invariant distribution. Since h is a bounded measurable function, it is harmonic. ∎

Remarks 1. The mean value property and $h \in L^1_{\text{loc}}$ imply h is C^∞ since $h * f$ is C^∞ if $f \in C_0^\infty$. If $\int f = 1$ and f is rotationally invariant with $\operatorname{supp} f \subset B_\delta$, then $(h * f)(y) = f(y)$ by the mean value property.

2. One can also base the proof on the Dirichlet ideas used to discuss hitting probabilities earlier.

3. Results of this genre go back to Kakutani [145, 146]. They are the starting point of "probabilistic potential theory" developed by Doob, Hunt, and others and summarized in Meyer's book [186]. They are central to some rather deep results of Burkholder, *et. al.* [28] on H^p spaces; see the readable lecture notes of Petersen [203]. Via these notions, Brownian motion can be used to prove some rather deep results in complex analysis [42].

The following result has already been proven as part of Theorem 7.11, but since we shall recover parts of that theorem from Theorem 7.20, we give an independent proof.

Lemma 7.21 Let K be a ball. Then $h(\mathbf{y}, K) \to 1$ as $\mathbf{y} \to K$.

Proof Let r be the radius of the ball and R the distance from y to the center of the ball. By the scaling relations $b(s) \doteq ab(sa^{-2})$, the hitting probability can only be a function $f(r/R)$ of r/R. We want to show that $f(x) \to 1$ as $x \uparrow 1$. Let K^α be the ball of radius α about the point $(-\alpha, 0, \ldots, 0)$. Let $\mathbf{y}_0 = (1, 0, 0, \ldots)$. Then

$$h(\mathbf{y}_0, K^\alpha) = f(\alpha/(1 + \alpha))$$

so

$$\begin{aligned} \lim_{x \uparrow 1} f(x) &= \lim_{\alpha \to \infty} h(\mathbf{y}_0, K^\alpha) \\ &= h(y, \bigcup K^\alpha) \qquad (\text{since } K_\alpha \subset K_\beta \text{ for } \alpha < \beta) \\ &= E(b_1(s) < -1, \text{ some } s) = 1 \end{aligned}$$

by the law of the iterated logarithm. ∎

Aside This lemma and Theorem 7.20 immediately imply the $v = 2$ case of Theorem 7.11, i.e., that $h(\mathbf{y}, K) = 1$ for $v = 2$ and K a sphere. For h is harmonic and rotationally invariant so $h(\mathbf{y}) = a + b \ln|\mathbf{y}|$. Thus $0 \leq h \leq 1$ implies $h(\mathbf{y}) = a$ and the lemma implies that $a = 1$. By a little more argument, one can recover all of Theorem 7.11 and also the fact (used in the Example following Theorem 7.11) that a fixed point not equal to zero is hit with probability zero if $v \geq 2$. The argument is as follows: fix $r < R$ and define, for $r < |\mathbf{y}| < R$:

$$g(\mathbf{y}) = E(|\mathbf{y} + b(t)| \quad \text{is first equal to } r \text{ before it is equal to } R)$$

i.e., $g(y)$ is the probability that a Wiener path starting at y hits the sphere $|\mathbf{x}| = r$ before it hits the sphere $|\mathbf{x}| = R$. By the law of the iterated logarithm, it eventually hits $|x| = R$. As in the proof of Theorem 7.20, $g(y)$ is harmonic in the region $r < |\mathbf{y}| < R$. Moreover, we claim that $g(y) \to 1$ (respectively, 0) as $|\mathbf{y}| \to r$ (respectively, R). To prove this let

$$q(\varepsilon) = E\left(\max_{0 \le t \le \varepsilon} |b_i(t)| \le \varepsilon^{1/3} \quad (i = 1, \dots, v);\ \min_{0 \le t \le \varepsilon} b_1(t) \le -\varepsilon^{2/3}\right)$$

Using (7.6) and (7.6′), one easily sees that $q(\varepsilon) \to 1$ as $\varepsilon \downarrow 0$ so that $g(r + \frac{1}{2}\varepsilon^{2/3}, 0, \dots, 0) \to 1$ as $\varepsilon \to 0$ by a geometric argument. We can therefore conclude that

$$g(y) = \begin{cases} (R - y)/(R - r), & v = 1 \\ [\ln R - \ln y]/(\ln R - \ln r), & v = 2 \\ [y^{-(v-2)} - R^{-(v-2)}]/[r^{-(v-2)} - R^{-(v-2)}], & v \ge 3 \end{cases}$$

If we take $R \to \infty$ with r fixed (and use the law of the iterated logarithm to note that if a path does not hit the sphere of radius r, it will always hit the path of radius R "first" and use the continuity of paths to note that if it hits the sphere of radius r, it will hit before the sphere of radius R for R large), we recover Theorem 7.11. On the other hand, if we first take $r \to 0$ and then $R \to \infty$, we see that a path starting at $y \ne 0$ will hit zero at some time with probability zero if $v \ge 2$ and with probability one, if $v = 1$.

We now return to the proof of Theorem 7.19(a).

Lemma 7.22 If K is a union of closed balls, then $h(\mathbf{y}, K) \to 1$ as $\mathbf{y} \to K$.

Proof Let $\mathbf{y}_n \to \mathbf{y}_0 \in K$. Since $\mathbf{y}_0 \in K$, it is in some ball $K_0 \subset K$. Then

$$\underline{\lim}\, h(\mathbf{y}_n, K) \ge \lim h(\mathbf{y}_n, K_0) = 1$$

by Lemma 7.21. ∎

The important idea of [145] is to use the notion of Newtonian capacity. We develop the ideas in $v = 3$ dimensions.

Definition Let $K \subset \mathbb{R}^3$ be compact and let $\mathcal{M}_{+,1}(K)$ be the probability measures on K. Then $C(K)$, **the Newtonian capacity of K**, is defined by

$$C(K) = \left[\inf\left\{\int |\mathbf{x} - \mathbf{y}|^{-1}\, d\mu(\mathbf{x})\, d\mu(\mathbf{y}) \,\middle|\, \mu \in \mathcal{M}_{+,1}(K)\right\}\right]^{-1}$$

Note that $C(K)$ may be zero, if the integral is always infinite, e.g., if $K = \{0\}$.

Theorem 7.23 For $\nu = 3$

$$h(\mathbf{y}, K) \geq C(K)[\max\{|\mathbf{x} - \mathbf{y}| \,|\, \mathbf{x} \in K\}]^{-1} \tag{7.21}$$

Remarks 1. Suitably generalized this is true for any $\nu \geq 3$.

2. In particular, this says that if $C(K) \neq 0$, then $h(\mathbf{y}, K) \neq 0$ for all y. There is a converse to this; namely, that if $C(K) = 0$, then $h(\mathbf{y}, K) = 0$. The latter follows from the fact [124, 164] that if $C(K) = 0$, then any positive function harmonic on $\mathbb{R}^\nu \backslash K$ with $f \to 0$ at infinity is either identically zero or unbounded. Since h is bounded, it must be zero.

Proof Let $\{\mathbf{y}_n\}$ be a countable dense set in K. Let K_ε be the union of the ε-balls about the points $\mathbf{y}_n$. We first claim that $K \subset K_\varepsilon$ and

$$h(\mathbf{y}, K) = \lim_{\varepsilon \downarrow 0} h(\mathbf{y}, K_\varepsilon) \tag{7.22}$$

For clearly, $h(\mathbf{y}, K) \leq h(\mathbf{y}, K_\varepsilon)$. If $\mathbf{b}(s_n) + \mathbf{y} \in K_{1/n}$, then since $|\mathbf{b}(s)| \to \infty$ as $s \to \infty$ (Theorem 7.12), the s_n are bounded so by passing to a subsequence, $s_n \to s$. It follows that $b(s) \in K$; i.e.,

$$\bigcap_\varepsilon \{\mathbf{b}(s) + \mathbf{y} \in K_\varepsilon \,|\, \text{some } s\} = \{\mathbf{b}(s) + \mathbf{y} \in K \,|\, \text{some } s\}$$

from which (7.19) follows. (Note: (7.22) fails in two dimensions if $K = \{0\}$.)

Clearly, $C(K_\varepsilon) \geq C(K)$. Thus, (7.21) for K_ε implies it for K; i.e., we can suppose that K is a countable union of ε-balls. Henceforth suppose that K has this form.

Next, we need a fact from potential theory [124, 164]: There is a probability measure $d\mu$ on K so that the function $\phi_\mu(\mathbf{x}) = \int |\mathbf{x} - \mathbf{y}|^{-1}\, d\mu$ has the value $C(K)^{-1}$ on K. (Intuitively, one gets μ by minimizing the energy

$$\int d\mu(\mathbf{x})\, d\mu(\mathbf{y}) |\mathbf{x} - \mathbf{y}|^{-1}$$

subject to $\mu \geq 0$, $\int d\mu = 1$.) It follows that $h(\mathbf{y}, K) = C(K)\phi_\mu(\mathbf{y})$ since both sides are harmonic on $\mathbb{R}^3 \backslash K$ going to zero at infinity and one at ∂K and thus (7.21) comes from $\phi_\mu(\mathbf{x}) \geq [\max(|\mathbf{x} - \mathbf{y}| \,|\, \mathbf{y} \in K)]^{-1}$. ∎

Theorem 7.24 Let $\nu = 3$. Fix $a, c > 0$. Then $K(\mathbf{b}) = \{\mathbf{b}(t) \,|\, a \leq t \leq c\}$ has strictly positive capacity with probability one.

Proof It suffices to find a probability measure $d\mu_\mathbf{b}$ on $K(\mathbf{b})$ so that $\int d\mu_\mathbf{b}(x)\, d\mu_\mathbf{b}(y) |x - y|^{-1} < \infty$. Define

$$d\mu_\mathbf{b}(A) = |c - a|^{-1} |\{s \,|\, \mathbf{b}(s) \in A;\, a \leq s \leq c\}|$$

Then

$$\int d\mu_{\mathbf{b}}(x)\, d\mu_{\mathbf{b}}(y)|x - y|^{-1} = |c - a|^{-2} \int_a^c ds \int_a^c dt\, |\mathbf{b}(s) - \mathbf{b}(t)|^{-1}$$

so, by Fubini's theorem, it suffices to show that

$$|c - a|^{-2} \int_a^c ds \int_a^c dt\, E(|\mathbf{b}(s) - \mathbf{b}(t)|^{-1}) < \infty \tag{7.23}$$

But $\mathbf{b}(s) - \mathbf{b}(t) \doteq |s - t|^{1/2}\mathbf{b}(1)$ and $E(|\mathbf{b}(1)|^{-1}) < \infty$ trivially (since $v > 1$). Thus (7.23) holds since

$$\int_a^c ds \int_a^c dt\, |s - t|^{-1/2} < \infty \quad \blacksquare$$

Proof of Theorem 7.19(a) We first prove that

$$\alpha \equiv E(\mathbf{b}(t) = \mathbf{b}(s)|\text{some } 0 \le t \le 1; 2 \le s \le \infty) > 0$$

For let $K(\mathbf{b}) = \{\mathbf{b}(s)|0 \le t \le 1\}$. Then

$$\alpha = E(h(\mathbf{b}(2), K(\mathbf{b})))$$

since the function $\mathbf{b}(s + 2) - \mathbf{b}(2)$ is a Brownian motion independent of $\{\mathbf{b}(t)|0 \le t \le 2\}$. For almost every $\mathbf{b}$, $h(\mathbf{b}(2), K(\mathbf{b})) > 0$. Thus $\alpha > 0$.

Next we note that

$$\alpha_T = E(\mathbf{b}(t) = \mathbf{b}(s)|\text{some } 0 \le t \le 1; 2 \le s \le T) > 0$$

for some $T < \infty$ since $\alpha_T \to \alpha$ as $T \to \infty$.

Now let

$$q_n(\mathbf{b}) = \begin{cases} 1, & \text{if } \mathbf{b}(t) = \mathbf{b}(s); \quad \text{some } nT \le t \le nT + 1, \\ & \qquad\qquad\qquad nT + 2 \le s \le (n + 1)T \\ 0, & \text{otherwise} \end{cases}$$

Then the $q_n(b)$ are independent random variables since $\mathbf{b}(t + nT) - \mathbf{b}(nT)$ is independent of the earlier $\mathbf{b}$'s. So, by the second Borel–Cantelli lemma (or alternately by the strong law of large numbers)

$$\sum q_n(\mathbf{b}) = \infty$$

almost everywhere in b. $\blacksquare$

Remark By scaling, for $v = 3$,

$$E(\mathbf{b}(t) = \mathbf{b}(s)|0 \le t \le \alpha; 2\alpha \le s \le T\alpha)$$

is independent of α. Thus picking $\alpha_n = T^{-1}2^{-n}$ and $T_n = 1 - 2^{-(n-1)}$, we see that

$$(\mathbf{b}(t) = \mathbf{b}(s) \mid T_n \leq t \leq T_n + \alpha_n;\ T_n + 2\alpha_n \leq s \leq T_{n+1})$$

are independent, identically distributed events. This shows that the path has infinitely many self-crossings for $0 \leq s, t \leq 1$. Then, by scaling and translation covariance there are infinitely many double points in any interval $a \leq s, t \leq b$. In particular, any point on the path is a limit of double points! By scaling again there are double points with $s - t \geq \alpha$ for any α. Brownian paths are very complicated indeed!

III

Bound State Problems

8. The Birman–Schwinger Kernel and Lieb's Formula

From the earliest days of quantum theory, the semiclassical (WKB) limit has been an important notion. Many rigorous developments of this limiting procedure have relied on the method of Dirichlet–Neumann bracketing, see, e.g., [252]. Kac in his original Feynman–Kac paper [138] suggested that Wiener integrals would be an ideal tool for this problem and his suggestion was implemented by Ray [212] in a suitable problem.

More recently, the idea has arisen that semiclassical results are not only the answers as $\hbar \to 0$, but that they, or some multiple of them, might bound the relevant quantum quantity for all values of $\hbar$. This idea occurred first in the bounds of Golden [111], Thompson [275] and Symanzik [270] for partition functions (extended by Lieb [170] to certain spin systems), then in certain beautiful bounds [40, 171, 224] on the number of bound states and most deeply in the realization of Lieb–Thirring [175] that these bounds were critical for the stability of matter. These works relied on a variety of methods; we want to concentrate on that of Lieb [171] who exploited Wiener integral methods. Indeed, the central tool will be a formula he derived. Apparently, unaware of the technical details of his work, Kac [142] derived some special cases of his formula and applied it to the $\hbar \to 0$ limit. In this section, we will present this formula; in the next, we will show (following Symanzik [270] and Lieb [171]) that classical bounds come from this formula and/or Feynman–Kac together with Jensen's inequality and in Section 10 (following Kac [138, 142]), that $\hbar \to 0$ limits come from these formulas and Tauberian theorems.

As a preliminary to Lieb's formula, we must introduce the Birman–Schwinger kernel. To avoid disagreeing with all other conventions (except

[142]), we let $N_\alpha(V)$ denote the number of bound states (eigenvalues) of $-\Delta + V$ of energy less than $-\alpha$. We will only take $\alpha \geq 0$ and will mainly consider the case $V \equiv -W \leq 0$. Since $N_\alpha(V) \leq N_\alpha(-V_-)$ for any V, this assumption will not affect bounds; it will affect limit theorems but the machinery could be extended to accommodate the general case.

As usual, $H_0 = -\frac{1}{2}\Delta$, so that $N_\alpha(V)$ is the number of bound states of $H_0 + \frac{1}{2}V$ of energy less than $-\frac{1}{2}\alpha$. Now, for any $V(\leq 0)$ which is H_0-form compact, one can easily [257] show that ($\gamma > 0$)

$$(H_0 + \lambda V)\phi = -\gamma\phi \tag{8.1}$$

if and only if ($\psi = |V|^{1/2}\phi$)

$$K_\gamma \psi = \lambda^{-1}\psi \tag{8.2}$$

where K_γ is the Birman [15]–Schwinger [233] "kernel":

$$K_\gamma = |V|^{1/2}(H_0 + \gamma)^{-1}|V|^{1/2} \tag{8.3}$$

Theorem 8.1 (The Birman–Schwinger principle) Let V be H_0-form compact.

(a) Let $\alpha > 0$. Then

$$N_\alpha(V) = \#\{\text{eigenvalues of } K_{(1/2)\alpha} > 2\}$$

(b) If $\alpha = 0$ and $\nu \geq 3$ (so that $K_0 \equiv \lim_{\alpha \downarrow 0} K_\alpha$ exists), then

$$N_0(V) = \#\{\text{eigenvalues of } K_0 > 2\}$$

Proof (a) Let $e_i(\lambda)$ denote the eigenvalues of $H_0 + \lambda V$ suitably ordered with $e_i(\lambda) = 0$ if $N_0(2\lambda V) \leq i - 1$. Then, the $e_i(\cdot)$ are continuous and *strictly* monotone decreasing in the region $\{\lambda | e_i(\lambda) < 0\}$. Thus, for $\alpha > 0$

$$\begin{aligned} N_\alpha(V) &= \#\{i | e_i(\lambda) < -\tfrac{1}{2}\alpha; \lambda = \tfrac{1}{2}\} \\ &= \#\{\lambda | e_i(\lambda) = -\tfrac{1}{2}\alpha; \text{ some } \lambda < \tfrac{1}{2}\} \\ &= \#\{\lambda | K_{(1/2)\alpha} \text{ has eigenvalue } \lambda^{-1} \text{ with } \lambda < \tfrac{1}{2}\} \\ &= \#\{\text{eigenvalues of } K_{(1/2)\alpha} > 2\} \end{aligned}$$

where we use the equivalence of (8.1) and (8.2) in the third inequality.

(b) Clearly $N_0(V) = \lim_{\alpha \downarrow 0} N_\alpha(V)$. But since K_α is monotone increasing as $\alpha \downarrow 0$,

$$\#\{\text{eigenvalues of } K_0 > 2\} = \lim(\#\{\text{eigenvalues of } K_\alpha > 2\}). \blacksquare$$

Now consider the operator ($\gamma > 0$):

$$A(\lambda) = W^{1/2}(H_0 + \lambda W + \gamma)^{-1}W^{1/2}$$

where $W \equiv -V \geq 0$. On the one hand,

$$(H_0 + \gamma)^{-1} = (H_0 + \lambda W + \gamma)^{-1} + \lambda(H_0 + \lambda W + \gamma)^{-1}W(H_0 + \gamma)^{-1}$$

so that

$$A(\lambda) = K_\gamma - \lambda A(\lambda)K_\gamma$$

or

$$A(\lambda) = K_\gamma/(1 + \lambda K_\gamma)$$

On the other hand, if W is in C_0^∞, then, by Theorem 6.6, $A(\lambda)$ has an integral kernel:

$$\begin{aligned} A(\lambda; \mathbf{x}, \mathbf{y}) &= W^{1/2}(\mathbf{x})W^{1/2}(\mathbf{y}) \int_0^\infty dt\, e^{-\gamma t} e^{-t(H_0 + \lambda W)}(\mathbf{x}, \mathbf{y}) \\ &= W^{1/2}(\mathbf{x})W^{1/2}(\mathbf{y}) \int_0^\infty dt\, e^{-\gamma t} \int d\mu_{0,\mathbf{x},\mathbf{y};t} \exp\left[-\lambda \int_0^t W(\boldsymbol{\omega}(s))\, ds\right] \end{aligned}$$

We would like a more general formula:

$$F(K_\gamma)(x, y) = W^{1/2}(\mathbf{x})W^{1/2}(\mathbf{y}) \int_0^\infty dt\, e^{-\gamma t} \int d\mu_{0,\mathbf{x},\mathbf{y};t}\, g\left(\int_0^t W(\boldsymbol{\omega}(s))\, ds\right) \tag{8.4}$$

The transformation $g \mapsto F$ should be linear and, by the above, should take $g(y) = e^{-\lambda y}$ into $F(x) = x(1 + \lambda x)^{-1}$. We thus try

$$F(x) = x \int_0^\infty e^{-y} g(xy)\, dy \tag{8.5}$$

or in terms of

$$f(y) = g(y)y \tag{8.6}$$

$$F(x) = \int_0^\infty e^{-y} f(xy)\, dy/y \tag{8.7}$$

Theorem 8.2 (Lieb's formula [171]) Let $W \in L^q(\mathbb{R}^\nu) + L^p(\mathbb{R}^\nu)$ with $q = \nu/2$ ($\nu \geq 3$), $q > 1$ ($\nu = 2$), $q = 1$ ($\nu = 1$), and $q < p < \infty$, with $W \geq 0$. Let $\gamma \geq 0$ or $\gamma = 0$, $\nu \geq 3$. Let f be a nonnegative lower semicontinuous function on $[0, \infty)$ with $f(0) = 0$ and let f, g, F be related by (8.5)–(8.7). Then (both sides may be simultaneously infinite):

$$\operatorname{Tr}(F(K_\gamma)) = \int_0^\infty dt\, e^{-t\gamma} \int d\mathbf{x}\, W(\mathbf{x}) \int d\mu_{0,\mathbf{x},\mathbf{x};t}\, g\left(\int_0^t W(\boldsymbol{\omega}(s))\, ds\right) \tag{8.8}$$

$$= \int_0^\infty \frac{dt}{t} e^{-t\gamma} \int d\mathbf{x} \int d\mu_{0,\mathbf{x},\mathbf{x};t}\, f\left(\int_0^t W(\boldsymbol{\omega}(s))\, ds\right) \tag{8.9}$$

Proof We first note that since

$$F(x) = \int_0^\infty e^{-zx^{-1}} g(z)\, dz$$

f positive implies F is monotone increasing. Thus either formula for $\gamma > 0$ implies the formula for $\gamma = 0$ by application of the monotone convergence theorem on both sides (K_γ increases as γ does, so its eigenvalues increase, and so $\mathrm{Tr}(F(K_\gamma))$ is a sum of increasing functions). Similarly, since any lower semicontinuous function is a monotone limit of continuous functions, we can suppose that f is continuous with support compact in $(0, \infty)$ to get (8.9). As a final preliminary we note that (8.8) and (8.9) are equivalent: for with respect to the measure $d\nu = d\mathbf{x}\, d\mu_{0,\mathbf{x},\mathbf{x};t}$, the variables $\boldsymbol{\omega}(s_1), \ldots, \boldsymbol{\omega}(s_n)$ are identically distributed to $\boldsymbol{\omega}(s_1 \text{ "+" } a), \ldots, \boldsymbol{\omega}(s_n \text{ "+" } a)$ where $s \text{ "+" } a$ means addition mod t. Thus

$$\int dx \int d\mu_{0,\mathbf{x},\mathbf{x};t}\, g\left(\int_0^t W(\boldsymbol{\omega}(s))\, ds\right) W(\boldsymbol{\omega}(u)) \equiv \beta(u)$$

is independent of u so $t^{-1} \int_0^t \beta(u) = \beta(0)$ which says (8.8) equals (8.9).

Thus we need only prove (8.8) for $\gamma > 0$, and continuous f supported in some $[a, b]$, $0 < a < b < \infty$. Suppose in addition that W is in C_0^∞. We have verified (8.4) for $g(y) = e^{-\lambda y}$. Since (by Stone–Weierstrass) sums of such functions are dense in the continuous functions on $[0, \infty)$ vanishing at infinity, (8.4) holds for the g in question. As in the proof of Theorem 6.6, the right-hand side of (8.4) is continuous in $\mathbf{x}$ and $\mathbf{y}$, so by a general result, $\mathrm{Tr}(F(K_\gamma))$ can be evaluated by setting $\mathbf{x} = \mathbf{y}$ and integrating ($F(K_\gamma)$ is a positive operator with continous kernel; see [216, 259]).

Next suppose that $W \in L^\infty$ with compact support. Then we can find $W_n \in C_0^\infty$ and a bounded set S so that $W_n \le a\chi_S$, $W_n \to W$ pointwise almost everywhere and in $L^{\nu/2}$ ($\nu \ge 3$; otherwise take L^2). Then $K_\gamma(W_n) \to K_\gamma(W)$ in norm, so that $F(K_\gamma(W_n)) \to F(K_\gamma(W))$ in norm. It follows (see [259]) that the traces converge if they are always finite (including for $F(K_\gamma(a\chi_S))$. Thus we will obtain (8.8) if we can show that the right sides converge. But for each t, and almost every $\mathbf{x}$ and $\boldsymbol{\omega}$, the integrands converge so it suffices to obtain a uniform L^1 bound on the integrand. Write $g(x) = x^m h(x)$ and let $d\alpha = d\mathbf{x}\, d\mu_{0,\mathbf{x},\mathbf{x};t}$ as above. Then ($\chi = \chi_S$):

$$W_n(\mathbf{x}) g\left(\int_0^t W_n(\boldsymbol{\omega}(s))\, ds\right) \le \|h\|_\infty a^m \chi(\mathbf{x}) \int_0^t ds_1 \cdots \int_0^t ds_m \prod_{i=1}^m \chi(\boldsymbol{\omega}(s_i))$$

so we only need that

$$\int_0^\infty dt\, e^{-\gamma t} \int d\alpha \int_0^t ds_1 \cdots \int_0^t ds_m\, \chi(\mathbf{x}) \cdots \chi(\boldsymbol{\omega}(s_m)) < \infty$$

But, by Hölder's inequality:

$$\int d\alpha\, \chi(\mathbf{x}) \cdots \chi(\boldsymbol{\omega}(s_m)) \leq \int d\alpha\, \chi^m(\mathbf{x})$$

so the above inequality is equivalent to

$$\int_0^\infty dt\, e^{-\gamma t} t^{-\nu/2} t^m < \infty$$

which is obvious if we take m large enough. This establishes (8.8) in case W is in L^∞ with compact support.

Finally, let W obey the hypotheses of the theorem. Choose W_n in L^∞ with compact support so that $W_n \uparrow W$. Since the eigenvalues of $K_\gamma(W_n)$ are monotone and F is monotone, the left sides of (8.8) converge. The integrands on the right converge pointwise, so it suffices to prove domination by an L^1-function. Let $h \geq g$ so that h has support in $[c, \infty)$ $(c > 0)$, h bounded and h monotone increasing. Then we only need

$$i \equiv \int_0^\infty dt\, e^{-\gamma t} \int dx \int d\mu_{0,\mathbf{x},\mathbf{x};t}\, h\left(\int_0^t W(\boldsymbol{\omega}(s))\, ds\right) < \infty$$

But (8.8) with g replaced by h and F by

$$H(x) = x \int_0^\infty e^{-y} h(xy)\, dy$$

holds by approximating with W_n's (here the integrals converge by using the monotone convergence theorem). Thus

$$i = \operatorname{Tr}(H(K_\gamma))$$

Notice that since $h(x) = x^m q(x)$ with $q \in L^\infty$

$$H(x) \leq x^{m+1} \|q\|_\infty \Gamma(m+1)$$

so $i < \infty$ follows if $\operatorname{Tr}(K_\gamma^m) < \infty$ for some m. But the hypothesis on W implies that $\operatorname{Tr}(K_\gamma^m) < \infty$ so long as $m > p$ [216]. ∎

Remark By a further approximation argument, one can easily extend (8.8), (8.9) to arbitrary measurable f's in L^1_{loc} with the property that

$$\varlimsup_{x \downarrow 0} \frac{|f(x)|}{|x|^\alpha} < \infty \qquad \text{for some} \quad \alpha > p$$

and

$$\varlimsup_{x \to \infty} \frac{|f(x)|}{|x|} < \infty \qquad (\gamma = 0)$$

or

$$\int |f(x)| e^{-\beta x} \, dx < \infty \qquad \text{for all} \quad \beta \qquad (\gamma > 0)$$

9. Phase Space Bounds

The main point here is that of Symanzik [270] and Lieb [171] that to get classical phase space bounds from path integrals, one need only apply Jensen's inequality:

Proposition 9.1 (Jensen's inequality) Let f be convex on $\mathbb{R}$ (**convex** means $f(\theta a + (1-\theta)b) \le \theta f(a) + (1-\theta) f(b)$; $0 < \theta < 1$, $a, b \in \mathbb{R}$). Then, for any probability measure, ν, on $\mathbb{R}$:

$$f\left(\int x \, d\nu(x)\right) \le \int f(x) \, d\nu(x) \tag{9.1}$$

so long as $\int |x| \, d\nu(x) < \infty$. If ν is supported on $[\alpha, \infty)$ for some $\alpha > -\infty$ and $f(\infty) \equiv \lim_{y \to \infty} f(y)$, then this last condition may be dropped.

Proof We first note that by convexity, for any y, we can find a with

$$f(x) - f(y) \ge a(x - y)$$

for all x; for example, convexity implies that $(f(x) - f(y))(x - y)^{-1}$ is monotone decreasing as $x \downarrow y$, so that $df(y + 0)/dx$ exists and convexity implies that one can take this value for a. Take $y = \int x \, d\nu(x)$ and integrate the above inequality with respect to $d\nu(x)$. Then since $\int (x - y) \, d\nu(x) = 0$:

$$\int f(x) \, d\nu(x) \ge f(y)$$

The case where $\int x \, d\nu(x) = \infty$ is handled by a simple limiting argument. ∎

By letting $d\nu$ be the probability distribution for X, we see that

$$f(E(X)) \le E(f(X)) \tag{9.2}$$

for any random variable X which is either bounded from below or has $E(|X|) < \infty$.

The basic philosophy of how to get phase space bounds is illustrated by the following theorem.

Theorem 9.2 Let V be in $L^1_{\text{loc}}(\mathbb{R}^\nu)$ and bounded from below. Let $H(\hbar) = -\hbar^2/2m\Delta + V$. Then

$$\operatorname{Tr}(\exp(-\alpha H(\hbar))) \le \int \frac{d^\nu p\, d^\nu x}{(2\pi\hbar)^\nu}\, e^{-\alpha(p^2/2m + V(\mathbf{x}))} \tag{9.3}$$

Proof By absorbing the α into V and/or $\hbar$ and doing the explicit p integration on the right-hand side, (9.3) is equivalent to

$$\operatorname{Tr}(\exp(-t(H_0 + V))) \le \int (2\pi t)^{-\nu/2} \exp(-tV(\mathbf{x}))\, d^\nu x \tag{9.4}$$

Suppose temporarily the following

$$\operatorname{Tr}(\exp(-t(H_0 + V))) = \int d^\nu x \int d\mu_{0,\mathbf{x},\mathbf{x};t} \exp\left(-\int_0^t V(\boldsymbol{\omega}(s))\, ds\right) \tag{9.5}$$

which is formally "obvious" from the Feynman–Kac formula. Since e^{-x} is convex and ds/t on $[0, t]$ is a probability measure, (9.2) implies that

$$\exp\left(-\int_0^t V(\boldsymbol{\omega}(s))\, ds\right) \le \int_0^t \exp(-tV(\boldsymbol{\omega}(s)))\, \frac{ds}{t}$$

so, using Fubini's theorem:

$$\begin{aligned}\text{r.h.s. of (9.5)} &\le \int d\mu_{0,\mathbf{0},\mathbf{0};t} \int_0^t \frac{ds}{t} \int d^\nu x \exp(-tV(\mathbf{x} + \boldsymbol{\omega}(s))) \\ &= \int d\mu_{0,\mathbf{0},\mathbf{0};t} \int_0^t \frac{ds}{t} \int d^\nu x \exp(-tV(\mathbf{x})) \\ &= \text{r.h.s. of (9.4)}\end{aligned}$$

where we used the translation invariance of $d^\nu x$ in the first equality and the fact that the resulting integrand is independent of s and $\boldsymbol{\omega}$ in the last step.

Thus all we need is to prove (9.5). We prove the more general:

$$\operatorname{Tr}(f e^{-t(H_0+V)} f) = \int d^\nu x\, f^2(\mathbf{x}) \int d\mu_{0,\mathbf{x},\mathbf{x};t} \exp\left(-\int_0^t V(\boldsymbol{\omega}(s))\, ds\right) \tag{9.6}$$

For $f \in C_0^\infty$ and $V \in C_0^\infty$, this follows from Theorem 6.6. For $V \in L^\infty$ with compact support and $f \in C_0^\infty$, we then obtain (9.6) by a limiting argument of the type used in Theorem 8.2 and then using the monotone convergence theorems for integrals and forms for arbitrary $V \in L^1_{\text{loc}}$, $V \ge a > -\infty$, and $f \in C_0^\infty$. Using monotone convergence again, we can obtain the result as $f \to 1$. ▮

Remarks 1. (9.3) is a celebrated inequality of Golden [111], Thompson [275], and Symanzik [270] obtained in [111, 275] from the inequality $\mathrm{Tr}(e^{A+B}) \le \mathrm{Tr}(e^{A/2}e^{B}e^{A/2})$. The above proof is that of Symanzik [270].

2. By simple limiting arguments the result extends to any V with $V_- - \Delta$ form bounded with relative bound zero.

The most beautiful phase space bound is on $N(V) \equiv N_0(V)$. In a semi-classical picture,

$$N_{sc}(V) \equiv (2\pi)^{-\nu}\tau_\nu \int |V_-(\mathbf{x})|^{\nu/2}\, d^\nu x$$

$$= |\{(\mathbf{p}, \mathbf{x}) | p^2 + V(\mathbf{x}) < 0\}|/(2\pi)^\nu$$

where $\tau_\nu \equiv$ volume of unit ball in $\mathbb{R}^\nu$.

Theorem 9.3 Let $\nu \ge 3$ and let $V \in L^{\nu/2}(\mathbb{R}^\nu)$. Then

$$N(V) \le a_\nu \int |V_-(\mathbf{x})|^{\nu/2}\, d^\nu x \tag{9.7}$$

for some universal constant a_ν.

Proof Since $N(V) \le N(-V_-)$, we can suppose $V = -W \le 0$. By the Birman–Schwinger principle for any f, F related by (8.7)

$$N(V) \le F(2)^{-1}\, \mathrm{Tr}(F(K_{\gamma=0})) \qquad (\text{since } F \text{ is monotone})$$

$$= F(2)^{-1} \int_0^\infty t^{-1}\, dt \int dx \int d\mu_{0,\mathbf{x},\mathbf{x};t}\, f\left(\int_0^t tW(\boldsymbol{\omega}(s))\, \frac{ds}{t}\right)$$

by Lieb's formula (8.9). If moreover f is convex, then $f(\int_0^t tW(\omega(s))\, ds/t) \le \int_0^t (ds/t)\, f(tW(\omega(s)))$. As in the proof of Theorem 9.2, we can now interchange the ds and $d\mu(\boldsymbol{\omega})$ integration with the x integration, eliminate $\boldsymbol{\omega}(s)$, and then trivially do the s and $\boldsymbol{\omega}$ integrations. This result is

$$N(V) \le F(2)^{-1} \int_0^\infty (2\pi)^{-\nu/2} t^{-1-\nu/2}\, dt \int d^\nu x\, f(tW(\mathbf{x})) = \tilde{a}_\nu \int d^\nu x\, W(\mathbf{x})^{\nu/2}$$

where

$$\tilde{a}_\nu = (2\pi)^{-\nu/2} \int_0^\infty s^{-1-\nu/2} f(s)\, ds/F(2)$$

(for change variables from t to $s = tW(\mathbf{x})$). Notice that for $\nu \ge 3$, $\tilde{a}_\nu < \infty$ for we can take $f \equiv 0$ near $s = 0$ and $f(s) \le Cs$ near infinity. But for $\nu = 1, 2,$

the integral diverges at infinity since $\lim_{s\to\infty} f(s)/s > 0$ so long as $f \geq 0$, f not zero identically. ∎

(9.7) was proven independently by Cwickel [40], Lieb [171], and Rosenbljum [224]. Lieb [171], whose proof is given above gets the best value for the constant a_ν among the three. Namely, if *we define*

$$a_\nu = \sup\left\{N(V)\left[\int |V(\mathbf{x})|^{\nu/2}\, d^\nu x\right]^{-1} \,\middle|\, V \in L^{\nu/2}\right\} \tag{9.8}$$

Then we have shown that

$$a_\nu \leq \inf\left\{(2\pi)^{-\nu/2}\int_0^\infty s^{-1-\nu/2} f(s)\, ds \,\middle|\, f \text{ is convex and } 1 = \int_0^\infty e^{-y} f(2y)\,\frac{dy}{y}\right\}$$

In particular, by minimizing over f's of the form:

$$f(s) = \begin{cases} 0, & 0 < s < s_0 \\ \alpha(s - s_0), & s_0 < s \end{cases}$$

Lieb finds [176]:

$$a_3 \leq 0.116 \equiv a_3(L) \tag{9.9}$$

There are two natural lower bounds on the precise value of a_ν. We will prove later (Section 10) that $N(\lambda V)/N_{\mathrm{sc}}(\lambda V) \to 1$ as $\lambda \to \infty$. This immediately implies that

$$a_\nu \geq \frac{\tau_\nu}{(2\pi)^\nu} \equiv a_\nu(c) \tag{9.10}$$

the "classical value."

Another lower bound on a_ν was found by Glaser *et al.* [103]: If $N(V) < 1$ then $-\Delta + V \geq 0$, so taking expectation values in the vector $\phi = |V|^{(\nu-2)/4}$ we find that ($p = 2\nu/(\nu - 2)$)

$$N(-|\phi|^{4/(\nu-2)}) < 1 \Rightarrow \int |\phi|^p\, d^\nu x \leq \int |\nabla\phi|^2\, d^\nu x$$

But, $\int |\phi|^p\, d^\nu x = \int |V|^{\nu/2}\, d^\nu x$ so

$$\int |\phi|^p\, d^\nu x < a_\nu^{-1} \Rightarrow \int |\phi|^p\, d^\nu x \leq \int |\nabla\phi|^2\, d^\nu x$$

or (using the freedom to scale ϕ), for any ϕ,

$$\left(\int |\phi|^p\, d^\nu x\right)^{1/p} \leq a_\nu^{1/\nu}\left(\int |\nabla\phi|^2\, d^\nu x\right)^{1/2}$$

Thus a_ν yields a constant in Sobolev's inequality; i.e., if

$$c_\nu \equiv \sup\{\|\phi\|_p \mid \|\nabla\phi\|_2 = 1; p^{-1} = \tfrac{1}{2} - 1/\nu\}$$

is the best constant in Sobolev's inequality; then

$$a_\nu \geq c_\nu^\nu \equiv a_\nu(s) \tag{9.11}$$

The bounds (9.10) and (9.11) correspond to the extreme cases $N(V) \sim 0$ or $N(V) \sim \infty$. Lieb and Thirring [176] made the following natural conjecture:

$$a_\nu \equiv \max(a_\nu(c), a_\nu(s)) = \begin{cases} a_\nu(c) & (\nu \geq 8) \\ a_\nu(s) & (\nu \leq 7) \end{cases}$$

The $(\nu \geq 8)$, $(\nu \leq 7)$ results follow from looking at the precise values of c_ν and τ_ν all of which are explicitly known. The value $a_3(L)$ in (9.9) is fairly close to what should be the exact value since

$$a_3(L) = 1.49 a_3(s)$$

(or put differently $0.077 \leq a_3 \leq 0.116$).

Recently, Glaser *et al.* [102] have shown by explicit examples that the Lieb–Thirring conjecture is false for $\nu \geq 8$; i.e., a_ν is strictly larger than $a_\nu(c)$ [which is the larger of $a_\nu(c)$ and $a_\nu(s)$ for $\nu \geq 8$]; in their *examples* it is never larger by more than 36%. On the basis of their work, they make the weaker conjecture

$$a_\nu = a_\nu(s) \qquad (\nu \leq 7)$$

$$\lim_{\nu \to \infty} \frac{a_\nu}{a_\nu(c)} = 1$$

Glaser *et al.* also prove that for $\nu = 4$ and V *spherically symmetric*

$$N(V) \leq a_\nu(s) \int |V(x)|^2 \, d^\nu x$$

(strongly supporting the Lieb–Thirring conjecture for $\nu \leq 7$). Additional information on best constants in the moment inequalities can be found in [0].

* * *

For further applications of Brownian motion to bound state problems, see [260a].

* * *

We want to explain, following Lieb–Thirring [175], how (9.7) leads to a proof of the stability of matter. Since the modification that electron spin makes in the Pauli principle is easily accommodated, we will ignore its effect. *We emphasize that the remainder of this section is an aside not used in the later sections.* Fix N and M and consider the operator on $L^2(\mathbb{R}^{3N})$. Let

$$H_N(\mathbf{R}_1, \ldots, \mathbf{R}_M) = -\sum_{i=1}^{N} \Delta_i - \sum_{i=1}^{N} \sum_{j=1}^{M} |\mathbf{r}_i - \mathbf{R}_j|^{-1} + \sum_{1 \le i < j \le N} |\mathbf{r}_i - \mathbf{r}_j|^{-1} + \sum_{1 \le i < j \le M} |\mathbf{R}_i - \mathbf{R}_j|^{-1}$$

where $\mathbf{R}_1, \ldots, \mathbf{R}_M$ are parameters in $\mathbb{R}^3$, and a point in $\mathbb{R}^{3N}$ is written $(\mathbf{r}_1, \ldots, \mathbf{r}_N)$, $\mathbf{r}_i \in \mathbb{R}^3$. H_N leaves invariant the subspace L_a^2 of functions $\psi(\mathbf{r}_1, \ldots, \mathbf{r}_N)$ antisymmetric under interchange of the $\mathbf{r}_i$'s, so we define

$$E(N, M) = \inf\{\inf \operatorname{spec}(H_N(\mathbf{R}_j) \upharpoonright L_a^2) \mid \text{all } \mathbf{R}_j\}$$

i.e., the ground state energy of N electrons and M infinitely heavy protons. **Stability of matter** says that

$$E(N, M) \ge -c(N + M)$$

This was first proven by Dyson–Lenard [68] whose proof was difficult and whose value for c extremely large ($\sim 10^{17}$ in units where $c = 1$ is presumably correct). The proof we are about to give due to Lieb–Thirring [175] is not only a considerable simplification, but the constant is only about 30 times too large.

It depends on some facts in the Thomas–Fermi theory. Define a functional on ρ's in $L^{5/3}$ with $\rho \ge 0$ by

$$\mathscr{E}_M^{(\alpha)}(\rho; \mathbf{R}_1, \ldots, \mathbf{R}_M) = \alpha \int \rho^{5/3}(\mathbf{x})\, d^3x - \int W(\mathbf{x})\rho(\mathbf{x})\, d^3x + \frac{1}{2} \int \rho(\mathbf{x})\rho(\mathbf{y})|\mathbf{x} - \mathbf{y}|^{-1}\, d^3x\, d^3y + \sum_{1 \le i < j \le M} |\mathbf{R}_i - \mathbf{R}_j|^{-1}$$

with $W(\mathbf{x}) = \sum_{i=1}^{M} |\mathbf{x} - \mathbf{R}_i|^{-1}$. The terms in $\mathscr{E}$ "approximate" $(\Psi, H_N \Psi)$ in a suitable semiclassical approximation. We will prove below that

$$\mathscr{E}_M^{(\alpha)}(\rho; \mathbf{R}_1, \ldots, \mathbf{R}_M) \ge -d(\alpha)M \tag{9.12}$$

where $d(\alpha)$ is an α-dependent constant; this stability of matter in the Thomas–Fermi approximation comes from an elementary effect: there is no binding of "molecules" in the Thomas–Fermi theory.

Let Ψ be a normalized antisymmetric function on $\mathbb{R}^{3N}$ and let

$$\rho_\Psi(\mathbf{x}) = N \int |\Psi(\mathbf{x}, \mathbf{x}_2, \ldots, \mathbf{x}_N)|^2 \, dx_2 \cdots dx_N \tag{9.13}$$

We are heading towards a proof of the fact that, for a suitable α and q:

$$(\Psi, H_N(\mathbf{R}_1, \ldots, \mathbf{R}_M)\Psi) \geq \mathscr{E}_M^{(\alpha)}(\rho_\Psi; \mathbf{R}_1, \ldots, \mathbf{R}_M) - qN \tag{9.14}$$

whence stability of matter will follow from (9.12). (9.14) follows from

$$\left(\Psi, \left(-\sum_{i=1}^{N} \Delta_i\right)\Psi\right) \geq \alpha_0 \int \rho_\Psi^{5/3}(\mathbf{x}) \, d^3x \tag{9.15}$$

$$\left(\Psi, \sum_{i<j} |\mathbf{r}_i - \mathbf{r}_j|^{-1}\Psi\right) \geq -\frac{1}{2}\alpha_0 \int \rho_\Psi^{5/3}(\mathbf{x}) \, d^3x - d(\tfrac{1}{2}\alpha_0)N + \frac{1}{2}\int |\mathbf{x} - \mathbf{y}|^{-1}\rho_\Psi(\mathbf{x})\rho_\Psi(\mathbf{y}) \, dx \, dy \tag{9.16}$$

$$\left(\Psi, \left(\sum_{i=1}^{N}\sum_{j=1}^{M} |\mathbf{r}_i - \mathbf{R}_j|^{-1}\right)\Psi\right) = \int W(\mathbf{x})\rho_\Psi(\mathbf{x}) \, dx \tag{9.17}$$

[for one can take $\alpha = \frac{1}{2}\alpha_0$ and $q = d(\frac{1}{2}\alpha_0)$ in (9.14)]. Equation (9.17) is obvious. The inequality (9.16) comes from (9.12) by the following clever device [175]: Take $\rho = \rho_\Psi$ and take $N = M$ and notice that

$$\begin{aligned}\int \mathscr{E}_N^{(\alpha)}(\rho_\Psi; \mathbf{R}_1, \ldots, \mathbf{R}_N)|\Psi(\mathbf{R}_1, \ldots, \mathbf{R}_N|^2 \, dR_1 \cdots dR_N \\ = \alpha \int \rho_\Psi^{5/3}(\mathbf{x}) \, d^3x - \int \rho_\Psi(\mathbf{x})|\mathbf{x} - \mathbf{y}|^{-1}\rho_\Psi(\mathbf{y}) \, d^3x \, d^3y \\ + \frac{1}{2}\int \rho_\Psi(\mathbf{x})\rho_\Psi(\mathbf{y})|\mathbf{x} - \mathbf{y}|^{-1} \, d^3x \, d^3y + \text{l.h.s of (9.16)}\end{aligned} \tag{9.18}$$

But, by (9.12), the left-hand side of (9.18) is greater than or equal to $-d(\alpha)N$.

This leaves the proof of (9.15) and it is here that the semiclassical bound (9.7) enters:

Theorem 9.4 ([175]) Let $e_j(V)$ denote the negative eigenvalues of $-\Delta + V$. Let $v \geq 3$. Then, for any $k > 0$ and $V \in L^{(1/2)v+k}$:

$$\sum |e_j(V)|^k \leq b_v(k) \int d^v x |V_-(\mathbf{x})|^{k+(1/2)v} \tag{9.19}$$

with $b_v(k) = ka_v \int_0^1 y^{k-1}(1 - y)^{v/2} \, dy$ (which is expressible in terms of Γ-functions).

Proof By a limiting argument, it suffices to prove (9.19) for $V \in C_0^\infty$ with $V \leq 0$. It is not hard to see that

$$N_\alpha(V) \leq N_0((V + \alpha)_-)$$

so that (9.7) implies that

$$N_\alpha(V) \leq a_\nu \int |(V + \alpha)_-|^{\nu/2}\, d^\nu x \tag{9.20}$$

In terms of Stieltjes integrals:

$$\begin{aligned} \sum |e_j(V)|^k &= -\int \alpha^k\, dN_\alpha \\ &= k \int \alpha^{k-1} N_\alpha\, d\alpha \qquad \text{(integrating by parts)} \\ &\leq k a_\nu \int d^\nu x \int_0^{-V(x)} \alpha^{k-1} |V(x) + \alpha|^{\nu/2}\, d\alpha \end{aligned}$$

by (9.20). We obtain (9.19) by scaling (i.e., change variables from α to $y = \alpha/[-V(x)]$). ∎

Remark There are also bounds of the genre of (9.19) for $\nu = 2$, $k > 0$ and $\nu = 1$, $k > \frac{1}{2}$ (see [176]):

Theorem 9.5 ([175]) Let $\Psi(\mathbf{r}_1, \ldots, \mathbf{r}_N) \in L^2(\mathbb{R}^{\nu N})$ be normalized and antisymmetric with respect to its $\mathbf{r}_i \in \mathbb{R}^\nu$ variables. Let ρ_Ψ be given by (9.13). Then ($H_0 = \sum_{i=1}^N - \Delta_i$)

$$(\Psi, H_0 \Psi) \geq l_\nu \int \rho_\Psi^{(\nu+2)/\nu}\, d^\nu x \tag{9.21}$$

where $l_\nu = b_\nu(1)^{-2/\nu}(2/2 + \nu)^{(2+\nu)/2}(\nu/2)$.

Proof Let $V(\mathbf{x}) = \rho_\Psi(\mathbf{x})^{2/\nu}$. Let $H_N = \sum_{i=1}^N (-\Delta_i - \lambda V(\mathbf{x}_i))$. Then, one can write down the eigenfunctions of H_N explicitly in terms of eigenfunctions of $-\Delta - \lambda V(\mathbf{x})$ and so determine that $H_N \restriction L_a^2 \geq -\sum_{j=1}^N |e_j(-\lambda V)|$, so by (9.19)

$$(\Psi, H_N \Psi) \geq -b_\nu(1)\lambda^{1+(1/2)\nu} \int \rho_\Psi^{(\nu+2)/\nu}(\mathbf{x})\, d^\nu x$$

or

$$(\Psi, H_0\Psi) \geq [\lambda - b_\nu(1)\lambda^{1+(1/2)\nu}] \int \rho_\Psi^{(\nu+2)/\nu}(\mathbf{x})\, d^\nu x$$

(9.21) follows from a maximization with respect to λ. ∎

We summarize what we have accomplished [modulo the proof of (9.12)]:

Theorem 9.6 stability of matter ([68, 175]) There is a universal constant c so that for any normalized antisymmetric Ψ in $L^2(\mathbb{R}^{3N})$ and any $\mathbf{R}_1, \ldots, \mathbf{R}_M \in \mathbb{R}^3$

$$\left(\Psi, \left[-\sum_{i=1}^{N} \Delta_i - \sum |\mathbf{r}_i - \mathbf{R}_j|^{-1} + \sum_{i<j} |\mathbf{r}_i - \mathbf{r}_j|^{-1} + \sum_{i<j} |\mathbf{R}_i - \mathbf{R}_j|^{-1}\right]\Psi\right)$$

$$\geq -c(N + M) \tag{9.22}$$

For further discussion, including how to go from (9.22) to a proof of non-collapse as $N, M \to \infty$, see Lieb [172].

It is very easy to see that

$$E(N, N) \leq -dN$$

for a suitable $d > 0$ and thus $-E(N, N)$ grows like N (indeed, $\lim_{N\to\infty} E(N, N)/N$ exists and is strictly negative; see [172]). If the antisymmetry restriction is removed, then the behavior is very different. Namely, let $E_0(N, M)$ be defined analogously to $E(N, M)$, but with L_a^2 replaced by L^2. Then, various methods, including those used above, show that

$$E_0(N, N) \geq -c' N^{5/3}$$

and Lieb [172a] has proven that

$$E_0(N, N) \leq -d' N^{5/3}$$

Therefore, if electrons were not fermions, bulk matter would collapse and we would not be here! We note that if the protons are given a finite mass, then the analog of $E_0(N, N)$ is believed to behave like $N^{7/5}$ rather than $N^{5/3}$, but a lower bound of this form has not been proven.

* * *

All that remains is a proof of (9.12). We sketch the argument; for details, see Lieb–Simon [173] whose proof of (9.12) uses heavily ideas of Teller

[274]. By a scaling argument, the α in the definition of $\mathscr{E}_M^{(\alpha)}$ can be absorbed, so we consider only the case $\alpha = \frac{3}{5}$; i.e.,

$$\mathscr{E}_M(\rho; \mathbf{R}_1, \ldots, \mathbf{R}_M) = \frac{3}{5}\int \rho^{5/3}(\mathbf{x})\, d^3x - \int W(\mathbf{x})\rho(\mathbf{x})\, d^3x$$
$$+ \frac{1}{2}\int \rho(\mathbf{x})\rho(\mathbf{y})|\mathbf{x} - \mathbf{y}|^{-1}\, d^3x\, d^3y + \sum_{i<j} |\mathbf{R}_i - \mathbf{R}_j|^{-1}$$
$$W(\mathbf{x}) = \sum |\mathbf{x} - \mathbf{R}_i|^{-1} \tag{9.23}$$

We will consider trial ρ's in

$$\mathscr{T} = \left\{\rho \geq 0 \,\Big|\, \int \rho^{5/3}(\mathbf{x})\, d^3x < \infty, \int \rho(\mathbf{x})|\mathbf{x} - \mathbf{y}|^{-1}\rho(\mathbf{y})\, d^3x\, d^3y < \infty\right\}$$

It is useful to consider the norms $\|\cdot\|_{5/3}$ and $\|\rho\|_+ = [\int \rho(\mathbf{x})|\mathbf{x} - \mathbf{y}|^{-1}\rho(\mathbf{y})]^{1/2}$ on $\mathscr{T}$.

Lemma 9.7 Fix $\mathbf{R}_1, \ldots, \mathbf{R}_M$. Then, there are linear functions l_1 and l_2 on $\mathscr{T}$ with

$$|l_1(\rho)| \leq aM\|\rho\|_{5/3}; \qquad |l_2(\rho)| \leq bM\|\rho\|_+$$

(a and b independent of M and $\mathbf{R}_i$) so that

$$\int W(\mathbf{x})\rho(\mathbf{x})\, d^3x = l_1(\rho) + l_2(\rho)$$

Proof It is easy to see that one need only consider the case $W(\mathbf{x}) = |\mathbf{x}|^{-1}$ (i.e., $M = 1$, $\mathbf{R}_1 = \mathbf{0}$). Let

$$W_2(\mathbf{x}) = \int_{|\mathbf{y}| \leq 1} |\mathbf{x} - \mathbf{y}|^{-1} \frac{3}{4\pi}\, d^3y$$

i.e., the potential due to a charge of one uniformly spread in a sphere of radius one. Then $W_2(\mathbf{x}) = |\mathbf{x}|^{-1}$ for $|\mathbf{x}| \geq 1$ and $W_2(\mathbf{x}) \leq |\mathbf{x}|^{-1}$ for $|\mathbf{x}| < 1$. Thus $W_1 \equiv W - W_2$ is in $L^{5/2}$ so $l_1(\rho) = \int W_1(\mathbf{x})\rho(\mathbf{x})\, d^3x$ is a bounded functional on $L^{5/3}$. By the positive definiteness of $|\mathbf{x} - \mathbf{y}|^{-1}$ and the Schwarz inequality:

$$\int W_2(\mathbf{x})\rho(\mathbf{x})\, d^3x \leq \|\rho\|_+ \left[\int_{\substack{|\mathbf{x}| \leq 1 \\ |\mathbf{y}| \leq 1}} |\mathbf{x} - \mathbf{y}|^{-1} \left(\frac{3}{4\pi}\right)^2 d^3x\, d^3y\right]^{1/2} \blacksquare$$

Theorem 9.8 Fix $\mathbf{R}_1, \ldots, \mathbf{R}_M$. Then

$$E(\mathbf{R}_1, \ldots, \mathbf{R}_M) = \inf_{\rho \in \mathscr{T}} \mathscr{E}_M(\rho; \mathbf{R}_1, \ldots, \mathbf{R}_M) \tag{9.24}$$

is finite (and bounded from below independently of $\mathbf{R}_i$ but depending on M). Moreover, there is a unique ρ in $\mathscr{T}$ for which the infimum in (9.24) is realized. This ρ, denoted by $\rho(\mathbf{x}; \mathbf{R}_1, \ldots, \mathbf{R}_M)$, obeys:

$$\phi(\mathbf{x}; \mathbf{R}_1, \ldots, \mathbf{R}_M) \equiv W(\mathbf{x}) - \int \rho(\mathbf{y})|\mathbf{x} - \mathbf{y}|^{-1} d^3y \geq 0$$

$$\rho = \phi^{3/2} \tag{9.25}$$

Proof By the lemma

$$\mathscr{E}_M(\rho) \geq \tfrac{3}{5}\|\rho\|_{5/3}^{5/3} - aM\|\rho\|_{5/3} + \tfrac{1}{2}\|\rho\|_+^2 - bM\|\rho\|_+$$

is clearly bounded from below. Moreover, $\mathscr{E}_M \to +\infty$ as either $\|\rho\|_{5/3}$ or $\|\rho\|_+ \to \infty$ so if we choose a sequence ρ_n with $\mathscr{E}_M(\rho_n) \to E(\mathbf{R}_1, \ldots, \mathbf{R}_M)$, then $\sup\|\rho_n\|_{5/3} < \infty$ and $\sup\|\rho_n\|_+ < \infty$. Thus, by the Banach–Alaoglu theorem, the infimum in (9.24) is taken on a set S which is compact in the topology of weak convergence, i.e., the topology in which $\rho_\alpha \to \rho$ if and only if $l(\rho_\alpha) \to l(\rho)$ for all l's continuous in *either* $\|\cdot\|_+$ or $\|\cdot\|_{5/3}$. Suppose that $\rho_\alpha \to \rho$ in this topology. By the lemma, $\int \rho_\alpha(\mathbf{x})W(\mathbf{x})\, d^3x \to \int \rho(\mathbf{x})W(\mathbf{x})\, d^3x$. Moreover, by the Hahn–Banach theorem, $\underline{\lim}\|\rho_\alpha\|_+ \geq \|\rho\|_+$ and $\underline{\lim}\|\rho_\alpha\|_{5/3} \geq \|\rho\|_{5/3}$. It follows that $\underline{\lim}\, \mathscr{E}_M(\rho_\alpha) \geq \mathscr{E}_M(\rho)$; i.e., $\mathscr{E}_M$ is lower semicontinuous. Since every lower semicontinuous function on a compact set takes its infimum, there is a minimizing ρ in $\mathscr{T}$. Its uniqueness follows from the strict convexity of $\mathscr{E}$; i.e.,

$$\mathscr{E}_M(\theta\rho_1 + (1 - \theta)\rho_2) < \theta\mathscr{E}_M(\rho_1) + (1 - \theta)\mathscr{E}_M(\rho_2); \quad 0 < \theta < 1; \quad \rho_1 \neq \rho_2$$

(which is easy to check).

Since $\rho(\mathbf{x}; \mathbf{R}_i)$ minimizes $\mathscr{E}_M$, it must obey the Euler–Lagrange equations $\partial\mathscr{E}_M/\partial\rho(\mathbf{x}) = 0$ at points where $\rho(\mathbf{x}) \neq 0$ ($\partial\mathscr{E}_M/\partial\rho(x) \geq 0$ at points where $\rho(\mathbf{x}) = 0$). Thus

$$\rho \equiv [\max(\phi, 0)]^{3/2}$$

where ϕ is given by (9.25). Thus, all that remains is the proof that $\phi \geq 0$.

Notice that the lemma shows that

$$\sup_{\mathbf{y}} \left| \int |\mathbf{x} - \mathbf{y}|^{-1} \rho_0(\mathbf{x})\, d^3x \right| \leq C[\|\rho_0\|_{5/3} + \|\rho_0\|_+]$$

for any ρ_0. This implies that $\phi - W$ is bounded. It is also continuous since near any $\mathbf{x}_0 \neq \mathbf{R}_i$ it is the sum of a harmonic function and the convolution of

$|\mathbf{x}|^{-1}$ with a $\tilde{\rho} \in L^\infty$ having compact support. Moreover, if we write $\rho = \rho_1 + \rho_0$ where ρ_1 has compact support and ρ_0 has small $\|\cdot\|_+$ and $\|\cdot\|_{5/3}$ norm, then we see that $\phi - W \to 0$ as $|\mathbf{x}| \to \infty$ (for $\int \rho_1(\mathbf{y})|\mathbf{x} - \mathbf{y}|^{-1} d^3y \to 0$ as $\mathbf{x} \to \infty$ since $\rho_1 \in L^1$). Thus $\phi \to 0$ at infinity. Clearly $\phi \to +\infty$ as $r \to R_i$. Now, let

$$S = \{\mathbf{x} \mid \phi(\mathbf{x}) < 0\}$$

which is open by continuity and disjoint from the $\mathbf{R}_i$. Since $\Delta\phi = 4\pi\rho$ (distributional sense) and $\rho = [\max(\phi, 0)]^{3/2}$, ϕ is harmonic on S, and therefore it takes its minimum value at $\partial S \cup \{\infty\}$. But clearly $\phi = 0$ on ∂S and $\phi \to 0$ at infinity, so the minimum value on S is zero. This is clearly impossible unless $S = \varnothing$. ▮

Remarks 1. The beautiful argument that $\phi \geq 0$ which is reused below is due to Teller [274].

2. By more work [173], one can show that $\phi > 0$, and that ϕ, ρ are C^∞ away from the $\mathbf{R}_i$ and that $\int \rho(\mathbf{x})\, dx = M$.

Now we generalize the problem slightly by adding parameters $z_i \geq 0$, and let

$$\mathscr{E}_M(\rho; \mathbf{R}_i, z_i) \equiv \frac{3}{5}\int \rho^{5/3}(\mathbf{x})\, d^3x - \int W_z(\mathbf{x})\rho(\mathbf{x})\, d^3x + \frac{1}{2}\int \rho(\mathbf{x})\rho(\mathbf{y})|\mathbf{x} - \mathbf{y}|^{-1}\, d^3x\, d^3y + \sum_{i<j} z_i z_j |\mathbf{R}_i - \mathbf{R}_j|^{-1}$$

with $W_z(\mathbf{x}) = \sum_{i=1}^M z_i|\mathbf{x} - \mathbf{R}_i|^{-1}$. As above $\mathscr{E}_M(\rho; \mathbf{R}_i, z_i)$ has a minimum $E(\mathbf{R}_i; z_i)$ and a minimizing $\rho(\mathbf{x}; \mathbf{R}_i, z_i)$ and corresponding $\phi(\mathbf{x}; \mathbf{R}_i, z_i)$.

Lemma 9.9 $E(\mathbf{R}_i, z_i)_{i \leq n} \to E(\mathbf{R}_i, z_i)_{i \leq n-1}$ as $z_n \to 0$. Moreover, in the region $z_i > 0$:

$$\frac{\partial E}{\partial z_i}(\mathbf{R}_j; z_j) = \lim_{\mathbf{x} \to \mathbf{R}_i} [\phi(\mathbf{x}; \mathbf{R}_j, z_j) - z_i|\mathbf{x} - \mathbf{R}_i|^{-1}]$$

Proof The first assertion follows easily from the bounds in Lemma 9.7. Formally the second assertion comes from writing

$$\frac{\partial E}{\partial z_i} = \frac{\partial}{\partial z_i}(\mathscr{E}(\rho(\mathbf{x}; \mathbf{R}_i, z_i); \mathbf{x}_i, z_i))$$

and noting that the $\partial\rho/\partial z_i$ terms are multiplied by $\partial\mathscr{E}/\partial\rho = 0$ since ρ is a minimum. The terms from $\partial\mathscr{E}/\partial z_i$ give precisely $-\int |\mathbf{x} - \mathbf{R}_i|\rho(\mathbf{x})\, d^3x + \sum_{j \leq i} z_j|\mathbf{R}_j - \mathbf{R}_i|^{-1}$ which is the limit in question. The rigorous points of this formal proof may be found in [173]. ▮

Lemma 9.10 (Teller's lemma) For fixed $\mathbf{x}$, $\mathbf{R}_1, \ldots, \mathbf{R}_M$, $\phi(\mathbf{x}; \mathbf{R}_i, z_i)$ is monotone increasing in z_i.

Proof Let $\bar{z}_i \geq z_i \geq 0$. Let

$$S = \{x \mid \phi(x; R_i, \bar{z}_i) < \phi(x; R_i, z_i)\}$$

S is open and avoids those R_i with $\bar{z}_i > z_i$. On S, the function $\psi \equiv \phi(\cdot, \bar{z}_i) - \phi(\cdot, z_i)$ obeys:

$$\begin{aligned} \Delta\psi &= 4\pi[\rho(\cdot, \bar{z}_i) - \rho(\cdot, z_i)] \\ &= 4\pi[\phi(\cdot, \bar{z}_i)^{3/2} - \phi(\cdot, z_i)^{3/2}] < 0 \end{aligned}$$

so ψ is superharmonic and thus it takes its minimum on $\partial S \cup \{\infty\}$. Since $\psi = 0$ on ∂S and $\psi \to 0$ at infinity, this minimum is zero so $S = \varnothing$. ∎

Theorem 9.11 (Teller's theorem)

$$\begin{aligned} E(\mathbf{R}_1, \ldots, \mathbf{R}_{N+M}; z_1, \ldots, z_{N+M}) \geq{} & E(\mathbf{R}_1, \ldots, \mathbf{R}_N; z_1, \ldots, z_N) \\ & + E(\mathbf{R}_{N+1}, \ldots, \mathbf{R}_{N+M}; z_{N+1}, \ldots, z_{N+M}) \end{aligned} \tag{9.26}$$

Proof Let $\Delta E =$ l.h.s. $-$ r.h.s. of (9.26). By the continuity as $z_i \to 0$, we need only prove $\partial \Delta E / \partial z_i \geq 0$ in the region $z_j > 0$. But, by Lemma 9.9, the derivative is a difference of ϕ's which is nonnegative by Teller's lemma. ∎

Remark Teller's theorem says that molecules do not bind in the Thomas–Fermi theory. For a physical "explanation" of this, see [274, 173].

Corollary 9.12 [Equation (9.12)]

$$E(\mathbf{R}_1, \ldots, \mathbf{R}_M; z_i = 1) \geq -dM$$

Proof By Teller's theorem, $E(\mathbf{R}_1, \ldots, \mathbf{R}_M; z_i = 1) \geq \sum_{i=1}^M E(\mathbf{R}_i, z_i = 1) = -dM$ since $E(\mathbf{R}, z = 1)$ is independent of $\mathbf{R}$ and finite. ∎

10. The Classical Limit

In the last section, we obtained classical bounds by managing to replace $\mathbf{x} + \boldsymbol{\omega}(s)$ (with respect to $d\mu_{0,\mathbf{0},\mathbf{0},t}$) by $\mathbf{x}$. In the limit as $t \to 0$, one expects $\boldsymbol{\omega}(s)$ to go to zero so that classical limits should be connected to $t \to 0$ limits. This idea of Kac [138, 142] will dominate this section. For another approach to the classical limit, see [252]. For a discussion of the classical limit for the pointwise solution of the heat equation, see the end of Section 18. The simplest result to prove is the folllowing.

Theorem 10.1 Let V be a continuous function on $\mathbb{R}^\nu$ with $\exp(-\beta V) \in L^1$. Let

$$Z_Q(\hbar) = \mathrm{Tr}\left[\exp\left(-\beta\left[-\frac{\hbar^2}{2m}\Delta + V\right]\right)\right]$$

$$Z_c(\hbar) = \int \frac{d^\nu p\, d^\nu x}{(2\pi\hbar)^\nu} \exp\left(-\beta\left[\frac{p^2}{2m} + V(\mathbf{x})\right]\right)$$

Then

$$\lim_{\hbar \downarrow 0} \frac{Z_Q(\hbar)}{Z_c(\hbar)} = 1 \tag{10.1}$$

Proof Let $W = \beta V$. By doing the p integration explicitly, (10.1) is seen to be equivalent to:

$$t^{\nu/2}\mathrm{Tr}[\exp(-t\,[H_0 + t^{-1}W])] \to (2\pi)^{-\nu/2} \int \exp(-W(\mathbf{x}))\, d^\nu x$$

as $t \downarrow 0$, where $H_0 = -\frac{1}{2}\Delta$ as usual. By (9.5), we need to show that

$$f(t) \equiv \int d^\nu x \int d\mu_{0,\mathbf{0},\mathbf{0},t} \exp\left(-t^{-1}\int_0^t W(\mathbf{x} + \boldsymbol{\omega}(s))\, ds\right)$$

obeys

$$f(t)t^{\nu/2} \to (2\pi)^{-\nu/2} \int \exp(-W(\mathbf{x}))\, d^\nu x \equiv f_0 \tag{10.2}$$

By Theorem 9.2, $f(t)t^{\nu/2} \le f_0$ for all t so $\overline{\lim}\, f(t)t^{\nu/2} \le f_0$.

Given δ, let $\Delta_{\mathbf{x}}^\delta$ be the hypercube of side δ centered at $\mathbf{x}$. Let $W_\delta(\mathbf{x}) = \max(W(\mathbf{y}) \mid y \in \Delta_x^\delta)$ and let $Q_\nu(\delta, t) = E(\mathbf{b}(s) \in \Delta_{\mathbf{0}}^\delta,$ all $0 \le s \le t \mid \mathbf{b}(t) = 0)$. Then, taking the contribution of those paths which stay in Δ_0^δ,

$$f(t) \ge (2\pi t)^{-\nu/2} Q_\nu(\delta, t) \int \exp(-W_\delta(x))\, dx$$

so

$$\underline{\lim}\, t^{\nu/2} f(t) \ge [\underline{\lim}\, Q_\nu(\delta, t)](2\pi)^{-\nu/2} \int \exp(-W_\delta(x))\, dx$$

Now $Q_\nu(\delta, t) = Q_1(\delta, t)^\nu$ by the independence of the components and as in the proof of Lemma 7.10,

$$Q_1(\delta, t) = P_{D;\,\delta}(0, 0; t)$$

where P_D has Dirichlet conditions at $x = \pm\delta$. By the method of images (see Proposition 7.17), we have that for each fixed δ, $Q_1(\delta, t) \to 1$ as $t \downarrow 0$, so $\underline{\lim}\, t^{\nu/2} f(t) \geq (2\pi)^{-\nu/2} \int \exp(-W_\delta)\, d^\nu x$. Now let $\delta \to 0$. Since $\exp(-W_\delta) \leq \exp(-W)$ and $W_\delta(x) \to W(x)$ pointwise (by continuity), $\int \exp(-W_\delta)\, d^\nu x \to \int \exp(-W)\, d^\nu x$. ∎

Remarks 1. Results of this genre, proven by different means, go back at least as far as Berezin [12]; see also Combes *et al.* [37]. The method of Lieb [170] can also be extended to prove this theorem [260b].

2. One can replace the continuity assumption by one that V is bounded from below and locally L^1 if one uses the fact that for any L^1-function W (which will be V restricted to a finite region)

$$\left(\int_{|\mathbf{y}| \leq \delta} dy\right)^{-1} \int_{|\mathbf{x}-\mathbf{y}| \leq \delta} W(\mathbf{x} + \mathbf{y})\, dy \to W(\mathbf{x})$$

as $\delta \to 0$ for almost everywhere x.

* * *

The remaining classical limit results will involve information on the growth of the number of eigenvalues of $H_0 + V$ which are less than E as $E \to \infty$, or of $H_0 + \lambda V$ which are less than zero as $\lambda \to \infty$. The latter case is the number of eigenvalues of $K_{\alpha=0}^{-1}$ less than $\lambda \to \infty$. In both cases, we will want to relate the divergence of $\dim E_{(0,a)}(A)$ as $a \to \infty$ for some operator to the divergence of $\mathrm{Tr}(e^{-tA})$ as $t \downarrow 0$ since it is the latter that we will compute using (9.5) or using Lieb's formula. More generally, we want to relate the divergence of $\int_0^a d\mu$ as $a \to \infty$ to that of $\int_0^\infty e^{-tx}\, d\mu(x)$ as $t \downarrow 0$. One direction of this relation is easy:

Theorem 10.2 (Abelian theorem) Let $d\mu$ be a (positive) Borel measure on $[0, \infty)$ and suppose that for some $\gamma \geq 0$, $C \geq 0$: $a^{-\gamma}\mu[0, a) \to C$ as $a \to \infty$. Then

$$\lim_{t \downarrow 0} t^\gamma \int e^{-tx}\, d\mu(x) = C\Gamma(\gamma + 1)$$

where $\Gamma(a) = \int_0^\infty e^{-x} x^{a-1}\, dx$.

Proof Let $G(a) = (a + 1)^{-\gamma} \int_0^a d\mu \equiv (a + 1)^{-\gamma}F(a)$. By hypothesis, $G(a) \to C$ as $a \to \infty$ and $g_0 = \sup_a |G(a)| < \infty$. Then, by a Stieltjes integration by parts,

$$t^\gamma \int e^{-tx}\, d\mu(x) = t^{\gamma+1} \int e^{-tx}F(x)\, dx$$

$$= t^{\gamma+1} \int e^{-tx}(x + 1)^\gamma G(x)\, dx$$

$$= \int e^{-y}(y + t)^\gamma G\left(\frac{y}{t}\right) dy$$

For $t \le 1$, $e^{-y}(y + t)^\gamma G(y/t) \le e^{-y}(y + 1)^\gamma g_0$ is in L^1; as $t \to 0$, $(y + t)^\gamma G(y/t) \to Cy^\gamma$ for y fixed, so the result follows from the dominated convergence theorem. ∎

The converse direction of this last theorem is much deeper. It depends critically on the positivity of $d\mu$ (for example, if $d\mu(x) = \sum_{n=0}^{\infty} (-1)^n \delta(x - n)$, then $\int_0^a d\mu$ does not have a limit but $\int e^{-tx}\, d\mu = 1/(1 + e^{-t}) \to \frac{1}{2}$ as $t \to 0$). Fortunately, a beautiful argument of Karamata [148] exists which makes the proof fairly easy:

Theorem 10.3 (Tauberian theorem) Let μ be a (positive) Borel measure on $[0, \infty)$ and suppose that $\int e^{-tx}\, d\mu < \infty$ for all $t > 0$ and that for some $\gamma \ge 0$, $D \ge 0$:

$$\lim_{t \downarrow 0} t^\gamma \int e^{-tx}\, d\mu(x) = D$$

Then

$$\lim_{a \to \infty} a^{-\gamma}\mu[0, a) = \frac{D}{\Gamma(\gamma + 1)} \tag{10.3}$$

Proof (version of Karamata [148] due to M. Aizenman, unpublished) If $\gamma = 0$, the result follows from the monotone convergence theorem, so suppose henceforth that $\gamma > 0$. Let $d\mu_t$ be the measure given by $\mu_t(A) = t^\gamma \mu(t^{-1}A)$ and let $dv = x^{\gamma-1}\, dx$ (so that $v_t \equiv v$). (10.3) says that

$$\lim_{t \to 0} \mu_t[0, 1) = cv([0, 1)) \tag{10.4}$$

where $c = D/\Gamma(\gamma)$. Suppose that we can show that

$$\lim_{t \to 0} \int f(x)\, d\mu_t(x) = c \int f(x)\, dv(x) \tag{10.4'}$$

for all $f \in C_0^\infty[0, \infty)$. Then using the fact that $\nu(\{1\}) = 0$, one easily sees that (10.4) holds (see, e.g., Proposition 17.2). By hypothesis

$$\lim_{t \to 0} \int e^{-x}\, d\mu_t(x) = c \int e^{-x}\, d\nu(x)$$

so the measures $e^{-x}\, d\mu_t$ are uniformly bounded. Thus (10.4′) follows from

$$\lim_{t \to 0} \int e^{-x} g(x)\, d\mu_t(x) = c \int g(x) e^{-x}\, d\nu(x) \tag{10.4″}$$

for a dense (in $\|\cdot\|_\infty$) set of g in $C_\infty[0, \infty)$, the continuous functions going to zero at infinity. But (10.4″) holds for $g(x) = e^{-nx}$ by hypothesis. Since polynomials in e^{-x} are dense in $C_\infty[0, \infty)$ by the Stone–Weierstrass theorem, we have proven (10.4″) for the required set. ∎

Remark The above results are called "Abelian" and "Tauberian" because of the earliest prototype results. If $d\mu(x) = \sum_{n=0}^\infty a_n \delta(x - n)$ and we let $\lambda = e^{-t}$, then $\lim_{t \downarrow 0} \int e^{-tx}\, d\mu(x) = \lim_{\lambda \uparrow 1} \sum_{n=0}^\infty a_n \lambda^n$ in which case Theorem 10.2 with $\gamma = 0$ just asserts that if a_n is absolutely summable then its "Abelian sum" is its ordinary sum. This is a famous result of Abel. Tauber was the first to consider converses in this case. Results of the genre of Theorem 10.3 were first obtained by Hardy–Littlewood but only with considerable effort.

There is one more result related to the above which we will need below:

Proposition 10.4 Let $d\mu$ be a positive measure on $[0, \infty)$. Suppose that for some $\gamma \geq 0$, $C \geq 0$

$$\lim_{a \to \infty} a^{-\gamma} \int_0^a d\mu(x) = C$$

Then

$$\lim_{a \to \infty} a^{-\gamma - 1} \int_0^a x\, d\mu(x) = \frac{\gamma}{\gamma + 1} C$$

Proof Let $F(a) = \int_0^a d\mu(x)$ and $G(a) = \int_0^a x\, d\mu(x)$. Then

$$G(a) = \int_0^a x\, dF = aF(a) - \int_0^a F(y)\, dy$$

Letting $H(a) = (a + 1)^{-\gamma}F(a)$, we see that (as in Abel's theorem)

$$a^{-\gamma-1}\int_0^a F(y)\,dy = \int_0^1 H(ya)(y + a^{-1})^\gamma\,dy \to C(\gamma + 1)^{-1}$$

as $a \to \infty$ (by the dominated convergence theorem). ∎

* * *

We can now return to the "classical" limit problems. Results of the following genre go back to Titchmarsh [276]; this kind of proof goes back to Ray [212].

Theorem 10.5 Let V be a continuous function on $\mathbb{R}^\nu$, going to infinity at infinity, so that for some $\gamma > 0$ and some $C > 0$

$$\lim_{E\to\infty}(2\pi)^{-\nu}E^{-\gamma}|\{(\mathbf{p}, \mathbf{x}) \mid \tfrac{1}{2}p^2 + V(\mathbf{x}) \le E\}| = C$$

Suppose, moreover, that

$$\lim_{E\to\infty}(2\pi)^{-\nu}E^{-\gamma}|\{(\mathbf{p}, \mathbf{x}) \mid \tfrac{1}{2}p^2 + V_\delta(\mathbf{x}) \le E\}| \equiv C_\delta \tag{10.5}$$

exists and $\lim_{\delta\downarrow 0} C_\delta = C$ where $V_\delta(\mathbf{x}) = \max\{V(\mathbf{y}) \mid \mathbf{y} \in \Delta_\mathbf{x}^\delta$, the hypercube of side δ about $\mathbf{x}\}$. Let $H = -\frac{1}{2}\Delta + V$ and let $n(E)$ be the number of eigenvalues of H less than E. Then

$$\lim_{E\to\infty} E^{-\gamma}n(E) = C$$

Remark For reasonable V's (e.g., polynomials), the quantity in (10.5) is δ-independent and equals C.

Proof By the Tauberian theorem, Theorem 10.3, we need only show that

$$\lim_{t\downarrow 0} t^\gamma\,\mathrm{Tr}(e^{-tH}) = C\Gamma(\gamma + 1) \tag{10.6}$$

Now, as in the proof of Theorem 10.1,

$$\mathrm{Tr}(e^{-tH}) \le (2\pi)^{-\nu}\int \exp[-t(\tfrac{1}{2}p^2 + V(\mathbf{x}))]\,d^\nu x\,d^\nu p$$

$$\mathrm{Tr}(e^{-tH}) \ge Q_\nu(\delta, t)(2\pi)^{-\nu}\int \exp[-t(\tfrac{1}{2}p^2 + V_\delta(\mathbf{x}))]\,d^\nu x\,d^\nu p$$

so that (10.6) follows by using the hypotheses and the Abelian theorem, Theorem 10.2. ∎

Remarks 1. Interestingly enough, Theorems 10.1 and 10.5 are intimately related. Both involve the behavior of $\mathrm{Tr}(e^{-t(H_0+\lambda V)})$ as $t \downarrow 0$. In one case, λ is fixed, and in the other, it is taken proportional to t^{-1}.

2. Using Proposition 10.4, one can immediately control the divergence of the sum of the jth powers of the eigenvalues of H which are smaller than E.

We can also recover the famous result of Weyl [285]:

Theorem 10.6 Let Ω be an arbitrary open set in $\mathbb{R}^\nu$ with $|\partial\Omega| = 0$. Let $H^{\Omega}_{0,\mathrm{D}} = -\frac{1}{2}$ (the Laplacian with Dirichlet boundary conditions in $\partial\Omega$). Let $n_\Omega(E)$ denote the number of eigenvalues of $H^{\Omega}_{0,\mathrm{D}}$ which are less than E. If the Lebesgue measure, $|\Omega|$, is finite, then

$$\lim_{E\to\infty} E^{-\nu/2} n_\Omega(E) = 2^{-\nu/2}\pi^{-\nu}\tau_\nu|\Omega|$$

where τ_ν is the volume of the unit ball in $\mathbb{R}^\nu$.

Remark See Section 21 for the definition of $H^{\Omega}_{0,\mathrm{D}}$ for general Ω. We assume here the relevant Feynman–Kac formulas proven there and in Section 22.

Proof Let $P_\Omega(\mathbf{x}, \mathbf{y}; t)$ be the integral kernel for $e^{-tH_{0,\mathrm{D}}}$ which is continuous on $\Omega \times \Omega$ so that

$$\mathrm{Tr}(e^{-tH_{0,\mathrm{D}}}) = \int_\Omega P_\Omega(\mathbf{x}, \mathbf{x}; t)\, d^\nu x$$

By a Tauberian theorem we need only show that as $t \downarrow 0$,

$$t^{\nu/2} \int P_\Omega(\mathbf{x}, \mathbf{x}; t)\, d^\nu x \to (2\pi)^{-\nu/2}|\Omega|$$

because

$$\tau_\nu = \pi^{\nu/2}\left[\Gamma\left(\frac{\nu}{2} + 1\right)\right]^{-1}$$

Now let P_0 be the P for $\Omega = \mathbb{R}^\nu$. Then, since (see Section 21)

$$\begin{aligned} P_\Omega(\mathbf{x}, \mathbf{x}; t) &= E\{\mathbf{x} + \mathbf{b}(s) \in \Omega; 0 \le s \le t \,|\, \mathbf{b}(t) = 0\} P_0(\mathbf{x}, \mathbf{x}; t) \\ &\le P_0(\mathbf{x}, \mathbf{x}; t) \end{aligned} \tag{10.7}$$

and $t^{\nu/2} P_0(\mathbf{x}, \mathbf{x}; t) \to (2\pi)^{-\nu/2}$, we need only prove that for almost every $\mathbf{x} \in \Omega$: $t^{\nu/2} P_\Omega(\mathbf{x}, \mathbf{x}; t) \to (2\pi)^{-\nu/2}$ and apply the dominated convergence theorem. But clearly, by (10.7), if $\Delta^\delta_{\mathbf{x}} \subset \Omega$:

$$Q_\nu(\delta, t) P_0(\mathbf{x}, \mathbf{x}; t) \le P_\Omega(\mathbf{x}, \mathbf{x}; t) \le P_0(\mathbf{x}, \mathbf{x}; t)$$

so the result is proven since $Q_\nu \to 1$ for any $\delta > 0$. ∎

Remarks 1. Actually, Wiener integrals, per se, are irrelevant to this proof; all that is relevant is the Tauberian theorem and the inequality for $(\Delta_{\mathbf{x}}^{\delta} \subset \Omega)$:

$$P_{\Delta_{\mathbf{x}}^{\delta}}(\mathbf{x}, \mathbf{x}; t) \leq P_{\Omega}(\mathbf{x}, \mathbf{x}; t) \leq P_0(\mathbf{x}, \mathbf{x}; t)$$

which can be proven using potential theory. In this form the proof is just that of Kac [140].

2. In the above, we established that $P_{\Omega}(\mathbf{x}, \mathbf{x}; t) \sim (2\pi t)^{-\nu/2}$ as $t \downarrow 0$. By related means, Kac [140] obtained the first few terms in an infinite asymptotic series $P_{\Omega}(\mathbf{x}, \mathbf{x}; t) \sim (2\pi t)^{-\nu/2} \sum_{n=0}^{\infty} a_n(\mathbf{x})t^n$. By very different means, Kannai [147] has obtained asymptotic series

$$P_{\Omega}(\mathbf{x}, \mathbf{y}; t) \sim (2\pi t)^{-\nu/2} \exp(-|\mathbf{x} - \mathbf{y}|^2/2t) \sum_{n=0}^{\infty} b_n(\mathbf{x}, \mathbf{y})t^n$$

(with $b_0 = 1$) for very general $\mathbf{x}$ and $\mathbf{y}$ (e.g., if Ω is convex, all $\mathbf{x}$ and $\mathbf{y}$).

3. There is a version of Weyl's theorem for unbounded regions due to Majda–Ralston [179a]: it involves the asymptotics of the S-matrix for acoustical scattering. The proof that Jensen–Kato [136a] gave of their result relates it directly to the ideas above; while they do not use path integrals for their estimates, one can.

Theorem 10.7 Let $V \leq 0$ be in $L^{\nu/2}(\mathbb{R}^\nu)$ with $\nu \geq 3$. Then

$$\lim_{\lambda \to \infty} \lambda^{-\nu/2} N(\lambda V) = (2\pi)^{-\nu}\tau_\nu \int |V(\mathbf{x})|^{\nu/2}\, d^\nu x$$

Proof We use Lieb's formula, (8.8), with $g(y) = 1$ (respectively, $g(y) = 0$) if $y > \beta$ (respectively, $y \leq \beta$) so that $F(x) = x \int_0^\infty e^{-y} g(xy)\, dy = xe^{-\beta/x}$. Thus letting μ_j denote the eigenvalues of $K_{y=0}$, we see that

$$\sum_{j=1}^{\infty} \mu_j e^{-\beta/\mu_j} = \int d^\nu x |V(\mathbf{x})| \int_0^\infty \frac{dt}{(2\pi t)^{\nu/2}} E\left(\int_0^t |V(\mathbf{x} + \mathbf{b}(s))|\, ds > \beta \,\middle|\, \mathbf{b}(t) = 0\right) \tag{10.8}$$

Change variables by $t = \beta\tau$, $s = \sigma\beta$ and find

$$\sum_{j=1}^{\infty} \mu_j e^{-\beta/\mu_j} = (2\pi)^{-\nu/2} \beta^{-(1/2)\nu + 1} \int d^\nu x |V(\mathbf{x})| \int_0^\infty \frac{d\tau}{\tau^{\nu/2}}$$

$$\times E\left(\int_0^\tau |V(\mathbf{x} + \mathbf{b}(\sigma\beta))|\, d\sigma > 1 \,\middle|\, \mathbf{b}(\beta\tau) = 0\right)$$

As $\beta \to 0$, $E(\int_0^\tau \cdots \mid \mathbf{b}(\beta\tau) = 0)$ is easily seen to go to zero (respectively, one) if $V(x)\tau < 1$ (respectively, $V(x)\tau > 1$) if V is continuous with $|\{\mathbf{y} \mid V(\mathbf{y}) = a\}| = 0$ for all a. Thus for $V \in C_0^\infty$ (in this case, the reader can provide the necessary justification of the interchange of $\int d^\nu x\, d\tau$ and $\lim_{\beta \downarrow 0}$), we see that

$$\lim_{\beta \downarrow 0} \beta^{(1/2)\nu - 1} \sum_{j=0}^{\infty} \mu_j e^{-\beta/\mu_j} = (2\pi)^{-\nu/2} \int d^\nu x |V(\mathbf{x})| \int_{1/V(\mathbf{x})}^{\infty} \frac{d\tau}{\tau^{\nu/2}}$$

$$= \left(\frac{2}{\nu - 2}\right)(2\pi)^{-(1/2)\nu} \int |V(\mathbf{x})|^{(1/2)\nu}\, d^\nu x$$

Applying a Tauberian theorem to the measure

$$d\alpha(x) = \sum \mu_j \delta(x - \mu_j^{-1})$$

we find that as $\lambda \to \infty$

$$\alpha(0, \lambda) \sim \lambda^{(1/2)\nu - 1}\left[\Gamma\left(\frac{\nu}{2}\right)\right]^{-1}\left(\frac{\nu}{2} - 1\right)^{-1}(2\pi)^{-\nu/2} \int |V(\mathbf{x})|^{\nu/2}\, d^\nu x$$

so applying Proposition 10.4:

$$\#\{\mu_j^{-1} < \lambda\} \sim \lambda^{\nu/2}\left[\Gamma\left(\frac{\nu}{2}\right)\right]^{-1}\left(\frac{\nu}{2}\right)^{-1}(2\pi)^{-\nu/2} \int |V(\mathbf{x})|^{\nu/2}\, d^\nu x$$

$$= \lambda^{\nu/2}\left[\Gamma\left(\frac{\nu}{2} + 1\right)\right]^{-1}(2\pi)^{-\nu/2} \int |V(\mathbf{x})|^{\nu/2}\, d^\nu x$$

$$= \lambda^{\nu/2}\tau_\nu 2^{-\nu/2}\pi^{-\nu} \int |V(\mathbf{x})|^{\nu/2}\, d^\nu x$$

But, by the Birman–Schwinger principle (Theorem 8.1)

$$N(\lambda V) = \#\{\mu_j^{-1} < \tfrac{1}{2}\lambda\}$$

which proves the result for nice V's. A limiting argument using Theorem 9.2 ([249]) handles the general case. ∎

Remarks 1. An identical proof shows that ($\nu = 1, 2$ is allowed for $\alpha > 0$)

$$N_{\lambda\alpha}(\lambda V) \sim \lambda^{\nu/2}(2\pi)^{-\nu}\tau_\nu \int |V(\mathbf{x}) + \alpha|^{\nu/2}\, d^\nu x$$

From this formula and $\sum |e_j(V)|^k = k \int \alpha^{k-1} N_\alpha(V)\, d\alpha$, one easily obtains limit theorems for $\sum |e_j(\lambda V)|^k$ as $\lambda \to \infty$.

2. In the next section we will need the $\gamma \neq 0$ analog of (10.8). Namely, if $\mu_j(\gamma)$ are the eigenvalues of K_γ and v *is arbitrary* ($\gamma > 0$), then

$$\sum_{j=1}^{\infty} \mu_j(\gamma)\exp(-1/\mu_j(\gamma)) = \int dx\, |V(\mathbf{x})| \int_0^\infty \frac{dt}{(2\pi t)^{v/2}} e^{-t\gamma} \times E\left(\int_0^t |V(\mathbf{x} + \mathbf{b}(s))|\, ds > 1 \,\middle|\, \mathbf{b}(t) = 0 \right) \quad (10.9)$$

3. Results of the genre of this theorem using rather different methods go back to Birman–Borzov [16], Martin [181], and Tamura [273]. The above proof is due to Kac [142].

4. The limit here should be classical since $N(\lambda V)$ is the number of bound states of $-\hbar^2\Delta + V$ with $\hbar = \lambda^{-1/2}$ so $\lambda \to \infty$ is the same as $\hbar \to 0$.

11. Recurrence and Weak Coupling

There is a striking difference between quantum mechanics in $v \geq 3$ and $v \leq 2$ dimensions. If $V \in L^{v/2}$, then for $v \geq 3$, $N(\lambda V) = 0$ for λ sufficiently small (this follows from Theorem 9.3, if we note that $N < 1$ implies $N = 0$). On the other hand, if $v = 1, 2$ and $V \leq 0$ ($V \not\equiv 0$), then $N(\lambda V) > 0$ for all $\lambda > 0$. There are a variety of ways of seeing this:

(a) If V is a spherical square well, one can solve $(-\Delta + \lambda V)u = Eu$ explicitly in terms of Bessel functions.

(b) (P. Lax, unpublished) One can construct rather simple trial functions u with $(u, (-\Delta + \lambda V)u) < 0$ in case V is a spherical square well.

(c) (Simon [251]) An elementary trial function argument shows that $\|K_\gamma\| \to \infty$ as $\gamma \to 0$. Since K_γ is compact, positive, and self-adjoint, its norm is an eigenvalue. Thus for any fixed λ, K_γ has at least one eigenvalue larger than $2\lambda^{-1}$ for all small γ.

(d) (Simon [248]; see also [17, 161]) One can develop a theory which gives explicit "series" for the actual weak coupling eigenvalue and which shows at the same time that such an eigenvalue exists.

Given all these relatively simple proofs, the considerations below have something of the character of using a sledgehammer to crack a peanut but the swing is still somewhat illuminating.

First, we mention some poetry of Kac [143] which has not yet been made into a rigorous proof. Let V be in C_0^∞, negative and strictly negative on some open set S, say, $V < -a$ on $S(a > 0)$. Then, if $H = -\frac{1}{2}\Delta + \lambda V$, and $f(x) \equiv 1$:

$$(e^{-tH}f)(0) = \int \exp\left(-\lambda \int_0^t V(\mathbf{b}(s))\, ds\right) Db$$

$$\geq \int \exp(a\lambda |\{s \leq t \mid \mathbf{b}(s) \in S\}|)\, Db$$

Recurrence (Theorem 7.13) tells us that $|\{s \leq t \mid \mathbf{b}(s) \in S\}| \to \infty$ as $t \to \infty$ almost everywhere in b, for $\nu = 1$ or 2 and thus in those dimensions $(e^{-tH}f)(0) \to \infty$ as $t \to \infty$. Thus, *as a map from* L^∞ *to* L^∞, $\|e^{-tH}\|$ diverges as $t \to \infty$ no matter how small λ is. The occurrence of a negative bound state is equivalent to $\|e^{-tH}\|$ diverging *as a map from* L^2 *to* L^2. Unfortunately, being unbounded as a map from L^∞ does not imply that the semigroup is unbounded on L^2; see [260a] for further discussion.

A proof that $\|K_\gamma\| \to \infty$ as $\gamma \downarrow 0$ can be based on (10.9) as follows below.

Theorem 11.1 Let $\nu = 1$ or 2. Let $V = -1$ on some cube C, 0 off C. Let $\mu_j(\gamma)$ be the jth eigenvalue of $K(\gamma)$, counting from the largest downwards. Then $\mu_1(\gamma) \to \infty$ as $\gamma \downarrow 0$. In particular, $H_0 + \lambda V$ has at least one bound state for any $\lambda > 0$.

Step 1 ($\sum_{j=2}^\infty \mu_j(\gamma) e^{-1/\mu_j(\gamma)}$ is bounded as $\gamma \downarrow 0$) Let H_1 (respectively, H_2) be $-\frac{1}{2}\Delta$ on $L^2(C)$ [respectively, $L^2(\mathbb{R}^\nu \backslash C)$] with Neumann boundary conditions on ∂C. Then (see, e.g., [217])

$$(H_0 + \gamma)^{-1} \leq (H_1 + H_2 + \gamma)^{-1} = (H_1 + \gamma)^{-1} \oplus (H_2 + \gamma)^{-1}$$

It follows that $K_\gamma \leq (H_1 + \gamma)^{-1} \oplus 0$. Thus, if e_n is the nth eigenvalue of H_1, listed in increasing order, $\mu_j(\gamma) \leq (e_j + \gamma)^{-1}$. Since only $e_1 = 0$ and $e_j \sim Cj^{1/\nu}$ we conclude that

$$\lim_{\gamma \downarrow 0} \sum_{j=2}^{\infty} \mu_j(\gamma) e^{-1/\mu_j(\gamma)} < \infty \tag{11.1}$$

The strategy now is to use (10.9) to show that $\sum_{j=1}^\infty \cdots \to \infty$ so that $\mu_1(\gamma) \to \infty$.

Step 2 (independence of finite times from infinite time) Let A be an event depending only on $\{\mathbf{b}(s) \mid 0 \leq s \leq t_0\}$. Then

$$E(A \mid \mathbf{b}(t) = 0) \to E(A) \qquad \text{as} \quad t \to \infty \tag{11.2}$$

For

$$E(A) = \int E(A \mid \mathbf{b}(t_0) = \mathbf{x})(2\pi t_0)^{-\nu/2} \exp(-\tfrac{1}{2}x^2 t_0^{-1})\, d^\nu x$$

and for $t > t_0$:

$$E(A \mid \mathbf{b}(t) = 0)$$
$$= \frac{\int E(A \mid \mathbf{b}(t_0) = \mathbf{x})(2\pi t_0)^{-\nu/2} \exp[-\tfrac{1}{2}x^2(t_0^{-1} + (t - t_0)^{-1})]\, d^\nu x}{\int (2\pi t_0)^{-\nu/2} \exp[-\tfrac{1}{2}x^2(t_0^{-1} + (t - t_0)^{-1})]\, d^\nu x}$$

on account of the fact that $\mathbf{b}(t) - \mathbf{b}(t_0)$ is independent of $\{\mathbf{b}(s) \mid 0 \leq s \leq t_0)$. (11.2) now follows from the monotone convergence theorem.

Step 3 (recurrence for paths conditioned on $\mathbf{b}(t) = 0$) We claim that in $\nu = 1$ or 2 dimensions, for each $\mathbf{x}$:

$$E\left(\int_0^t |V(\mathbf{x} + \mathbf{b}(s))|\, ds > 1 \,\middle|\, \mathbf{b}(t) = 0\right) \to 1 \tag{11.3}$$

as $t \to \infty$. Clearly for $t > t_0$

$$E\left(\int_0^t |V(\mathbf{x} + \mathbf{b}(s))|\, ds > 1 \,\middle|\, \mathbf{b}(t) = 0\right)$$
$$\geq E\left(\int_0^{t_0} |V(\mathbf{x} + \mathbf{b}(s))|\, ds > 1 \,\middle|\, \mathbf{b}(t) = 0\right)$$

Fix t_0, take $t \to \infty$ and use (11.2):

$$\lim_{t\to\infty} E\left(\int_0^t |V(\mathbf{x} + \mathbf{b}(s))|\, ds > 1 \,\middle|\, \mathbf{b}(t) = 0\right) \geq E\left(\int_0^{t_0} |V(\mathbf{x} + \mathbf{b}(s))|\, ds > 1\right)$$

Now take $t_0 \to \infty$ and use Theorem 7.13 to get (11.3).

Step 4 (completion of proof) Looking at (10.9) and using (11.3), one immediately sees that for $\nu = 1, 2$

$$\sum_{j=1}^{\infty} \mu_j(\gamma) \exp\left(\frac{-1}{\mu_j(\gamma)}\right) \to \infty$$

as $\gamma \downarrow 0$. Thus $\mu_1(\gamma) \to \infty$. ∎

Remarks 1. The proof actually shows much more. Namely

$$\mu_1(\gamma) \gtrsim \begin{cases} \log(\gamma^{-1}) & (\nu = 2) \\ \gamma^{-1/2} & (\nu = 1) \end{cases}$$

This is actually the precise behavior (see [248]).

2. The reader might think that $\nu = 1, 2$ enters critically in the above proof in Step 3. This is actually wrong, for while (11.3) will not hold if $\nu \geq 3$, the

lim–inf will have a nonzero value, and that was all we really needed in Step 4. The crucial place that $\nu = 1, 2$ enters is in the $t^{-\nu/2}$ at infinity. Thus, in some sense, the above proof does not so much derive weak coupling bound states from recurrence but rather shows they both come from a common cause (compare Proposition 7.17).

* * *

Although we have given a path integral proof that for $\nu \geq 3$ and V "nice" one has that $N(\lambda V) = 0$ for λ small, it is not a very direct proof via recurrence. Suppose that $W(\mathbf{x}) = -V(\mathbf{x})$ is the characteristic function of $\{\mathbf{x} \mid |\mathbf{x}| \leq 1\}$. We have already remarked in our discussion of recurrence that (see the Remark following Proposition 7.18)

$$E\left(\int_0^\infty W(\mathbf{b}(s))\, ds\right) = E(|\{s \mid |\mathbf{b}(s)| \leq 1\}|) < \infty$$

and one sees by a similar argument that

$$\sup_{\mathbf{x}} E\left(\int_0^\infty W(\mathbf{b}(s) + \mathbf{x})\, ds\right) < \infty \tag{11.4}$$

On the other hand, suppose we know that

$$a \equiv \sup_{\mathbf{x}} E\left(\exp\left[\lambda \int_0^\infty W(\mathbf{b}(s) + \mathbf{x})\, ds\right]\right) < \infty \tag{11.5}$$

Then

$$\begin{aligned}\|\exp[-t(H_0 + \lambda V)]f\|_\infty &= \sup_{\mathbf{x}} \left| E\left(\exp\left[\lambda \int_0^t W(\mathbf{b}(s) + \mathbf{x})\, ds\right] f(\mathbf{x} + \mathbf{b}(t))\right) \right| \\ &\leq \|f\|_\infty \sup_{\mathbf{x}} E\left(\exp\left[\lambda \int_0^t W(\mathbf{b}(s) + \mathbf{x})\, ds\right]\right) \\ &\leq a\|f\|_\infty\end{aligned}$$

Thus $e^{-t(H_0 + \lambda V)}$ is bounded by a as a map from L^∞ to L^∞. By duality and interpolation it is bounded by a, independently of t, as a map from L^2 to L^2. We conclude that $H_0 + \lambda V \geq 0$ if (11.5) holds, so $N(\lambda V) = 0$.

At first sight it seems unlikely that (11.4) implies (11.5) since $E(\exp(\lambda X)) < \infty$ is much stronger than $E(X) < \infty$. That makes the following result of Portenko [204a] especially striking:

Theorem 11.2 ([204a]) Let $W \geq 0$ be measurable. If

$$\gamma \equiv \sup_{\mathbf{x}} E\left(\int_0^\infty W(\mathbf{b}(s) + \mathbf{x})\, ds\right) < 1$$

then

$$\sup_{\mathbf{x}} E\left(\exp\left[\int_0^\infty W(\mathbf{b}(s) + \mathbf{x})\, ds\right]\right) \le (1 - \gamma)^{-1} < \infty$$

Proof Expanding the exponential, it suffices to show that

$$\int_{0 < s_1 < \cdots < s_n < \infty} E(W(\mathbf{b}(s_1) + \mathbf{x}) \cdots W(\mathbf{b}(s_n) + \mathbf{x}))\, ds_1 \cdots ds_n \le \gamma^n \quad (11.6)$$

since we can use the monotone convergence theorem to interchange E and the expansion. Fix $0 < s_1 < \cdots < s_{n-1}$. Let $\tilde{\mathbf{b}}(t) = \mathbf{b}(s_{n-1} + t) - \mathbf{b}(s_{n-1})$. Then

$$\int_{s_{n-1}}^\infty E(W(\mathbf{b}(s_1) + \mathbf{x}) \cdots W(\mathbf{b}(s_n) + \mathbf{x}))\, ds_n$$

$$= \int_0^\infty E(W(\mathbf{b}(s_1) + \mathbf{x}) \cdots W(\mathbf{b}(s_{n-1}) + \mathbf{x}) W(\tilde{\mathbf{b}}(t) + \mathbf{b}(s_{n-1}) + \mathbf{x}))\, dt$$

$$\le \int_0^\infty E(\cdots \tilde{E}(W(\tilde{\mathbf{b}}(t) + \mathbf{b}(s_{n-1}) + \mathbf{x}))\, dt$$

$$\le \gamma E(W(\mathbf{b}(s_1) + \mathbf{x}) \cdots W(\mathbf{b}(s_{n-1} + \mathbf{x})))$$

since $\tilde{\mathbf{b}}$ is independent of $\mathbf{b}(s_1), \ldots, \mathbf{b}(s_{n-1})$. (11.6) now follows by induction. ∎

Remarks 1. If viewed properly, one sees all that was really used was that $\mathbf{b}$ has the Markov property.

2. This result was rediscovered by Berthier–Gaveau [13] from whom we learned it. The proof is somehow a probabilistic version of some of the ideas of Kato [151] (see also [247]).

3. The above results can be rephrased in L^p language as saying that if $(-\Delta)^{-1}W$ is a strict contraction on L^1, then $N(V) = 0$. This can be seen without recourse to path integrals as follows: By duality $W(-\Delta)^{-1}$ is a strict contraction on L^∞, so the Stein interpolation theorem applied to $F(z) = W^z(-\Delta)^{-1}W^{(1-z)}$ implies that $W^{1/2}(-\Delta)^{-1}W^{1/2}$ is a strict contraction on L^2. By the Birman–Schwinger principle $N(V) = 0$. We note however that Theorem 11.2 is stronger than $N(V) = 0$, for $N(V) = 0$ says that e^{-tH} is a contraction L^2; Theorem 11.2 says it is a contraction on L^∞.

4. This result will play a central role in Section 25.

IV

Inequalities

12. Correlation Inequalities

Most of the inequalities discussed in this section and the next do not absolutely require path integrals for their statement or proof. Indeed, we will prove them by replacing $\int_0^t V(\boldsymbol{\omega}(s))\,ds$ by $\sum_{j=1}^n (t/n)V(\boldsymbol{\omega}(jt/n))$ and looking at the joint distribution of the $\{\boldsymbol{\omega}(jt/n)\}_{j=1}^n$. Thus we will undo the Trotter product formula proof of the Feynman–Kac formula; clearly, we could just use the Trotter formula and state all inequalities in terms of the semigroups. Path integrals are notationally and, more importantly, conceptually clarifying for the results.

In this section, we will consider a variety of inequalities called "correlation inequalities" after their original occurrence in the statistical mechanics of lattice gases. There the earliest results were obtained by Griffiths [115]; a bibliography on the subject including the large number of inequalities and their often impressive application to lattice gases would run to hundreds of papers! The applicability of the most general of these inequalities to Euclidean quantum field theory was discovered by Guerra *et al.* [120]; in essence, the results of this section are the specialization of their idea and its development to zero space dimensional quantum fields, i.e., $P(\phi)_1$-processes!

Consider the following generalization of the $P(\phi)_1$-process of Section 5:

Fix f, $g \in L^2(\mathbb{R})$ positive functions and V which we suppose continuous and nonnegative. Then

$$Z^{-1}f(\omega(0))g(\omega(t))\exp\left(-\int_0^t V(\omega(s))\,ds\right)d\mu_0(\omega) \equiv d\mu(\omega) \tag{12.1}$$

$$Z = (f, \exp(-t(H_0 + V))g)$$

is a probability measure on the paths $\{\omega(s) \mid 0 \le s \le t\}$. If $V = \frac{1}{2}x^2 + P$ and $g = f = \Omega_P$, then $d\mu$ is just the restriction of $d\nu_P$ to $\Sigma_{(0,t)}$ (see Theorem 6.7).

Now fix n and let us approximate $d\mu(\omega)$ by

$$Z_n^{-1} f(\omega(0) g(\omega(t)) \exp\left(-\sum_{j=1}^{n} V\left(\omega\left(\frac{jt}{n}\right)\right)\frac{t}{n}\right) d\mu_0(\omega) \equiv d\mu_n(\omega)$$
$$Z_n = (f, (e^{-tH_0/n} e^{-tV/n})^n g) \tag{12.2}$$

and consider the joint distribution of $\{\omega(jt/n)\}_{j=0}^{n}$ with respect to $d\mu_n$. It has the form

$$Z_n^{-1}\left[\exp\left(J \sum_{j=1}^{n} x_j x_{j-1}\right) \prod_{j=0}^{n} d\nu_j(x_j)\right] \tag{12.3}$$

where

$$J = n/t, \qquad d\nu_0(x) = f(x) e^{-(1/2)Jx^2}\, dx$$
$$d\nu_j(x) = e^{-Jx^2} e^{-(t/n)V(x)}\, dx \qquad (j = 1, \ldots, n-1)$$
$$d\nu_n(x) = g(x) e^{-(1/2)Jx^2} e^{-(t/n)V(x)}\, dx$$

The reader who has seen the theory of lattice gases [129, 133, 226] will recognize (12.3) as the probability distribution of an Ising-type ferromagnet except that the $d\nu(x_i) = \frac{1}{2}[\delta(x_i - 1) + \delta(x_i + 1)]$ of the Ising model is replaced by a more general type of distribution.

The following result of Ginibre [101] generalizes the classic inequalities of Griffiths [115] as extended by Kelly–Sherman [156]; hence, called GKS (Griffiths–Kelly–Sherman) inequalities. Below, in dealing with measures like (12.3), we will expand exponentials without concern about convergence. *All the theorems below should have additional hypotheses* (usually $\int e^{ax^2}\, d\nu_j(x) < \infty$ for suitable a will do) *which justify the convergence of these series.* In applications, one can prove the additional hypotheses or can use a limiting argument.

Theorem 12.1 ([101]) Let $\mathscr{F}_1$ be the functions on $\mathbb{R}$ which are nonnegative and monotone on $[0, \infty)$ and either even or odd (denoted respectively, $\mathscr{F}_1^e$, $\mathscr{F}_1^o$). Let $\mathscr{F}_n$ be the functions on $\mathbb{R}^n$ of the form $f_1(x_1) \cdots f_n(x_n)$ with $f_i \in \mathscr{F}_1$. Let $d\mu$ be a probability measure of the form of (12.3) where $J \ge 0$, each $d\nu_j$ has the form $\exp(f_j(x))\, d\lambda_j(x)$ with $f_j \in \mathscr{F}_1$ and $d\lambda_j$ even. Then $(\langle \cdot \rangle \equiv \int \cdot\, d\mu)$

$$\text{(GKS I)} \quad \langle f \rangle \ge 0 \tag{12.4}$$

$$\text{(GKS II)} \quad \langle fg \rangle \ge \langle f \rangle \langle g \rangle \tag{12.5}$$

for all $f, g \in \mathscr{F}_{n+1}$.

Proof (*GKS I*) We first note that for any $f_1, \ldots, f_n \in \mathscr{F}_1$ and any even measure $d\lambda$ on $\mathbb{R}$

$$\int \prod_{j=1}^{n} f_j \, d\lambda \geq 0 \tag{12.6}$$

For the integral is zero if there are an odd number of factors from $\mathscr{F}_1^o$ since then Πf_i is odd and $d\lambda$ is even. If there are an even number of factors from $\mathscr{F}_1^o$, then $\prod_{j=1}^n f_j$ is even and nonnegative for $x \geq 0$ and so for all x, so (12.6) holds. If $g = g_1(x_1) \cdots g_{n+1}(x_{n+1}) \in \mathscr{F}_{n+1}$, then

$$Z\langle g \rangle = \int \left[\prod_{j=1}^{n} g_j(x_j) \right] \left[\prod_{j=1}^{n} \exp(f_j(x_j)) \right] \left[\prod_{j=1}^{n} \exp(Jx_j x_{j-1}) \right] \prod_{j=1}^{n} d\lambda_j(x_j)$$

Expanding each exponential and noticing $x \in \mathscr{F}_1$, the integral is a sum of factors of the form of (12.6) and thus positive.

(*GKS II*) Let $d\mu(x)$ be the measure on $\mathbb{R}^{n+1}$ and consider "the duplicate system," i.e., $d\mu(x)\, d\mu(y)$ on $(\mathbb{R}^{n+1})^2$. Then

$$\langle fg \rangle - \langle f \rangle \langle g \rangle = \frac{1}{2} \int (f(x) - f(y))(g(x) - g(y)) \, d\mu(x) \, d\mu(y).$$

Let

$$dQ(x) = \prod d\lambda_j(x_j)$$

Then

$$2Z^2(\langle fg \rangle - \langle f \rangle \langle g \rangle) = \int (f(x) - f(y))(g(x) - g(y)) \exp(J(\textstyle\sum x_i x_j + y_i y_j))$$
$$\times \exp(\textstyle\sum f_i(x_i) + f_i(y_i)) \, dQ(x) \, dQ(y)$$

Expanding the exponentials, we see that it suffices to prove that

$$\int \prod_{i=1}^{m} (F_i(x) + \varepsilon_i F_i(y)) \, dQ(x) \, dQ(y) \geq 0 \tag{12.7}$$

for each choice of $\varepsilon_i = \pm 1$ and each $F_i \in \mathscr{F}_{n+1}$ (since $x_i x_j \in \mathscr{F}_{n+1}$). Now

$$ab \pm a'b' = \tfrac{1}{2}(a + a')(b \pm b') + \tfrac{1}{2}(a - a')(b \mp b')$$

By repeated use of this equality we can reduce (12.7) to the special case where each $F_i(x)$ has the form $f(x_j)$ with $f \in \mathscr{F}_1$. But then (12.7) breaks up into a product of one-dimensional integrals; i.e., we only need to prove that for $d\lambda$ even on $\mathbb{R}$ and $f_1, \ldots, f_m \in \mathscr{F}_1$:

$$\int \prod_{i=1}^{m} (f_i(x) + \varepsilon_i f_i(y)) \, d\lambda(x) \, d\lambda(y) \geq 0 \tag{12.8}$$

Now, since each f_i is even or odd

$$f_i(\sigma x) = \sigma^{\pi(i)} f_i(x)$$

for $\sigma = \pm 1$, and $\pi(i) = 1$ or 2. Using $d\lambda(x) = d\lambda(-x)$, we can write the left-hand side of (12.8) as

$$\sum_{\sigma_1 = \pm 1, \sigma_2 = \pm 1} \int_{x, y \geq 0} \prod (f_i(\sigma_1 x) + \varepsilon_i f_i(\sigma_2 y))\, d\lambda(x)\, d\lambda(y)$$

$$= \sum_{\sigma_3 = \pm 1} [1 + (-1)^{\Sigma \pi(i)}] \int_{x, y \geq 0} \prod [f_i(x) + \varepsilon_i \sigma_3^{\pi(i)} f_i(y)]\, d\lambda(x)\, d\lambda(y)$$

where $\sigma_3 = \sigma_1 \sigma_2$. Thus it suffices to prove (12.8) where the integral is over the region $x, y \geq 0$. But this integral is $(1 + \prod_{j=1}^m \varepsilon_i)$ times the integral over the region $x \geq y \geq 0$ (consider the $x \leftrightarrow y$ interchange) and thus we need only prove (12.8) in case the integral is over the region $x \geq y \geq 0$. But in that case $f_i(x) - f_i(y) \geq 0$ since f is monotone, and $f_i(x) + f_i(y) \geq 0$ since f is positive on $[0, \infty)$. ∎

By taking limits, we have some inequalities for path integrals:

Theorem 12.1′ ([120]) Suppose that $V = V_1 + V_2$ where V_1 is even and $-V_2 \in \mathscr{F}_1^{\circ}$ (i.e., V_2 odd, negative, and monotone decreasing on $[0, \infty)$). Let f, g be positive even functions. Let $A = a(\omega(t_1), \ldots, \omega(t_n))$ with $a \in \mathscr{F}_n$ and similarly for B. Let $\langle \cdot \rangle = \int \cdot\, d\mu(\omega)$ where $d\mu(\omega)$ is given by (12.1). Then, if $\langle A^2 \rangle < \infty$, $\langle B^2 \rangle < \infty$,

$$\langle A \rangle \geq 0 \qquad (12.4')$$

$$\langle AB \rangle - \langle A \rangle \langle B \rangle \geq 0 \qquad (12.5')$$

In particular, this holds for any $P(\phi)_1$-process with $P(x) = P_e(x) + P_o(x)$ with P_e even and P_o odd, negative and monotone decreasing for $x \geq 0$.

Remarks 1. To get the $P(\phi)_1$ result for the result from f, g even, we take $f, g = \Omega_0$ and then take $\omega(s) \to \omega(s - \frac{1}{2}t)$ and $t \to \infty$ using Theorem 6.9 to be sure we recover the $P(\phi)_1$-process.

2. To handle unbounded a's, we replace a by truncated functions and take limits.

Looking at the proof of (GKS I), one sees that the monotonicity of the functions was never used. Thus, we have the following.

Theorem 12.1″ ([173a]) Let V be a function on $\mathbb{R}^\nu$ so that for $x_2, \ldots, x_\nu$ fixed and $x_1 \geq 0$ we have that

$$V(\mathbf{x}) \leq V(-x_1, x_2, \ldots) \tag{12.9}$$

and let $\langle \cdot \rangle = \int \cdot \, d\mu(\boldsymbol{\omega})$ where $d\mu(\boldsymbol{\omega})$ is given by (12.1) extended to $\mathbb{R}^\nu$ with f, g even and positive. Then

$$\langle A \rangle \geq 0 \tag{12.4″}$$

if $A = a(\boldsymbol{\omega}(t_1))$ with $a(x_1, \ldots, x_\nu)$ a function which is odd in x_1 for $x_2, \ldots, x_\nu$ fixed and positive for $x_1 \geq 0$.

Proof After discretizing, fix the $x_2, \ldots, x_\nu$ coordinates in each variable. Expand in the x_1 couplings and get positivity by the proof of (GKS I). If one then integrates in the $x_2, \ldots, x_\nu$ variables, positivity results. ∎

Remark If V is unbounded below, one proves (12.4″) for an approximating sequence of V_n's bounded below and takes limits.

As an application we have the following result of [173a]:

Theorem 12.2 ([173a]) Let $e(R)$ be the ground state energy of $-\frac{1}{2}\Delta - z_1|\mathbf{r}|^{-1} - z_2|\mathbf{r} - R\mathbf{e}|^{-1}$ where $\mathbf{e}$ is a unit vector and $z_1, z_2 \geq 0$. Then $e(R)$ increases as R does in the region $R > 0$.

Proof Without loss take $\mathbf{e} = (-1, 0, 0)$. Then, if Ω_R is the ground state eigenvector:

$$\begin{aligned} \frac{de}{dR} &= (\Omega_R, [z_2(x_1 + R)|\mathbf{r} - R\mathbf{e}|^{-3}]\Omega_R) \\ &= z_2(\tilde{\Omega}_R, x_1|\mathbf{r}|^{-3}\tilde{\Omega}_R) \end{aligned}$$

where $\tilde{\Omega}_R$ is the ground state for $\tilde{H} = -\frac{1}{2}\Delta + \tilde{V}$ with

$$\tilde{V} = -z_1|\mathbf{r} + R\mathbf{e}|^{-1} - z_2|\mathbf{r}|^{-1}$$

and we have used translation covariance. Since $|\mathbf{r}|^{-1}$ is monotone, $\tilde{V}$ obeys (12.9), so by (12.4″) and a limiting argument $(\tilde{\Omega}_R, x_1|\mathbf{r}|^{-3}\tilde{\Omega}_R) \geq 0$. ∎

Now suppose that $W \in \mathscr{F}_1$ and we let $\langle \cdot \rangle_\lambda$ be of the form (12.1) with V replaced by $V(x) + \lambda W(x)$. Then

$$\frac{d}{d\lambda}\langle A \rangle_\lambda = -\int_0^t [\langle A W(\omega(s)) \rangle_\lambda - \langle A \rangle_\lambda \langle W(\omega(s)) \rangle_\lambda] \, ds$$

for the $\langle AW\rangle_\lambda$ term comes from the derivative of the $\int Ae\cdots$ term and $\langle A\rangle_\lambda\langle W\rangle_\lambda$ from the derivative of Z^{-1}. We have thus proven (taking suitable limits for $P(\phi)_1$ and unbounded W's):

Corollary 12.3 Let $\frac{1}{2}x^2 + P$, Q be functions which are bounded from below with $P = P_e + P_o$, $Q = Q_e + Q_o$ (P_e even, etc.) with $-P_o$, Q_e, Q_o, $-P_o - Q_o \in \mathscr{F}_1$. Suppose that the $R(\phi)_1$-process exists for $R = P + \lambda Q$, $\lambda \in [0, 1]$, and use $\langle\cdot\rangle_\lambda$ to denote its expectations. Then, for any $A = a(q(t_i))$; $a \in \mathscr{F}_n$,

$$\langle A\rangle_{\lambda_1} \leq \langle A\rangle_{\lambda_0} \qquad \text{if} \quad \lambda_1 \geq \lambda_0$$

Our first application of this result is taken from [120]:

Theorem 12.4 Let P, Q be even, $Q \in \mathscr{F}_1$. Let $E_1(P) = \inf \operatorname{spec}(L_0 + P)$ $E_2(P) = \inf \operatorname{spec}(L_0 + P \restriction \Omega_P^\perp)$. Then

$$E_2(P + \lambda Q) - E_1(P + \lambda Q)$$

is monotone increasing in λ.

Proof By adding an εx^2 to P, we can suppose that all $L = L_0 + \lambda Q$ have purely discrete spectrum in which case E_1 and E_2 are the first two eigenvalues. Let Ω_1, Ω_2 be the corresponding eigenvectors. We claim that $(x\Omega_1, \Omega_2) \neq 0$, for Ω_1 is nodeless and Ω_2 has exactly one node at $x = 0$. Thus, making an eigenvector expansion of $x\Omega_1$ and noticing $(\Omega_1, x\Omega_1) = 0$

$$\begin{aligned} -(E_2 - E_1) &= \lim_{t\to\infty} t^{-1} \ln[(x\Omega_1, e^{-t(L-E_1)}x\Omega_1)] \\ &= \lim_{t\to\infty} t^{-1} \ln\langle q(t)q(0)\rangle_\lambda \end{aligned}$$

is monotone decreasing in λ. ∎

Example If $E_i(a_0, a_1, \ldots, a_m)$ is the ith eigenvalue of $-(d^2/dx^2) + \sum_{j=0}^m a_j x^{2j}$, then in the region $a_m > 0$, $E_2(a) - E_1(a)$ is monotone in the a's.

Our second application is from [6]:

Theorem 12.5 Let V be an arbitrary even function on $\mathbb{R}$ with $V_- \in L^1$, $V_+ \in L^1_{\text{loc}}$. Let W be an even function monotone increasing on $(0, \infty)$. Suppose that $-(d^2/dx^2) + V + \lambda W$ has a normalized ground state (that is, an eigenvector corresponding to lowest eigenvalue), ψ_λ, for each $\lambda \in [0, 1]$. Then for each a

$$\int_{-a}^{a} |\psi_1(x)|^2 \, dx \geq \int_{-a}^{a} |\psi_0(x)|^2 \, dx$$

Proof Let $F(\lambda) = 1 - \int_{-a}^{a} |\psi_\lambda(x)|^2 \, dx$. We want to show that $F(\lambda)$ is monotone decreasing in λ. But

$$F(\lambda) = \langle G(q(0)) \rangle_\lambda$$

where $G(x) = 0$ (respectively, 1) for $|x| < a$ (respectively, $\geq a$) and $\langle \cdot \rangle_\lambda$ is the $P(\phi)_1$-process for $-d^2/dx^2 + V + \lambda W$. Since $G \in \mathscr{F}_1$, this follows from Corollary 12.3. ∎

Martin [182] has discovered an "elementary" proof of Theorem 12.5 that avoids path integrals; see also [289, 290].

There is one other family of inequalities that holds for all even $d\lambda$'s, namely the FKG (for Fortuin, Kasteleyn, and Ginibre [85]) inequalities. We state the following without proof (see, e.g., [258]).

Theorem 12.6 Let F be a nonnegative function on $\mathbb{R}_+^m = [0, \infty)^m$ obeying

$$F(\mathbf{x} \wedge \mathbf{y}) F(\mathbf{x} \vee \mathbf{y}) \geq F(\mathbf{x}) F(\mathbf{y}) \tag{12.10}$$

where $(\mathbf{x} \wedge \mathbf{y})_i = \min(x_i, y_i)$; $(\mathbf{x} \vee \mathbf{y})_i = \max(x_i, y_i)$. Let

$$\langle \cdot \rangle = \frac{\int \cdot \, F \, d^m x}{\int F \, d^m x}$$

Let f, g be functions on $\mathbb{R}_+^m$ which are monotone increasing in each x_j (with x_i for $i \neq j$ fixed). Then

$$\langle fg \rangle \geq \langle f \rangle \langle g \rangle$$

Let us note some cases where (12.10) holds:

Lemma 12.7 ([6])

(a) If $F(x) = \exp(-W(x))$ with W a C^2-function, then (12.10) holds if and only if

$$\frac{\partial^2 W}{\partial x_i \, \partial x_j} \leq 0, \qquad i \neq j \tag{12.11}$$

(b) Let dS_ν be the usual measure on unit sphere in $\mathbb{R}^\nu$ with $\nu = 1$ or 2, and let

$$F(\mathbf{x}) = \int \exp\left(\sum_{i \neq j} J_{ij} x_i x_j \boldsymbol{\sigma}_i \cdot \boldsymbol{\sigma}_j\right) \prod_{i=1}^{n} dS_\nu(\boldsymbol{\sigma}_i)$$

for $x \in \mathbb{R}_+^m$. Then F obeys (12.10) if $J_{ji} = J_{ij} \geq 0$.

Proof (a) (12.10) is equivalent to

$$W(\mathbf{x} \wedge \mathbf{y}) + W(\mathbf{x} \vee \mathbf{y}) - W(\mathbf{x}) - W(\mathbf{y}) \leq 0 \tag{12.12}$$

and this difference can be written as a sum of integrals of $\partial^2 W/\partial x_i\, \partial x_j$. Thus (12.11) implies (12.10). By taking $\mathbf{y} = \mathbf{x} + \varepsilon \mathbf{e}_i - \varepsilon \mathbf{e}_j$, (12.12) yields (12.11).

(b) Let $\langle \cdot \rangle_{\mathbf{x}} = \int \cdot \exp(\cdots)\pi\, dS / \int \exp(\cdots)\pi\, dS$ where the centered dot is a function of the $\boldsymbol{\sigma}$'s. Let $W = -\ln F$. Then

$$\frac{\partial W}{\partial x_i} = -\left\langle \sum_j J_{ij} x_j \boldsymbol{\sigma}_i \cdot \boldsymbol{\sigma}_j \right\rangle_{\mathbf{x}}$$

$$\frac{\partial^2 W}{\partial x_i\, \partial x_j} = -\langle J_{ij} \boldsymbol{\sigma}_i \cdot \boldsymbol{\sigma}_j \rangle_{\mathbf{x}} - \sum_{k,l} J_{il} J_{jk} x_l x_k \langle \boldsymbol{\sigma}_i \cdot \boldsymbol{\sigma}_l;\, \boldsymbol{\sigma}_j \cdot \boldsymbol{\sigma}_k \rangle_{T,\mathbf{x}}$$

where $\langle A;\, B \rangle_T = \langle AB \rangle - \langle A \rangle \langle B \rangle$. Thus (12.11) holds if $\langle \boldsymbol{\sigma}_i \cdot \boldsymbol{\sigma}_j \rangle_{\mathbf{x}} \geq 0$, $\langle \boldsymbol{\sigma}_i \cdot \boldsymbol{\sigma}_l;\, \boldsymbol{\sigma}_j \cdot \boldsymbol{\sigma}_k \rangle_{T,\mathbf{x}} \geq 0$. For $\nu = 1$, this is just the usual GKS inequality; for $\nu = 2$, it is an inequality of Ginibre [101]. ∎

As a typical application of this consider the following.

Proposition 12.8 ([6]) Let V be a function on $\mathbb{R}^3$ which is only a function of $\rho = (x^2 + y^2)^{1/2}$ and $|z|$, with $\partial^2 V/\partial \rho\, \partial z \leq 0$ in the region $\rho, z \geq 0$. Suppose that V is monotone in $\rho, |z|$, that G is another function of $\rho, |z|$ which is monotone, and let W be a function of ρ alone. Suppose that $-\Delta + W + \lambda V$ has a ground state for each $\lambda \in [0, 1]$, and let Ω_λ be the corresponding ground state eigenvector. Then $(\Omega_\lambda, G\Omega_\lambda)$ is monotone decreasing as λ increases.

Proof Pass to the path integral [$P(\phi)_1$-process] for $-\Delta + W + \lambda V$; call it $\langle \cdot \rangle_\lambda$. As in Corollary 12.3, it suffices to prove that

$$\langle G(q(0)) V(q(s)) \rangle \geq \langle G(q(0)) \rangle \langle V(q(s)) \rangle$$

and this follows if we can prove the FKG condition (12.10) for the measure obtained by first discretizing and then "integrating" over the angles in the

(x, y)-planes and signs of z's. The F that results is a product of three kinds of factors:

$$\exp(-\alpha W(\rho_i)), \qquad \exp(-\alpha V(\rho_i, |z_i|))$$

$$\int \exp(-\beta \sum (\mathbf{r}_i - \mathbf{r}_j)^2)\, d\phi_i\, d(\operatorname{sgn} z_i)$$

It is enough to check (12.10) for each factor: The W-factors trivially obey (12.10), the V-factors are all right by hypothesis and Lemma 12.7(a), and the final factors are all right by Lemma 12.7(b). ∎

An interesting application of this last result is to the Zeeman effect in hydrogen, i.e., hydrogen in a constant magnetic field. In the approximation of infinite nuclear mass, with suitable units and a magnetic field B_0 in the z-direction, the Hamiltonian is given by (1.2) which can be written:

$$H = -\Delta + (B^2/4)\rho^2 - BL_z - r^{-1}$$

where $L_z = (1/i)(\partial/\partial\phi)$ is the z-component of angular momentum. H commutes with L_z and one can ask what is the value of L_z in the ground state of H. Consider the Hamiltonians

$$H(m, \lambda) = -\Delta + m^2\rho^{-2} + (B^2/4)\rho^2 - mB - \lambda r^{-1}$$

Now $H \restriction$ (functions with $L_z = m$) is isomorphic under the natural association to $H(m, \lambda = 1) \restriction$ (functions with $L_z = 0$), so that

$$\begin{aligned} E(m) \equiv \inf(H \restriction L_z = m) &= \inf(H(m, \lambda = 1) \restriction L_z = 0) \\ &= \inf(H(m, \lambda = 1)) \end{aligned}$$

where the last equality comes from the fact that $H(m, \lambda)$ has a strictly positive ground state eigenvector and hence one with $L_z = 0$. Now think of m as a continuous parameter. Then

$$\frac{\partial E(m, \lambda)}{\partial m} = \langle 2m\rho^{-2}\rangle_{m,\lambda} - B$$

where $\langle\cdot\rangle_{m,\lambda}$ denotes expectations in the ground state of $H(m, \lambda)$. $V(\rho, |z|) = -(\rho^2 + z^2)^{-1/2}$ is easily seen to be monotone in ρ, z and to obey $\partial^2 V/\partial\rho\, \partial z \le 0$. Since $-\rho^{-2}$ is monotone in ρ, Proposition 12.8 implies that $\langle\rho^{-2}\rangle_{m,\lambda}$ is monotone increasing in λ. Thus for $m \ge 0$,

$$\frac{\partial E(m, \lambda = 1)}{\partial m} \ge \frac{\partial E(m, \lambda = 0)}{\partial m} = 0$$

where the last equality comes from the fact that $E(m, \lambda = 0) = 2B$ for all $m \geq 0$ [the ground state in the ρ-variables is a multiple of $\rho^m \exp(-\frac{1}{4}B\rho^2)$]. This shows that

$$E(m) \geq E(0) \tag{12.13}$$

for $m \geq 0$. Since $E(-m) = E(m) + 2mB$ for $m > 0$, (12.13) holds for all m. By an additional argument [6], the inequality can be shown to be strict. We thus have proven the following.

Theorem 12.9 ([6]) The ground state of the Hamiltonian (1.2) of a hydrogen atom in a magnetic field has $m = 0$ for any B.

Remark The above proof will work if $-r^{-1}$ is replaced by any function $V(r)$ with V negative, monotone increasing and concave on $[0, \infty)$. By an additional trick, one can avoid the hypothesis of concavity [6]. If V is not monotone, then the ground state need not have $m = 0$ (see [4] and references therein).

* * *

The above results (at least in one dimension) give one some control on differential operators $-d^2/dx^2 + V$ for general even V's. One could get control over such general V's because there was no restriction other than evenness on the "single spin distribution," $d\lambda$. One can get much greater control by placing stronger restrictions on V. The earliest result of this genre is the following obtained by Simon and Griffiths [261]:

Theorem 12.10 Let $\langle \cdot \rangle$ be the $P(\phi)_1$ expectation for $P(x) = ax^4 + bx^2$ ($a > 0$, b in $\mathbb{R}$). Then any multilinear inequality true for an arbitrary spin-$\frac{1}{2}$ ferromagnet is valid for $\langle \cdot \rangle$.

Remarks 1. By an arbitrary spin-$\frac{1}{2}$ ferromagnet, we mean the measure on $\{-1, 1\}^m$ given by

$$\langle f(\sigma) \rangle = \sum_{\substack{\sigma_\alpha = \pm 1 \\ \alpha = 1, \ldots, m}} f(\sigma_\alpha) \exp\left(\sum J_{\alpha\beta} \sigma_\alpha \sigma_\beta\right) / \text{Normalization}$$

with $J_{\alpha\beta} \geq 0$. By a multilinear inequality we mean one of the form

$$\sum_{P \in \mathscr{P}} d(P)(\sigma_{\alpha_1}, \ldots, \sigma_{\alpha_n})_P \geq 0$$

where $\mathscr{P}$ is the family of all partitions, P, of $\{1, \ldots, n\}$ into disjoint subsets

$$\Pi_1, \ldots, \Pi_k \qquad \text{and} \qquad (\sigma_{\alpha_1}, \ldots, \sigma_{\alpha_n})_P = \prod_{j=1}^{k} \left\langle \prod_{m \in \Pi_j} \sigma_{\alpha_m} \right\rangle$$

2. Examples of such inequalities are those of Newman [196]

$$\langle q(t_1)\cdots q(t_{2n})\rangle \le \sum \prod_{k=1}^{n} \langle q(t_{i_k})q(t_{j_k})\rangle$$

with the sum over all pairings of $\{1, \ldots, 2n\}$, the GHS inequality discussed below and certain inequalities of Lebowitz [165].

3. For the proof, see [261]. See [76, 62] for further discussion including multicomponent inequalities. The idea in [261] is the following: If σ_i are independent spins, then the distribution of $\sum_{i=1}^{N} \sigma_i/\sqrt{N}$ is

$$(2\pi)^{-1/2} \exp(-x^2/2)$$

in the limit of $N \to \infty$. If one couples the spins together with a Hamiltonian $(2N)^{-1}(\sum_{i=1}^{N} \sigma_i)^2$, then this leading behavior is precisely canceled and it turns out that $\sum_{i=1}^{N} \sigma_i/N^{3/4}$ approaches $c \exp(-ax^4)$ for a, c suitable. In this way one can approximate a "single" spin with $e^{-x^4}\,dx$ distribution with a sum of ferromagnetically coupled spin-$\frac{1}{2}$ spins. Thus ferromagnetically coupled $e^{-x^4}\,dx$ spins can be approximated by a larger array of spin-$\frac{1}{2}$ spins. Since multilinear inequalities extend to sums of spin-$\frac{1}{2}$ spins, the theorem is proven.

* * *

The most interesting applications of Theorem 12.10 involve the GHS (for Griffiths, Hurst, and Sherman [117]) inequalities which have been proven for many V's by Ellis, Monroe and Newman [74] whose treatment we follow below. Similar results have been obtained by Sylvester [269]. As a preliminary, we discuss the **Ursell functions** or **cumulants**. Given n random variables, $X_1, \ldots, X_n$, we define

$$u_n(X_1, \ldots, X_n) = \frac{\partial^n}{\partial h_1 \cdots \partial h_n} \ln\left\langle \exp\left(\sum_{i=1}^{n} h_i X_i\right)\right\rangle \Bigg|_{h_i=0} \tag{12.14}$$

Thus,

$$u_1(X_1) = \langle X_1\rangle$$

$$u_2(X_1, X_2) = \langle X_1X_2\rangle - \langle X_1\rangle\langle X_2\rangle$$

$$u_3(X_1, X_2, X_3) = \langle X_1X_2X_3\rangle - \langle X_1\rangle\langle X_2X_3\rangle - \langle X_2\rangle\langle X_1X_3\rangle - \langle X_3\rangle\langle X_1X_2\rangle + 2\langle X_1\rangle\langle X_2\rangle\langle X_3\rangle$$

and for X's with $\langle\prod X_i^{k_i}\rangle = 0$ for $\sum k_i$ odd,

$$u_4(X_1, X_2, X_3, X_4) = \langle X_1X_2X_3X_4\rangle - \langle X_1X_2\rangle\langle X_3X_4\rangle - \langle X_1X_4\rangle\langle X_2X_3\rangle - \langle X_1X_3\rangle\langle X_2X_4\rangle$$

There are explicit formulas for u_n but of more value is an axiomatic characterization of Percus [202]:

(P1) u_n is multilinear in $X_1, \ldots, X_n$ (in the sense described after Theorem 12.10).

(P2) $u_n(X_1, \ldots, X_n) = \langle X_1 \cdots X_n \rangle$ + sums of products of two or more factors

(P3) If the X's break up into two mutually independent sets, then $u_n = 0$.

Proposition 12.11 (Percus' lemma [202]) Fix n. Then u_n obeys (P1)–(P3). Moreover, any expression obeying (P1)–(P3) is identical to u_n.

Proof (P1) and (P2) are easy to check. (P3) follows if we note that if the X's break up into independent sets then $\ln\langle \exp \cdots \rangle$ is a sum of two functions each of which is independent of some h_i.

Conversely, given a function $\tilde{u}_n(X_1, \ldots, X_n)$ obeying (P1)–(P3), we can write

$$\tilde{u}_n(X_i) = \sum_{P \in \mathscr{P}} d(P) \langle X \rangle_P$$

by (P1). Here $d(P)$ is a number, $\mathscr{P}$ is the family of all partitions of $\{1, \ldots, n\}$

$$P = \left\{ \prod_1 \cdots \prod_k \right\} \qquad (k \equiv \#(P))$$

and

$$\langle X \rangle_P = \prod_{i=1}^{\#(P)} \left\langle \prod_{j \in \Pi_i} X_j \right\rangle$$

By (P2), $d(P)$ is determined for $\#(P) = 1$. We claim that (P3) then determines $d(P)$ inductively in $\#(P)$. For fix a partition P_o and let $P \lhd P_o$ indicate that P_o is a refinement of P. Let $P_o = \{\pi_i\}$. Then

$$\sum_{P \lhd P_o} d(P) = 0 \qquad (12.15)$$

for suppose that $\{X_j\}_{j \in \pi_i}$ are independent. Then $\sum_{P \lhd P_o} d(P)$ is coefficient of $\langle X \rangle_{P_o}$. But (12.15) determines $d(P_o)$ in terms of $d(P)$'s with $\#(P) < \#(P_o)$. ∎

As a typical application of Percus' lemma, we note the following formulas of Cartier [33]: Given random variables $X_1, \ldots, X_n$ on $(X, \mathscr{F}, \mu)$, take n independent copies of X (i.e., the n-fold Cartesian product of X with the product measure), let $X_i^{(j)}$ be the copy of X_i on the jth factor. Let $\omega = e^{2\pi i/n}$ be a primitive nth root of unity and let $\tilde{X}_i = \sum_{j=1}^n \omega^j X_i^{(j)}$. Then **(Cartier's formula)**

$$u_n(X_1, \dots, X_n) = n^{-1}E(\tilde{X}_1, \dots, \tilde{X}_n) \tag{12.16}$$

To prove (12.16), note that (P1) and (P2) are easy; (P3) follows if we note that under the measure preserving map $X_i^{(j)} \to X_i^{(j-1)}$ (with $X^{(0)} \equiv X^{(n)}$), $\tilde{X}_i \mapsto \omega\tilde{X}_i$. Thus

$$E(\tilde{X}_{j_1}, \dots, \tilde{X}_{j_l}) = \omega^l E(\tilde{X}_{j_1}, \dots, \tilde{X}_{j_l})$$

is zero if $l < n$.

Theorem 12.12 ([74, 269]) Let $d\mu$ be a measure on $\mathbb{R}^n$ of the form

$$d\mu(x) = Z^{-1} \exp\left(-\sum_{i=1}^n V_i(x_i) + \sum_{i=1}^n h_i z_i + \sum_{i,j} J_{ij}x_i x_j \right) d^n x \tag{12.17}$$

where Z is a normalizing factor (assumed finite). Suppose that $J_{ji} = J_{ij} \geq 0$; $h_i \geq 0$ and that each $V_i(x)$ is an even function of x, C^1 on $(-\infty, \infty)$ with V_i' convex in the region $(0, \infty)$. Then with respect to $d\mu$ **(GHS inequality)**

$$u_3(x_1, x_2, x_3) \leq 0 \tag{12.18}$$

If all $h_i = 0$, then **(Lebowitz inequality)**

$$u_4(x_1, \dots, x_4) \leq 0 \tag{12.19}$$

Remarks 1. If we take $V_i(x) = V(x)$ with $V(x) = a[(x^2 - 1)^2]$ and take $a \to \infty$,

$$e^{-V(x)}\, dx/\text{Normalization} \to \tfrac{1}{2}[\delta(x+1) + \delta(x-1)]$$

and one recovers the original GHS [117] inequality for the usual spin-$\frac{1}{2}$ Ising model.

2. Since $u_3 = 0$ if all h_i are zero and by (12.18), $u_3 \leq 0$ if $h_4 > 0$, $h_i = 0$ $(i \neq 4)$, we have $\partial u_3/\partial h_4 \leq 0$ at all $h_i = 0$. That is, (12.18) implies (12.19). (12.19) was originally obtained by Lebowitz [165] as one of a large number of new inequalities; it is a remark of Newman (unpublished) that it follows from (12.18).

3. At first sight, it may be surprising that for spin-$\frac{1}{2}$ models $u_1 \geq 0$, $u_2 \geq 0$ but $u_3 \leq 0$. There is a good physical reason for this: Typically, the magnetization of a magnet is as drawn in Figure 1, so that in the region $h > 0$, it is positive, monotone, and *concave*. Since $m = \langle \sigma_1 \rangle$, this says $\langle \sigma_1 \rangle \geq 0$, $\sum_i \partial\langle \sigma_1 \rangle/\partial h_i \geq 0$ but $\sum_{i,j} \partial^2\langle \sigma_1 \rangle/\partial h_i\, \partial h_j \leq 0$.

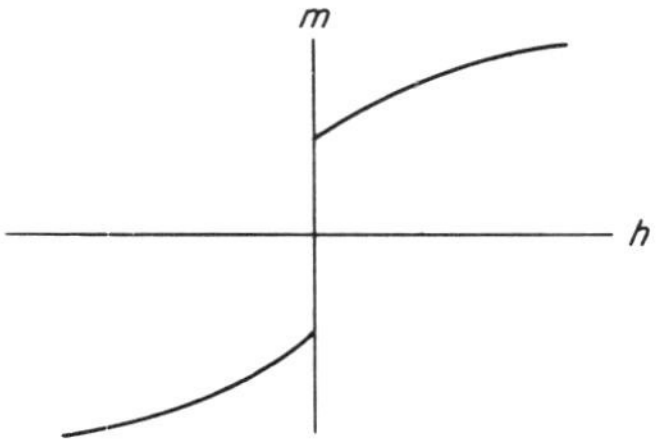

Figure 1. A typical magnetization curve.

Proof ([74]) GKS I was proven with one copy of $d\mu$, GKS II with two copies; now we will use four copies, i.e., $d\mu(x^{(1)}) \cdots d\mu(x^{(4)})$ with components $x_i^{(k)}$. Let B be the orthogonal matrix

$$B = \frac{1}{2}\begin{pmatrix} 1 & 1 & 1 & 1 \\ 1 & -1 & 1 & -1 \\ 1 & 1 & -1 & -1 \\ -1 & 1 & 1 & -1 \end{pmatrix} \tag{12.20}$$

and let $w_i^{(k)} = \sum b_{kl} x_i^{(l)}$. We first claim that

$$u_3(x_1, x_2, x_3) = -2E(w_1^{(2)} w_2^{(3)} w_3^{(4)}) \tag{12.21}$$

This follows, either by explicit expansion of the right-hand side of (12.21) into 64 terms or by using Percus' lemma: the right-hand side of (12.21) is multilinear, the leading coefficient is $\langle x_1 x_2 x_3 \rangle$, and (P3) is obeyed since $E(w_i^{(j)}) = 0$ for $j = 2, 3, 4$. Thus (12.18) and (by Remark 2) (12.19) follow if we show that

$$E\left(\prod_{i=1}^{n} F_i(\mathbf{w}_i)\right) \geq 0 \tag{12.22}$$

where $F_i(w^{(1)}, \ldots, w^{(4)})$ is a function positive in the region with all $w^{(k)} \geq 0$ and odd or even in each $w^{(k)}$.

Now B is an orthogonal matrix, so $\prod_{j=1}^{4} d^n x^{(j)} = \prod_{j=1}^{4} d^n w^{(j)}$ and

$$\sum_{k=1}^{4} x_i^{(k)} x_j^{(k)} = \sum_{k=1}^{4} w_i^{(k)} w_j^{(k)}$$

Thus,

$$Z^4 E\left(\prod_{i=1}^{n} F_i(\mathbf{w}_i)\right) = \int \exp\left(\sum_{i=1}^{n} 2h_i w_i^{(1)} + \sum_{i,j,k} J_{ij} w_i^{(k)} w_j^{(k)}\right)$$
$$\times \left[\prod_{i=1}^{n} F_i(\mathbf{w}_i)\right] \prod_{i=1}^{n} \exp\left(-\sum_{k=1}^{4} V_i(x_i^{(k)})\right) \prod_{k=1}^{4} d^n w^{(k)}$$

If we now expand the first exponential and note that products of the allowed F's are allowed and that $w_i^{(k)}$ is allowed, we see that (12.22) is implied by one-half of the lemma below. ∎

Lemma 12.13 Let $x^{(i)} = \sum_{j=1}^{4} B_{ji} w^{(j)}$ where B is the matrix (12.20). (Note: $B^{-1} = B^{\mathrm{T}}$.) Let $\mathscr{W}$ consist of all functions of $w^{(1)}, \ldots, w^{(4)}$ which are odd or even under change of sign of each $w^{(k)}$ (odd under some, even under others is allowed) positive if all $w^{(k)} \geq 0$. Let

$$d\mu(w) = \exp\left(-\sum_{i=1}^{4} V(x^{(i)})\right) \prod_{k=1}^{4} dw^{(k)}$$

where V is an even C^1-function. Then

$$\int F(w)\, d\mu(w) \geq 0 \tag{12.23}$$

for all $F \in \mathscr{W}$ if and only if V' is convex on $(0, \infty)$.

Proof Suppose that $F(w)$ is multiplied by $(-1)^{m_i}$ if the sign of $w^{(i)}$ is changed. $d\mu$ is left invariant under $w^{(i)} \to -w^{(i)}$ ($i = 1, 2, 3, 4$), since V is even, and also under reversal of two w's (e.g., $w^{(2)} \to -w^{(2)}$, $w^{(4)} \to -w^{(4)}$ corresponds to interchanging x_1 with x_2 and x_3 with x_4). Thus the integral in (12.23) is zero unless all m_i are even or all are odd. If all are even, then $F \geq 0$ so (12.23) is trivial. We are therefore restricted to consideration of all m_i odd. The invariance of $d\mu$ under reversal of two signs then yields the integral over the sixteen "quadrants" as two sets of identical contributions; i.e.,

$$\int F(w)\, d\mu(w) = 8 \int_{w_i \geq 0} F(w)[d\mu(w) - d\mu(-w^{(1)}, w^{(2)}, w^{(3)}, w^{(4)})]$$

For this integral to be positive for *all* $F \geq 0$ and $w^{(i)} \geq 0$, it is necessary and sufficient that the measure in $[\cdots]$ be positive. Thus, (12.23) is equivalent to the symmetric condition

$$\begin{aligned} &V(w^{(1)} + w^{(2)} + w^{(3)} + w^{(4)}) + V(w^{(1)} + w^{(2)} - w^{(3)} - w^{(4)}) + 2 \text{ others} \\ &\quad - V(w^{(1)} + w^{(2)} + w^{(3)} - w^{(4)}) - 3 \text{ others} \geq 0 \end{aligned} \tag{12.24}$$

for all $w^{(i)} \geq 0$.

Suppose that (12.24) holds. Fix a. Then (12.24) holds for $w^{(1)} = a + y$ ($|y| < \frac{1}{2}a$), and so it holds for $w^{(1)} = a$ and V replaced by V_δ, the convolution of V with a positive C^∞- function supported in $(-\frac{1}{2}a, \frac{1}{2}a)$. V_δ is C^∞ so we can choose $w^{(1)} = a$, $w^{(2)} = w^{(3)} = w^{(4)} = \varepsilon$ and expand in Taylor series; (12.24) says:

$$V_\delta(a + 3\varepsilon) + 3V_\delta(a - \varepsilon) - V_\delta(a - 3\varepsilon) - 3V_\delta(a + \varepsilon) \geq 0$$

so taking $\varepsilon \to 0$, $V_\delta'''(a) \geq 0$. Taking $\delta \to 0$, we see that the third distributional derivative of V is a measure on $(0, \infty)$, so V' is convex there.

Conversely suppose V is C^1 with V' convex on $(0, \infty)$. By symmetry, we need only check (12.24) for $w^{(1)} \geq w^{(4)}$. Then:

$$\text{l.h.s. of (12.24)} = w^{(4)} \int_{-1}^{1} G(w^{(1)} + rw^{(4)}, w^{(2)}, w^{(3)})\, dr$$

with

$$G(x, y, z) = V'(x + y + z) + V'(x - y - z) \\ - V'(x + y - z) - V'(x + z - y)$$

so we need only prove G is positive in the region $x \geq y \geq z \geq 0$ (G is symmetric). Now since V' is continuous on *all* of $\mathbb{R}$ and convex (respectively, concave) on $x > 0$ ($x < 0$), V'' exists for all but countably many x's and V' is the integral of V''. Thus

$$G(x, y, z) = z \int_{-1}^{1} [V''(x + rz + y) - V''(x + rz - y)]\, dr$$

Since $x \geq z$, we only need that $x, y \geq 0$ implies

$$V''(x + y) \geq V''(x - y)$$

and this follows from the fact that V'' is even and monotone on $(0, \infty)$. ∎

As usual, we immediately obtain inequalities for the $P(\phi)_1$-process:

Theorem 12.14 Let P have the form $Q - hx$ with $h \geq 0$ and Q a C^1 even function with Q' convex on $(0, \infty)$. Then

(a) $\langle q(t)q(s)\rangle - \langle q(t)\rangle\langle q(s)\rangle$ is monotone decreasing as h increases.
(b) If $h = 0$,

$$\langle q^2(t_1)q^2(t_2)\rangle - \langle q^2(t_1)\rangle\langle q^2(t_2)\rangle \leq 2\langle q(t_1)q(t_2)\rangle^2$$

Remark (a) comes by noting that $u_3 = \partial u_2/\partial h_3$ and (b) by using Lebowitz' inequality with $\sigma_1 = \sigma_2$, $\sigma_3 = \sigma_4$.

Corollary 12.15 ([261]) Under the above hypotheses on Q, let $E_i(h)$ be the ith eigenvalue of $\frac{1}{2}(d^2/dx^2) + Q(x) - hx$. Then $E_2(h) - E_1(h)$ is monotone increasing in h.

Proof If Ω_i is the ith eigenvector, then $(\Omega_2, q\Omega_1) \neq 0$ since Ω_2 has one node say at a, so $(\Omega_2, q\Omega_1) = (\Omega_2, (q - a)\Omega_1) \neq 0$. Thus, if $\psi = q\Omega_1 - (\Omega_1, q\Omega_1)\Omega_1$ and $\tilde{H} = H - E_1$:

$$-(E_2 - E_1) = \lim_{t \to \infty} t^{-1} \ln[(\psi, e^{-t\tilde{H}}\psi)]$$

$$= \lim_{t \to \infty} t^{-1} \ln[\langle q(t)q(0)\rangle - \langle q(t)\rangle\langle q(0)\rangle]$$

Now use Theorem 12.14(a). ▮

Corollary 12.16([107]; see also [264, 196]) Under the above hypotheses on Q and notation,

$$E_3(h) - E_2(h) \geq E_2(h) - E_1(h)$$

at $h = 0$.

Proof Since $\Omega_i(-q) = (-1)^{i+1}\Omega_i(q)$, $(\Omega_2, q^2\Omega_1) = 0$. On the other hand, since Ω_3 has nodes at $q = \pm\sqrt{a}$,

$$(\Omega_3, q^2\Omega_1) = (\Omega_3, (q^2 - a)\Omega_1) \neq 0$$

so, using Theorem 12.14(b):

$$-(E_3 - E_1) = \lim_{t \to \infty} t^{-1} \ln[\langle q^2(t)q^2(0)\rangle - \langle q^2(t)\rangle\langle q^2(0)\rangle]$$

$$\leq \lim_{t \to \infty} t^{-1} \ln[2\langle q(t)q(0)\rangle^2] = -2(E_2 - E_1)$$

so

$$E_3 - E_1 \geq 2(E_2 - E_1)$$ ▮

There is a final aspect of Theorem 12.12/Lemma 12.13 of interest:

Theorem 12.17 ([75]) Let V be an even C^1-function on $\mathbb{R}$ with $V'(x)$ convex on $[0, \infty)$. Suppose that $H = -d^2/dx^2 + V(x)$ has an eigenvalue at the bottom of its spectrum with eigenvector $e^{-f(x)}$. Then f is C^1 and f' is convex on $[0, \infty)$.

Proof Let $\langle\cdot\rangle$ be four independent copies of the path integral associated to H. By the generalized GHS inequality, (12.22), we have that

$$\langle F(\mathbf{w})\rangle \geq 0$$

for any $F \in \mathscr{W}$, the family in Lemma 12.13. Thus (12.23) holds with the "V" in $d\mu$ replaced by $2f$. Since V is (locally) C^1, a general argument implies that

f is (locally) C^1 (actually C^2 can be shown), so by the converse direction of Lemma 12.13, f' is convex on $[0, \infty)$. ∎

Remark V' continuous up to zero is not needed. All that is needed is $\lim_{x \downarrow 0} V'(0) \leq 0$, since one can then obtain V' as a limit of W's with $W'(0) = 0$ and W convex.

Example We want to describe an example with V even and C^∞, with $V^{(m)}(x) \geq 0$ for all $x \geq 0$ all integers $m \geq 0$ but for which the ground state $\psi = e^{-f}$ with $f^{(4)}$ negative for large x. V will just be x^4. If E is the energy of the lowest eigenvalue, then the *formal* WKB form for the asymptotics of ψ is

$$\psi \sim (V - E)^{-1/4} \exp\left(-\int^x \sqrt{V - E}\, dx\right)$$

and thus for $V = x^4$

$$f \sim \frac{1}{3}x^3 + \ln x + O(x^{-1})$$

and

$$f^{(4)} \sim -6x^{-4} + O(x^{-5})$$

The point is that using ODE methods, one can prove these asymptotic formulas.

13. Other Inequalities: Log Concavity, Symmetric Rearrangement, Conditioning, Hypercontractivity

A. Log Concavity

Definition A function $F\colon \mathbb{R}^n \to [0, \infty)$ is called **log concave** (respectively, **log convex**) if and only if

$$F(\lambda\mathbf{x} + (1 - \lambda)\mathbf{y}) \geq F(\mathbf{x})^\lambda F(\mathbf{y})^{1-\lambda} \qquad (\text{resp.}, \leq F(\mathbf{x})^\lambda F(\mathbf{y})^{1-\lambda})$$

for all $\mathbf{x}, \mathbf{y} \in \mathbb{R}^n$, $0 \leq \lambda \leq 1$.

Examples 1. If C is a convex set, its characteristic function is log concave.

2. If $\{a_{ij}\}_{i,j=1}^n$ is a positive definite matrix then $\exp(-\sum x_i x_j a_{ij})$ is log concave.

The following is an elementary consequence of Hölder's inequality.

Theorem 13.1 If $F:\mathbb{R}^{m+n} \to [0, \infty)$ with $F(\cdot, \mathbf{y})$ log convex for each fixed $\mathbf{y}$ in $\mathbb{R}^n$, then

$$G(\mathbf{x}) = \int_{\mathbb{R}^n} F(\mathbf{x}, \mathbf{y})\, d^n y$$

is log convex in $\mathbf{x}$.

By taking limits (or using Hölder directly on the path space), one sees that the map $V \mapsto \int dq \exp(-\int_0^t V(q(s))\, ds) = (\Omega_0, \exp(-t(L_0 + V))\Omega_0)$ is log convex so that using (1.9), we recover the result (which also follows from the Rayleigh–Ritz principle) that $V \mapsto \inf \operatorname{spec}(L_0 + V)$ is concave.

Much subtler is the following result proven by Prékopa [205] and then independently by Rinott [220] and Brascamp–Lieb [19].

Theorem 13.2 If $F:\mathbb{R}^{m+n} \to [0, \infty)$ is log concave, then

$$G(\mathbf{x}) = \int F(\mathbf{x}, \mathbf{y})\, d^n y$$

is log concave.

Proof ([19]) By induction, we need only consider the case $n = 1$. Moreover, since log concavity is an expression about G over lines, we need only consider the case $m = 1$. Fix $x_0, x' \in \mathbb{R}$ and $\lambda \in (0, 1)$. If $G(x_0)$ or $G(x')$ is zero, the inequality

$$G(\lambda x_0 + (1 - \lambda)x') \geq G(x_0)^\lambda G(x')^{1-\lambda}$$

is trivial. Moreover, by replacing F by $F(x, y)\exp(-\varepsilon y^2)$ we can suppose G is everywhere finite. Then replacing F by $e^{a+bx}F$ for a, b suitable we can suppose that

$$\sup_y F(x_0, y) = \sup_y F(x', y) \equiv z_0 \tag{13.1}$$

Fix $0 < z < z_0$ and let $C(z) = \{(x, y) \mid F(x, y) \geq z\}$. Log concavity says that $C(z)$ is convex and nonempty. Thus for $x = \lambda x_0 + (1 - \lambda)x'$

$$C(x, z) = \{y \mid (x, y) \in C(z)\}$$

is nonempty (by 13.1) and an interval $[a(x, z), b(x, z)]$ with a and $-b$ convex in x. In particular, $g(x, z) = \operatorname{meas}(C(x, z)) = b(x, z) - a(x, z)$ is concave; i.e.,

$$g(\lambda x_0 + (1 - \lambda)x', z) \geq \lambda g(x_0, z) + (1 - \lambda)g(x', z)$$

But

$$G(x) = -\int_0^\infty z\,dg(x, z) = \int_0^\infty g(x, z)\,dz \geq \int_0^{z_0} g(x, z)\,dz$$

with equality for $x = x_0$ or x'. Thus

$$\begin{aligned} G(\lambda x_0 + (1 - \lambda)x') &\geq \lambda G(x_0) + (1 - \lambda)G(x') \\ &\geq G(x_0)^\lambda G(x')^{1-\lambda} \quad \blacksquare \end{aligned}$$

Corollary 13.3 The convolution of two log concave functions is log concave.

Proof If F and G are log concave on $\mathbb{R}^n$, then $F(\mathbf{x} - \mathbf{y})G(\mathbf{y})$ is log concave on $\mathbb{R}^{2n}$. ▮

Most of the results below can be obtained by systematically using Corollary 13.3 and the Trotter formula. As usual, we give instead a path integral result. For convenience, we state things for the Wiener process. Similar results hold for *any* Gaussian process and for Wiener measure.

Theorem 13.4 Let $F(\mathbf{b}, \boldsymbol{\lambda})$ be a function on ν-dimensional Wiener paths depending on an additional parameter $\lambda \in \mathbb{R}^n$. Suppose that $F(\mathbf{b}, \boldsymbol{\lambda}) = \lim_{m\to\infty} F_m(\mathbf{b}, \boldsymbol{\lambda})$ in $L^1(Db)$ for each fixed λ for F_m obeying

$$F_m(\mathbf{b}, \boldsymbol{\lambda}) = G_m(\mathbf{b}(s_1^{(m)}), \ldots, \mathbf{b}(s_m^{(m)}), \lambda)$$

with G_m log concave on $\mathbb{R}^{n+m\nu}$. Then

$$H(\lambda) = E(F(\mathbf{b}, \lambda))$$

is log concave.

Proof $E(F_m(\mathbf{b}, \boldsymbol{\lambda})) = \int G_m(\mathbf{x}_1, \ldots, \mathbf{x}_m, \boldsymbol{\lambda})Q(\mathbf{x})\,d^{m\nu}x$ where $Q(\mathbf{x})$ is a Gaussian. Since G_m and Q are log concave, so is $G_m Q$ and thus so is the integral by Theorem 13.2. ▮

The following applications are from [19, 20].

Corollary 13.5 If V is convex on $\mathbb{R}^\nu$ and $H = -\frac{1}{2}\Delta + V$ has ground state $\Omega(\mathbf{x}) = e^{-f(\mathbf{x})}$, then f is convex.

Proof Let Ω_0 be the ground state of the harmonic oscillator, L_0. By a limiting argument using (1.10) it suffices to show that $(e^{-tH}\Omega_0)(\mathbf{x})$ is log concave for each t. But

$$(e^{-tH}\Omega_0)(\mathbf{x}) = E\left(\exp\left(-\int_0^t V(\mathbf{x} + \mathbf{b}(s))\, ds\right)\Omega_0(\mathbf{x} + \mathbf{b}(t))\right)$$

and $G(\mathbf{u}_1, \ldots, \mathbf{u}_m, \mathbf{x}) \equiv \exp(-(t/m)\sum_{j=1}^m V(\mathbf{x} + \mathbf{u}_j))\Omega_0(\mathbf{x} + \mathbf{u}_m)$ is log concave on $\mathbb{R}^{(m+1)\nu}$. Thus the result follows from Theorem 13.4. ∎

Remark Let S be an open convex set in $\mathbb{R}^\nu$ and let $V_n = n\,\mathrm{dist}(x, S)$. Then V_n is convex. Taking $n \to \infty$, we see that the ground state, ψ, of the Dirichlet operator $-\Delta_D^S$ is log concave; in particular, the level sets $\{x \mid \psi(x) = \alpha\}$ are the boundaries of convex sets $\{x \mid \psi(x) \geq \alpha\}$. This is a result of Brascamp–Lieb [20].

Corollary 13.6 Let $V(\mathbf{x}, \lambda)$ be a convex function on $\mathbb{R}^{\nu+1}$. Let $E(\lambda) = \inf \mathrm{spec}(H_0 + V(x, \lambda))$. Then $E(\lambda)$ is convex in λ.

Proof The proof follows from Theorem 13.4 extended to $d\mu_0$ and the formula (1.9):

$$-E(\lambda) = \lim_{t\to\infty} t^{-1} \ln \int \Omega_0(\boldsymbol{\omega}(0))\Omega_0(\boldsymbol{\omega}(t))\exp\left(-\int_0^t V(\boldsymbol{\omega}(s), \lambda)\, ds\right) d\mu_0(\omega) \quad ∎$$

This result is useful to obtain lower bounds on ground state energies which complement Rayleigh–Ritz results.

Example $V(x, \lambda) = \lambda x^2 + x^4 + \frac{1}{6}\lambda^2$ is jointly convex in the region $\lambda \geq 0$ (compute the matrix of second partial derivatives). In particular, if $a(\lambda) \equiv \inf \mathrm{spec}(-\frac{1}{2}\, d^2/dx^2 + \lambda x^2 + x^4)$, then for $0 \leq \lambda$:

$$-\tfrac{1}{3} \leq a''(\lambda) \leq 0$$

If $b(\lambda) = \inf \mathrm{spec}(-\frac{1}{2}\, d^2/dx^2 + x^2 + \lambda x^4)$, then [241] $b(\lambda) = \lambda^{1/3}a(\lambda^{-2/3})$ by a scaling argument, so we also obtain information about b.

Brunn–Minkowski inequalities for Gaussian measure have been extensively studied by probabilists; see [2a, 118a, 18a, 18b, 18c]. Our treatment follows that in [19] and begins with the classical Brunn–Minkowski theorem and its relation to Theorem 13.2.

Theorem 13.7 (Brunn–Minkowski inequality) Let C_0, C_1 be nonempty compact convex sets in $\mathbb{R}^n$ and let $C_\lambda = \lambda C_1 + (1 - \lambda)C_0$. Then ($|\cdot|$ indicates Lebesgue measure)

$$|C_\lambda|^{1/n} \geq \lambda|C_1|^{1/n} + (1 - \lambda)|C_0|^{1/n}$$

Proof By a limiting argument (replace C_i by $\{\mathbf{x} \mid \operatorname{dist}(\mathbf{x}, C_i) \le \varepsilon\}$) we may suppose that $|C_i| > 0$. Take

$$\tilde{C}_1 = |C_1|^{-1/n} C_1; \qquad \tilde{C}_0 = |C_0|^{-1/n} C_0$$

and suppose we know that $|\tilde{C}_\lambda| \ge 1$. Then

$$\begin{aligned} C_\lambda &= \lambda |C_1|^{1/n} \tilde{C}_1 + (1 - \lambda)|C_0|^{1/n} \tilde{C}_0 \\ &= [\lambda |C_1|^{1/n} + (1 - \lambda)|C_0|^{1/n}] \tilde{C}_{\tilde{\lambda}} \end{aligned}$$

for suitable $\tilde{\lambda}$. But then

$$|C_\lambda|^{1/n} = [\lambda |C_1|^{1/n} + (1 - \lambda)|C_0|^{1/n}] |\tilde{C}_{\tilde{\lambda}}|^{1/n}$$

proving the result.

We are thus reduced to the case $|C_0| = |C_1| = 1$. Let

$$D = \{(\mathbf{x}, \lambda) \mid \mathbf{x} \in C_\lambda, 0 \le \lambda \le 1\}$$

Then D is convex by construction, so its characteristic function, χ, is log concave. But then $|C_\lambda| = \int \chi(\mathbf{x}, \lambda)\, dx$ is log concave, so $|C_\lambda| \ge |C_0|^{1-\lambda} |C_1|^\lambda = 1$. ∎

Remarks 1. The Brunn–Minkowski inequality is known to be true for nonconvex C_i although a different proof is needed (see, e.g., [20]).

2. The isoperimetric inequality is one consequence of the Brunn–Minkowski inequality: For let C be an arbitrary convex set with smooth boundary and let B be the unit ball. Then the surface area, $s(C)$, of C is given by

$$\begin{aligned} s(C) &= \lim_{\varepsilon \to 0} \varepsilon^{-1} |\{\mathbf{x} \mid \operatorname{dist}(\mathbf{x}, C) \le \varepsilon\}| \\ &= \lim_{\varepsilon \to 0} \varepsilon^{-1} [|C + \varepsilon B| - |C|] \\ &= \lim_{\varepsilon \to 0} \varepsilon^{-1} [(1 + \varepsilon)^\nu |(1 + \varepsilon)^{-1} C + (1 + \varepsilon)^{-1} \varepsilon B| - |C|] \\ &\ge \lim_{\varepsilon \to 0} \varepsilon^{-1} [(1 + \varepsilon)^\nu [(1 + \varepsilon)^{-1} |C|^{1/\nu} + (1 + \varepsilon)^{-1} \varepsilon |B|^{1/\nu}]^\nu - |C|] \\ &= \lim_{\varepsilon \to 0} \varepsilon^{-1} \{[|C|^{1/\nu} + \varepsilon |B|^{1/\nu}]^\nu - |C|\} \\ &= \nu |C|^{(\nu - 1)/\nu} |B|^{1/\nu} \end{aligned}$$

where we have used the Brunn–Minkowski inequality. Since $s(B) = \nu |B|$, we see that if $|C| = |B|$, then $s(C) \ge s(B)$, which is the isoperimetric inequality.

Theorem 13.8 (abstract Gaussian Brunn–Minkowski [19]) Let $\mathbb{R}^\infty = \bigotimes_{n=1}^\infty (-\infty, \infty)$ with the measure $\bigotimes_{n=1}^\infty ((2\pi)^{-1/2} \exp(-\frac{1}{2}x_n^2)\, dx_n) = d\mu$. Let C_0, C_1 be measurable convex sets in $\mathbb{R}^\infty$ and let $C_\lambda = \lambda C_0 + (1 - \lambda)C_1$. Then $\mu(C_\lambda)$ is log concave in λ.

Proof Let $f(\lambda, \mathbf{x}) = 1$ (respectively, 0) if $\mathbf{x} \in C_\lambda$. Then $f(\lambda, \mathbf{x})$ is log concave so

$$f_1(\lambda, \mathbf{x}) = (2\pi)^{-1/2} \int_{y_i = x_i (i > 1)} f(\lambda, \mathbf{y}) e^{-1/2 y_1^2}\, dy_1$$

is log concave as in the proof of Theorem 13.2. Integrating over n-other variables and taking $n \to \infty$ (C_0, C_1 measurable implies convergence) we obtain the result. ∎

Remark Since Wiener measure is "isomorphic" to the last measure on $\mathbb{R}^\infty$ by a linear map, this theorem extends to, say, Wiener measure, realized on continuous functions.

Log concave functions can also be used to prove Theorem 12.2:

Alternate proof of Theorem 12.2 [173a] We will prove a stronger result: Namely, fix $\mathbf{R}_1, \ldots, \mathbf{R}_m$ and let $e(\lambda)$ be the ground state of

$$-\tfrac{1}{2}\Delta - \sum_{i=1}^m z_i |\mathbf{r} - \lambda \mathbf{R}_i|^{-1}$$

Then $e(\lambda)$ increases as λ increases. By the usual kinds of limiting arguments, it suffices to prove that

$$F(\lambda) = (\Omega, (e^{-tH_0/n} e^{-t\tilde{V}/n})^n \Omega)$$

is decreasing for Ω a Gaussian and $\tilde{V}$ a cutoff V arranged to be bounded below. Since $e^{+z_i |\mathbf{x}|^{-1}/n}$ is monotone decreasing, we can write the cutoff V's as an integral of characteristic functions of balls centered at zero [see (13.2) below], so it suffices to prove that

$$\tilde{F}(\lambda) = \left(\Omega, e^{-tH_0/n} \prod_{j=1}^m \chi_{j,1}(\mathbf{r} - \lambda \mathbf{R}_j) e^{-tH_0/n} \cdots \prod_{j=1}^m \chi_{j,n}(\mathbf{r} - \lambda \mathbf{R}_j)\Omega\right)$$

is monotone in λ if each $\chi_{j,i}$ ($j = 1, \ldots, m$; $i = 1, \ldots, n$) is the characteristic function of some ball centered at zero. Since $\chi_j(\mathbf{r} - \lambda \mathbf{R}_j)$ is log concave in r and λ and $e^{-tH_0/n}$ has a log concave integral kernel, $\tilde{F}(\lambda)$ is log concave. Since it is also clearly symmetric ($\tilde{F}(\lambda) = \tilde{F}(-\lambda)$) it is easily seen to be monotone decreasing. ∎

* * *

B. Symmetric Rearrangement

A function f on $\mathbb{R}^n$ is said to be **symmetric decreasing** if and only if f is a function, g, of $|x|$, and g is positive and monotone decreasing as $|x|$ increases. There is a particularly useful way of writing symmetric decreasing f's. Namely,

$$f = \int \chi_\lambda \, d\lambda \tag{13.2}$$

where χ_λ is the characteristic function of the ball about zero of radius $r(\lambda) = \max\{|\mathbf{x}| \mid f(\mathbf{x}) \geq \lambda\}$. The proof of (13.2) is immediate, it just says that

$$f(\mathbf{x}) = \int_0^{f(\mathbf{x})} d\lambda$$

It realizes f as a "sum" of characteristic functions of balls about zero. If χ_1 and χ_2 are two such functions, then $\chi_1 * \chi_2$ is seen to be symmetric decreasing by a simple geometric argument. Thus (13.2) implies the following lemma.

Lemma 13.9 If f and g are symmetric decreasing so are fg and $f * g$.

Theorem 13.10 If $-V(\mathbf{x})$ is symmetric decreasing and $\psi(\mathbf{x})$ is the ground state of $H = -\frac{1}{2}\Delta + V$, then $\psi(\mathbf{x})$ is symmetric decreasing.

Proof By (1.10), it suffices to show $e^{-tH}\Omega_0$ is symmetric decreasing where Ω_0 is the ground state of L_0. But, by the Trotter formula, it suffices that $(e^{-tH_0/n}e^{-tV/n})^n\Omega_0$ be symmetric decreasing, and this follows from Lemma 13.9 if we note that $e^{-tH_0/n}$ is convolution with a symmetric decreasing function. ∎

Remark Theorems 12.17, 13.5, and 13.10 say if V is a C^3 even function on $\mathbb{R}$ and e^{-f} is the ground state of $d^2/dx^2 + V(x)$, then for $k = 1, 2, 3$, we have that $V^{(k)} \geq 0$ on $(0, \infty)$ implies that $f^{(k)} \geq 0$. This does not hold for $k \geq 4$; see the example at the end of Section 12.

Given any nonnegative measurable f on $\mathbb{R}^\nu$ with the property that $|\{\mathbf{x} \mid f(\mathbf{x}) > \mu\}| < \infty$ for each $\mu > 0$, there is an essentially unique function f^* that is symmetric decreasing and obeys

$$|\{\mathbf{x} \mid f(\mathbf{x}) > \mu\}| = |\{\mathbf{x} \mid f^*(\mathbf{x}) > \mu\}|$$

It is called the **symmetric decreasing rearrangement** of f [if $\mu \mapsto |\{\mathbf{x} \mid f(\mathbf{x}) > \mu\}|$ is continuous, f^* is uniquely determined; otherwise it is only almost everywhere determined and can be fixed by demanding $f^*(\mathbf{x}) = \lim_{\lambda \uparrow 1} f^*(\lambda \mathbf{x})$, say]. The earliest results that sums increase, under symmetric rearrangement, go back to Hardy and Littlewood. The ideas were developed by among others, Hardy, Littlewood, and Polya and by Sobolev. The strongest version is the following.

Theorem 13.11 ([22]) Let $f_1, \ldots, f_k$ be positive measurable functions on $\mathbb{R}^\nu$ and let $f_1^*, \ldots, f_k^*$ be their symmetric decreasing rearrangements. Let $\mathbf{a}_1, \ldots, \mathbf{a}_k \in \mathbb{R}^n$ and define $l_i: \mathbb{R}^{n\nu} \to \mathbb{R}^\nu$ by $l_i(\mathbf{x}_1, \ldots, \mathbf{x}_n) = \sum_{j=1}^n a_{ij}\mathbf{x}_j (\mathbf{x}_j \in \mathbb{R}^\nu)$. Then

$$\int \prod_{i=1}^k f_i(l_i(\mathbf{x}))\, d^{n\nu}x \le \int \prod_{i=1}^k f_i^*(l_i(\mathbf{x}))\, d^{n\nu}x$$

For a proof, see [22]. The idea is to use (13.2) and its analog for nonsymmetric f to note that the f's and f^* can be taken to be characteristic functions in which case the inequality is a geometric statement. The analog of this statement with * replaced by symmetrization about a single plane ("Steiner symmetrization") is proven using Brunn–Minkowski and then * is realized as a limit of Steiner symmetrizations about different planes.

Typical of the applications of this inequality to Wiener integrals (really Trotter formulas) is the following result (it was proven in [178] assuming Theorem 13.11 as a conjecture and served as motivation for [22]).

Theorem 13.12 Let $V = -W$ be negative and let $V^* \equiv -W^*$. Let $E(V) \equiv \inf \operatorname{spec}(-\frac{1}{2}\Delta + V)$ and let $P_t(V) \equiv \exp(-t(-\frac{1}{2}\Delta + V))$. Then

$$E(V^*) \le E(V) \tag{13.3}$$

$$(f, P_t(V)g) \le (f^*, P_t(V^*)g^*) \tag{13.4}$$

Proof (13.3) follows from (13.4) and (1.9). (13.4) is easy to prove in a Trotter approximation using Theorem 13.11 and the fact that $\exp(\frac{1}{2}t\Delta)$ is convolution with a symmetric decreasing function. ∎

Remark One might conjecture that $N(V) \le N(V^*)$ on the basis of this. However, this is wrong (Lieb, private communication): If V is a sum of very shallow square wells in one or two dimensions very far apart, then $N(V)$ can be large while $N(V^*) = 1$. Even in three dimensions, two wells far apart, each of which just binds a state, will yield a V with $N(V) = 2$ and $N(V^*) = 1$ (M. Klaus, private communication).

* * *

C. Conditioning

We discuss briefly some inequalities due to Guerra *et al.* [120, 121], which are of some technical use in quantum field theory (e.g., [94, 96, 110, 121]). Let $\{X_i\}_{i\in I}$ be a family of objects which are Gaussian random variables with respect to two different probability measures μ and ν. We say that μ is obtained from ν **by conditioning** and write $\mu \prec \nu$ if and only if

$$\int |\sum z_j X_{i_j}|^2 \, d\mu \leq \int |\sum z_j X_{i_j}|^2 \, d\nu$$

for all $z_1, \ldots, z_n \in \mathbb{C}$; i.e., if and only if $a^\nu - a^\mu$ is a positive definite matrix where $a_{ij}^\nu = \int X_i X_j \, d\nu$.

Examples 1. If $X = \alpha(t)$, the Brownian bridge and $b(t)$, the Brownian motion, then $D\alpha \prec Db$. For $\alpha(t) \oplus tb(1) \doteq b(t)$ implies that

$$\int |\sum z_i b(t_i)|^2 \, Db = \int |\sum z_i \alpha(t_i)|^2 \, D\alpha + |\sum t_i z_i|^2$$

2. Let $L_0(\omega) = -\frac{1}{2} d^2/dx^2 + \frac{1}{2}\omega^2 x^2 - \frac{1}{2}\omega$. The corresponding path integral has covariance $(2\omega)^{-1} \exp(-\omega|t - s|)$. Since

$$(2\omega)^{-1} e^{-\omega|x-y|} = (2\pi)^{-1} \int (k^2 + \omega^2)^{-1} e^{ik(x-y)} \, dk$$

we see that as ω increases, Dq_ω decreases in conditioning sense.

The description above is an active one; i.e., we fix the X_i's (thought of as coordinates) and vary measures. It is useful to change the point of view to a "passive" one where we fix the measure as $d\nu$ and think of changing from X_i to $\tilde{X}_i$. Without loss, we can extend the indexing set for $\{X_i, d\nu\}$ to a Hilbert space and suppose that $\int X_i X_j \, d\nu = \int \phi(f_i)\phi(f_j) \, d\nu = (f_i, f_j)$ for suitable $f_i, f_j \in \mathscr{H}$. Now define an operator A from $\mathscr{H}$ to $\mathscr{H}$ by

$$\left(\sum_i z_i f_i, A\left(\sum_j w_j f_j\right)\right) = \int \sum \bar{z}_i w_j X_i X_j \, d\mu$$

Then $0 \leq A \leq 1$ since $\mu \prec \nu$. Let $B = A^{1/2}$. Let $\phi_1(f) = \phi(Bf)$, $\phi_2(f) = \phi((1 - A)^{1/2} f)$. Then $\tilde{X}_i = \phi_1(f_i)$ with the measure $d\nu$ is a "model" for $(X_i, d\mu)$. Consider the process $\phi_1 \oplus \phi_2$, i.e., the product of $d\nu \otimes d\nu$ with

$$(\phi_1 \oplus \phi_2)(f)(x, y) = \phi_1(f)(x) + \phi_2(f)(y)$$

It is not hard to see that $\phi_1 \oplus \phi_2$ is a model for (X_i, dv). Now Jensen's inequality says that

$$\int \exp\left(\int F(\phi_1 \oplus \phi_2)\, dv_2\right) dv_1 \le \int \exp(F(\phi_1 \oplus \phi_2))\, dv_1 \otimes dv_2$$

Thus given any function F of the X_i, there is another function G of the X_i given by

$$\int F(\phi_1 \oplus \phi_2)\, dv_2$$

so that

$$\int \exp(G)\, d\mu \le \int \exp(F)\, dv \tag{13.5}$$

Can one make the map $F \mapsto G$ more concrete? The answer is yes, and the point is that $\int \cdots dv_2$ is the same as taking a conditional expectation with respect to ϕ_2. Thus, by (3.9′),

$$F = {:}\exp\left(\sum_j \alpha_j X_j\right){:}_v \mapsto G = {:}\exp\left(\sum_j \alpha_j X_j\right){:}_\mu$$

where we use $:\!-\!:_v$ and $:\!-\!:_\mu$, because in the original active picture $:\!-\!:$ depends on covariance; e.g.,

$${:}\exp(X_1){:}_\lambda = \exp\left(X_1 - \frac{1}{2}\int X_1^2\, d\lambda\right)$$

for $\lambda = \mu, v$. A little more useful in applications than $:\exp \phi(f):$ are the Wick-ordered polynomials $:P(\phi):$ defined by the formal generating functional

$$\sum_{n=0}^{\infty} t^n{:}\phi^n{:}/n! = {:}\exp(t\phi){:} = \exp(t\phi - \tfrac{1}{2}t^2\langle \phi^2 \rangle)$$

for Gaussian ϕ's. Thus, e.g.,

$${:}\phi^4{:} = \phi^4 - 6\langle \phi^2 \rangle \phi^2 + 3\langle \phi^2 \rangle^2$$

Since the map $F \mapsto G$ is linear, it takes $:\phi^n:_v$ into $:\phi^n:_\mu$.

We thus have the following theorem.

Theorem 13.13 (conditioning inequality [120]) If $\mu \prec v$, for two Gaussian processes and if P is a polynomial, then

$$\int \exp\left(-\sum_{i=1}^{n} a_i {:}P(X_i){:}_\mu\right) d\mu \le \int \exp\left(-\sum_{i=1}^{n} a_i {:}P(X_i){:}_v\right) dv$$

Example If $:x^4:_\omega$ is the polynomial

$$:x^4:_\omega = x^4 - 3\omega^{-1}x^2 + \tfrac{3}{4}\omega^{-2}$$

then the above and Example 2 above imply that the ground state energy of $L_0(\omega) + \lambda{:}x^4{:}_\omega$, where

$$L_0(\omega) + \lambda{:}x^4{:}_\omega = -\lim_{t\to\infty} t^{-1} \ln \int \exp\left(-\int_0^t \lambda{:}q(s)^4{:}\, ds\right) Dq_\omega$$

increases as ω increases. Thus the ground state energy of

$$H(\omega, \lambda) = -\frac{1}{2}\frac{d^2}{dx^2} + \lambda x^4 + x^2\left(\frac{1}{2}\omega^2 - \frac{3\lambda}{\omega}\right) + \left(\frac{3\lambda}{\omega^2} - \frac{1}{2}\omega\right)$$

increases as ω does. This can be turned into an inequality on the derivative of the ground state energy of $-\frac{1}{2}\, d^2/dx^2 + \alpha x^2 + x^4$ under change of α.

* * *

D. Hypercontractivity

Finally we state an estimate which has been very useful in quantum field theory: it has not yet had any striking application to (one-dimensional) path integrals.

Theorem 13.14 (hypercontractive estimates) Let $E(\cdot \mid q(0))$ be a conditional expectation with respect to the oscillator process. Suppose that p and r are such that

$$(r - 1) \le e^{2t}(p - 1)$$

Then

$$\|E(f(q(t)) \mid q(0))\|_r \le \|f(q(t))\|_p$$

where $\|\cdot\|_p$ is the L^p-norm with respect to dq.

Remarks 1. This inequality for $r - 1 < e^{2t}(p - 1)$ with a constant $a(p, r, t)$ was first proven by Nelson [192]; Glimm [104] showed that for p, r fixed by choosing t large, one could take $a = 1$; this was significant for applications to systems with an infinity of variables such as field theory. The full result is due to Nelson [194].

2. For proofs see Nelson [194], Gross [118], Brascamp–Lieb [21], and Neveu [195b].

3. Generalizations to other $P(\phi)_1$-processes are due to Rosen [222], Eckmann [70], and Carmona [31].

4. Certain kinds of improvements to Hölder's inequality ("checkerboard estimates") in special cases come from this estimate and the Markov property; see [120].

V

Magnetic Fields and Stochastic Integrals

14. Itô's Integral

Our goal here is to define $\int f(s)\,db$ and more generally $\int g(b(s))\,db$. This integral cannot be viewed as a Stieltjes integral, for b is never of bounded variation (equivalently the curve $b(s)$ is not rectifiable):

Theorem 14.1 For one-dimensional Brownian motion, define for $\alpha > 0$ and n a positive integer

$$f(b; n, \alpha) = \sum_{k=1}^{2^n} \left| b\left(\frac{k}{2^n}\right) - b\left(\frac{k-1}{2^n}\right) \right|^{\alpha}$$

(a) If $\alpha < 2$, then $f(b; n, \alpha) \to \infty$ as $n \to \infty$ with probability one.
(b) If $\alpha = 2$, then $f(b; n, \alpha) \to 1$ as $n \to \infty$ with probability one.
(c) If $\alpha > 2$, then $f(b; n, \alpha) \to 0$ as $n \to \infty$ with probability one.
(d) The convergence in (b), (c) is also in any L^p-space with $p < \infty$.

Proof Let $c_\alpha = (2\pi)^{-1/2} \int |x|^\alpha e^{-(1/2)x^2}\,dx$ so $c_2 = 1$ and

$$E\left(\left| b\left(\frac{k}{2^n}\right) - b\left(\frac{k-1}{2^n}\right) \right|^{\alpha}\right) = 2^{-(1/2)n\alpha} c_\alpha \equiv d(n, \alpha)$$

By the indepencence of the $|b(k/2^n) - b((k-1)/2^n)|$ as k varies:

$$E\left(\left[\sum_{k=1}^{2^n}\left(\left|b\left(\frac{k}{2^n}\right) - b\left(\frac{k-1}{2^n}\right)\right|^\alpha - d(n,\alpha)\right)\right]^2\right)$$

$$= \sum_{k=1}^{2^n} E\left(\left[\left|b\left(\frac{k}{2^n}\right) - b\left(\frac{k-1}{2^n}\right)\right|^\alpha - d(n,\alpha)\right]^2\right)$$

$$\leq \sum_{k=1}^{2^n} E\left(\left|b\left(\frac{k}{2^n}\right) - b\left(\frac{k-1}{2^n}\right)\right|^{2\alpha}\right) = 2^n\, d(n, 2\alpha)$$

where we have used $E([X - E(X)]^2) \leq E(X^2)$. Thus

$$E(|f(b; n, \alpha) - 2^{n(1-(1/2)\alpha)}c_\alpha|^2) \leq 2^{n(1-\alpha)}c_{2\alpha} \tag{14.1}$$

If $\alpha < 2$, then for any fixed k, $\frac{1}{2}c_\alpha 2^{n(1-(1/2)\alpha)} > k$ for n sufficiently large. For such n:

$$\begin{aligned} E(f(b; n, \alpha) < k) &\leq E(f - 2^{n(1-(1/2)\alpha)}c_\alpha < -\tfrac{1}{2}c_\alpha 2^{n(1-(1/2)\alpha)}) \\ &\leq 4c_\alpha^{-2}c_{2\alpha}2^{-n} \end{aligned}$$

by (14.1). Thus, by the first Borel–Cantelli lemma, $f(b; n, \alpha) \geq k$ for n large. Since k is arbitrary, (a) is proven.

(b) and (c) have similar proofs. For (b), one replaces $E(f < k)$ by $E(|f - 1| > \varepsilon)$ and for (c) by $E(|f| > \varepsilon)$. L^p convergence in case (b) for $p = 2m$ (m an integer) is by direct calculation of $E(|f - 1|^{2m})$ (or one can use a suitable corollary of hypercontractive estimates). In case (c), we just note by the triangle inequality for $\|\cdot\|_p$:

$$\begin{aligned} E(|f(b; n, \alpha)|^p) &\leq 2^{np}E(|b(1/2^n)|^{p\alpha}) \\ &\leq 2^{np(1-(1/2)\alpha)} \end{aligned}$$

goes to zero. ∎

Remarks 1. By the extra devices used in the proof of the strong law of large numbers (Theorem 7.14), one can replace 2^n by n and take n to infinity.

2. The same result (except the 1 in (b) is replaced by ν) holds for ν-dimensional Brownian motion; indeed, all but (b) follow from $|b_1| \leq |\mathbf{b}| \leq \sum |b_i|$, and (b) comes from $|\mathbf{b}|^2 = \sum |b_i|^2$.

3. If "probability one" is interpreted to mean "except for a set of measure zero," the same result holds for Wiener measure. By local equivalence, similar results hold for the oscillator process and the Brownian bridge.

For suitable f's, *functions of s alone*, Payley *et al.* [199] succeeded in defining $\int_0^1 f(s)\,db(s) \equiv G(f)$ by the following device: If f is C^1 and $f(1) = f(0) = 0$, then *define*

$$G(f) = -\int_0^1 f'(s)b(s)\,ds$$

a formal integration by parts. With this definition

$$E(G(f)^2) = \int_0^1 ds \int_0^1 dt\ f'(s)f'(t)\min(t, s)$$

$$= \int_0^1 |f(s)|^2\,ds$$

since $(\partial^2/\partial s\,\partial t)\min(s, t) = \delta(s - t)$. Thus $G(f)$ extends to a linear map from $L^2(0, 1)$ to $L^2(C(0, 1), Db)$. We plan to extend this definition vastly using ideas of Itô [134]. What we learn from the above is that despite the fact that b is not smooth, one can define $\int f\,db$ for more f's than one might expect. Moreover, L^2 calculations and extensions in L^2 are useful. Our presentation below will actually settle for considerably weaker results than obtained by Itô, for we will settle for an L^2 definition of $\int f(b(s))\,db$ similar in flavor to that above; i.e., we will define the integral for nice f's and extend by L^2-continuity. Itô has a "pointwise" definition which is more complicated but yields more information about the result. For example, $F(b, t) = \int_0^t f(b(s))\,db$ can be shown (for any f for which it can be defined) to be continuous in t for almost every b. All we get is continuity in L^2-sense. For our purposes, this L^2 definition will suffice. See McKean [183] for the "pointwise" treatment.

The basic idea of Itô's definition is to take

$$\int_0^1 f(b(s))\,db = \lim_{n\to\infty} \sum_{m=1}^{2^n} f\left(b\left(\frac{m-1}{2^n}\right)\right)\left[b\left(\frac{m}{2^n}\right) - b\left(\frac{m-1}{2^n}\right)\right] \tag{14.2}$$

Before discussing the limit, we want to note some important and perhaps surprising aspects of this formula. First, the differentials stick out into the future; i.e., $f(b(s))$ is evaluated at the left endpoint. This is convenient because $f(b((m-1)/2^n))$ is independent of $b(m/2^n) - b((m-1)/2^n)$ while $f(b(m/2^n))$ is not. For this reason, we shall be able to define this integral for nonsmooth f's; if we had taken the differentials pointing into the past, it turns out that f' would have entered naturally. Secondly, *the ordinary formal rules of calculus do not hold* for the stochastic integral; e.g., $\int_0^1 b(s)\,db = \frac{1}{2}(b^2(1) - 1)$ not $\frac{1}{2}(b^2(1))$! [Note that $b(0) = 0$.] The computation of this integral is illuminating since it also illustrates the difference between differentials pointing into the past and the future. Define

$$I_+(n) = \sum_{m=1}^{2^n} b\left(\frac{m}{2^n}\right)\left[b\left(\frac{m}{2^n}\right) - b\left(\frac{m-1}{2^n}\right)\right]$$

$$I_-(n) = \sum_{m=1}^{2^n} b\left(\frac{m-1}{2^n}\right)\left[b\left(\frac{m}{2^n}\right) - b\left(\frac{m-1}{2^n}\right)\right]$$

We have that for any n

$$I_+(n) + I_-(n) = b(1)^2$$

since

$$\left[b\left(\frac{m}{2^n}\right) + b\left(\frac{m-1}{2^n}\right)\right]\left[b\left(\frac{m}{2^n}\right) - b\left(\frac{m-1}{2^n}\right)\right] = b^2\left(\frac{m}{2^n}\right) - b^2\left(\frac{m-1}{2^n}\right)$$

and the sum telescopes. On the other hand, by Theorem 14.1(b),

$$I_+(n) - I_-(n) \to 1$$

pointwise as $n \to \infty$. It follows that

$$\lim_{n\to\infty} I_\pm(n) = \tfrac{1}{2}(b(1)^2 \pm 1)$$

showing the difference between I_+ and I_- and their unintuitive answer. Of course $\frac{1}{2}(I_+ + I_-)$ gives the intuitive answer here but if we took $\int b^k\, db$, this prescription would not yield $(k+1)^{-1}b^{k+1}$.

Theorem 14.2 Let f be a C^1 function on $\mathbb{R}$ with f and f' bounded. Then, the limit on the right-hand side of (14.2) exists in L^2-sense and ($\|\cdot\|_2 = L^2$-norm)

$$\left\|\int_0^1 f(b(s))\, db\right\|_2^2 = E\left[\int_0^1 f(b(s))^2\, ds\right] \tag{14.3}$$

Proof Define

$$J_n(f) = \sum_{m=1}^{2^n} f\left(b\left(\frac{m-1}{2^n}\right)\right)\left[b\left(\frac{m}{2^n}\right) - b\left(\frac{m-1}{2^n}\right)\right] \equiv \sum_{m=1}^{2^n} q_{n,m}$$

Now for $i \neq j$, $E(q_{n,i}q_{n,j}) = 0$, since for $j > i$ there is a factor of $b(j/2^n) - b((j-1)/2^n)$ independent of the rest of $q_{n,i}q_{n,j}$, and $E(b(k/2^n)) = 0$. Moreover, since $f(b((m-1)/2^n))$ is independent of $b(m/2^n) - b((m-1)/2^n)$, we have

$$E(q_{n,m}^2) = E\left(\left[f\left(b\left(\frac{m-1}{2^n}\right)\right)\right]^2\right)\frac{1}{2^n}$$

Thus,

$$\|J_n(f)\|_2^2 = \sum_{m=1}^{2^n} E\left(\left[f\left(b\left(\frac{m-1}{2^n}\right)\right)\right]^2\right)\frac{1}{2^n} \tag{14.4}$$

Since $\sum_m f(b((m-1)/2^n))^2 1/2^n$ converges pointwise to $\int_0^1 f(b(s))^2\,ds$ (f and b are continuous) and the sum is dominated by $\|f\|_\infty^2$, (14.3) will follow if we show that $J_n(f)$ converges.

A calculation similar to the above shows

$$\|J_{n+1}(f) - J_n(f)\|_2^2 = \sum_{m=0}^{2^n-1} E\left(\left[f\left(b\left(\frac{2m+1}{2^{n+1}}\right)\right) - f\left(b\left(\frac{2m}{2^{n+1}}\right)\right)\right]^2\right)\frac{1}{2^{n+1}}$$

$$\leq \frac{1}{2^{n+1}} C \sum_{m=0}^{2^n-1} E\left(\left[b\left(\frac{2m+1}{2^{n+1}}\right) - b\left(\frac{2m}{2^{n+1}}\right)\right]^2\right) = \frac{C}{2^{n+2}}$$

where $C = \|f'\|_\infty^2$ so $|f(x) - f(y)|^2 \leq C|x-y|^2$. Thus $J_n(f)$ is Cauchy in L^2. ▮

We now extend the definition of $\int f(b(s))\,ds$ from C^1-functions to an arbitrary f with

$$E\left(\int_0^1 f(b(s))^2\,ds\right) = \int_0^1 \left(\int dx\, f(x)^2 (2\pi s)^{-1/2} e^{-x^2/2s}\right) ds$$

finite; e.g., f bounded near zero and L^2 will certainly suffice. This integral is in L^2 for any such f. We can actually go one step further: Let χ_a be the characteristic function of $(-a, a)$ and suppose that

$$E\left(\int_0^1 \chi_a(b(s)) f(b(s))^2\,ds\right) < \infty$$

as will be true if f is L^∞ near zero and locally L^2 elsewhere. Then for each a, we can define $\int_0^1 f\chi_a(b(s))\,ds$. We claim that for almost every b

$$\lim_{a\to\infty} \int_0^1 \chi_a(b(s)) f(b(s))\,ds \tag{14.5}$$

exists pointwise and is finite. For $b(s)$ is continuous on $[0, 1]$ and so bounded. Thus for any fixed b, (14.5) is independent of a for a large.

We also want to extend the scope of definition. First we can clearly use

ν-dimensional Brownian motion and define $\int_0^1 \mathbf{f}(\mathbf{b}(s)) \cdot d\mathbf{b}$, and we can easily define $\int_s^t$ in place of $\int_0^1$. More importantly for the applications we have in mind, we can define $\int_0^1 \mathbf{f}(\boldsymbol{\omega}(s)) \cdot d\boldsymbol{\omega}$ for the $d\mu_0$—Wiener paths. The construction is identical to the above but now

$$E\left(\int_0^1 |f(\boldsymbol{\omega}(s))|^2 \, ds\right) = \int |f(\mathbf{x})|^2 \, d^\nu x$$

since $\boldsymbol{\omega}(s)$ has the distribution of Lebesgue measure. Extending via the analog of (14.5), we see that *for any* $f \in L^2_{\mathrm{loc}}(\mathbb{R}^\nu)$, $\int_0^t \mathbf{f}(\boldsymbol{\omega}(s)) \, d\boldsymbol{\omega}$ can be defined for μ_0—almost every path. This is the integral we will need in Section 15. This random variable is called the **Itô integral**.

If one looks at the construction above, it is clear that we can allow f to be a function on $\mathbb{R}^\nu \times \mathbb{R}$ and define $\int_0^1 f(\mathbf{b}(s), s) \, d\mathbf{b}$ so long as $E(\int_0^1 |f(\mathbf{b}(s), s)|^2 \, ds) < \infty$. It is a simple exercise that $\int_0^1 f(s) \, db$ agrees with the Payley *et al.* [199] definition. In fact, one can deal with general **nonanticipatory functionals**, i.e., functions $f(b, s)$ with the property that $f(\cdot, s)$ is only a function of $\{b(t) \,|\, t \le s\}$, i.e., is $\mathscr{B}_s$-measurable; for example, if τ is a stopping time one can define $\int_0^\tau f(b(s), s) \, db$. The only subtle step in the construction for general nonanticipatory functionals is the density of "smooth functions" for which one can obtain the integral as a limit; in fact, it is more useful to deal initially with f's with $f(\cdot, s)$ constant for $k/2^n \le s < (k + 1)/2^n$ and take limits; see, e.g., [183].

For calculation and understanding of stochastic integrals a lemma of Itô [135] is particularly useful. We state it for the full Wiener integral since it is somewhat easier to give optimal conditions on f; a similar result holds for the Brownian path if, for example (to state an overly strong condition), f is C^2 globally. This is a special case of a more general result of Itô which will be featured prominently in Section 16.

Theorem 14.3 (Itô's lemma) Let $f(\mathbf{x}, t)$ be a function on $\mathbb{R}^\nu \times [0, t]$ and use ∇f, etc., for $\mathbf{x}$ derivatives and $\dot{f}$ for t derivatives. Suppose that f is in L^2_{loc} and that its distributional derivatives ∇f, Δf, and $\dot{f}$ lie in L^2_{loc}. Then

$$f(\boldsymbol{\omega}(t), t) - f(\boldsymbol{\omega}(0), 0) = \int_0^t (\nabla f)(\boldsymbol{\omega}(s), s) \cdot d\boldsymbol{\omega} + \frac{1}{2} \int_0^t (\Delta f)(\boldsymbol{\omega}(s), s) \, ds + \int_0^t \dot{f}(\boldsymbol{\omega}(s), s) \, ds \tag{14.6}$$

for almost every $\boldsymbol{\omega}$.

Proof By a limiting argument we can suppose that $f \in C_0^\infty$ and for notational simplicity that $t = 1$. Write

$$f(\boldsymbol{\omega}(1), 1) - f(\boldsymbol{\omega}(0), 0) = A_n + B_n + C_n$$

$$A_n = \sum_{m=1}^{2^n} (\nabla f)\left(\boldsymbol{\omega}\left(\frac{m-1}{2^n}\right), \frac{m-1}{2^n}\right) \cdot \delta\boldsymbol{\omega}_{n,m}$$

$$B_n = \sum_{m=1}^{2^n} f\left(\boldsymbol{\omega}\left(\frac{m}{2^n}\right), \frac{m-1}{2^n}\right) - f\left(\boldsymbol{\omega}\left(\frac{m-1}{2^n}\right), \frac{m-1}{2^n}\right)$$
$$- (\nabla f)\left(\boldsymbol{\omega}\left(\frac{m-1}{2^n}\right), \frac{m-1}{2^n}\right) \cdot \delta\boldsymbol{\omega}_{n,m}$$

$$C_n = \sum_{m=1}^{2^n} f\left(\boldsymbol{\omega}\left(\frac{m}{2^n}\right), \frac{m}{2^n}\right) - f\left(\boldsymbol{\omega}\left(\frac{m}{2^n}\right), \frac{m-1}{2^n}\right)$$

with $\delta\boldsymbol{\omega}_{n,m} = \boldsymbol{\omega}(m/2^n) - \boldsymbol{\omega}((m-1)/2^n)$. By construction

$$A_n \to \int_0^1 (\nabla f)(\boldsymbol{\omega}(s), s) \cdot d\boldsymbol{\omega}$$

in L^2 as $n \to \infty$. Using

$$\left| f(\mathbf{x}) - f(\mathbf{y}) - (\nabla f)(\mathbf{y}) \cdot (\mathbf{x} - \mathbf{y}) - \frac{1}{2} \sum \frac{\partial^2 f}{\partial y_i \, \partial y_j} (x - y)_i (x - y)_j \right|$$
$$\leq C |\mathbf{x} - \mathbf{y}|^3$$

and Theorem 14.1, we see that up to a term going to zero in L^2-norm, B_n equals

$$\tilde{B}_n = \frac{1}{2} \sum_{i,j} \frac{\partial^2 f}{\partial x_i \, \partial x_j} \left(\boldsymbol{\omega}\left(\frac{m-1}{2^n}\right), \frac{m-1}{2^n}\right) (\delta\boldsymbol{\omega}_{n,m})_i (\delta\boldsymbol{\omega}_{n,m})_j$$

As in the proof of Theorem 14.1, $\tilde{B}_n$ equals

$$\frac{1}{2} \sum_{i,j} \frac{\partial^2 f}{\partial x_i \, \partial x_j} \left(\boldsymbol{\omega}\left(\frac{m-1}{2^n}\right), \frac{m-1}{2^n}\right) \delta_{ij} \frac{1}{2^n}$$

plus a term going to zero in L^2. Thus $B_n \to \frac{1}{2} \int_0^t (\Delta f)(\boldsymbol{\omega}(s), s)\, ds$. Finally, using

$$|f(\mathbf{x}, t) - f(\mathbf{x}, s) - \dot{f}(\mathbf{x}, s)(t - s)| \leq C|t - s|^2$$

one easily sees that $C_n \to \int_0^t \dot{f}(\boldsymbol{\omega}(s), s)\, ds$. ∎

(14.6) is often written in infinitesimal form as the formal expression

$$df = \nabla f \, d\omega + (\tfrac{1}{2}\Delta f + \dot{f}) dt \tag{14.7}$$

The surprise is that *second derivatives of f appear in the first order differential.* (14.7) can be expressed even more succinctly by

$$d\omega_i \, d\omega_j = \delta_{ij} \, dt$$

(14.6) can be used for calculational purposes:

Example ($v = 1$) f's with $\frac{1}{2}f'' + \dot{f} = 0$ will have particularly simple stochastic integrals. For example, if

$$f(x, t) = \exp(\alpha x - \tfrac{1}{2}\alpha^2 t)$$

then $df = \nabla f \, d\omega$. Recognizing the Wick-ordered exponential $f(\omega(s), s) = {:}\exp(\alpha\omega(s)){:}$ and taking ω to b, we see that

$$\int_0^t {:}\exp(\alpha b(s)){:} \, db = \alpha^{-1}{:}\exp(\alpha b(t)){:} - \alpha^{-1} \tag{14.8}$$

Thus, it is the Wick-ordered exponential, rather than the ordinary exponential which stochastically integrates like an exponential! It is a curious historical coincidence that the work of Itô leading to (14.8) is approximately simultaneous to that of Wick on "normal ordering." If (14.8) is expanded in α, one finds that

$$\int_0^t {:}b^n(s){:} \, db = (n + 1)^{-1}{:}b^{n+1}(t){:}$$

If we take $n = 1$ we find that

$$\int_0^t b(s) \, db = \tfrac{1}{2}{:}b^2(t){:} = \tfrac{1}{2}(b^2(t) - t)$$

a result already obtained.

Itô's lemma provides directly the connection of Brownian motion and $-\frac{1}{2}\Delta = H_0$ and also a new proof of the Feynman–Kac formula (somewhat related to the second proof we gave). A key observation is that because the differentials point into the future, we have that $E(f(b(s)) \, db) = E(f(b(s))E(db) = 0$ since $E(db) = 0$; i.e., $E(\int_a^b f(b(s)) \, db) = 0$, more strongly

$$E\left(g \int_a^b f(b(s)) \, db\right) = 0 \tag{14.9}$$

if g is measurable with respect to $\{b(u) \mid u \leq a\}$.

Choose $f(\mathbf{x}, s) = (e^{-(t-s)H_0}g)(\mathbf{x})$, so $\dot{f} = H_0 f$. Thus $\dot{f} + \frac{1}{2}\Delta f = 0$ so that Itô's lemma implies that

$$f(\mathbf{b}(t), t) = f(\mathbf{b}(0), 0) + \int_0^t (\nabla f)(\mathbf{b}(u), u) \cdot d\mathbf{b}$$

Using (14.9), we see that

$$E(f(\mathbf{b}(t), t)) = E(f(\mathbf{b}(0), 0))$$

Since $f(\mathbf{b}(t), t) = g(\mathbf{b}(t))$ and $f(\mathbf{b}(0), 0) = (e^{-tH_0}g)(\mathbf{0})$, we have proven once again that

$$E(g(\mathbf{b}(t))) = (e^{-tH_0}g)(\mathbf{0})$$

The proof of the Feynman–Kac formula is similar but more complicated. Here, we take $H = H_0 + V$ and

$$f(\mathbf{x}, s) = (e^{-(t-s)H}g)(\mathbf{x})$$

Then, at least for sufficiently nice V's (say, $V \in C_0^\infty$),

$$f(\mathbf{b}(s), s) = f(\mathbf{b}(t), t) - \int_s^t V(\mathbf{b}(u))f(\mathbf{b}(u), u)\, du - \int_s^t (\nabla f)(\mathbf{b}(u), u) \cdot d\mathbf{b}$$

Notice that by (14.9) the last term has zero expectation. Iterating this equation once:

$$f(\mathbf{b}(0), 0) = f(\mathbf{b}(t), t) - \int_0^t V(\mathbf{b}(u))f(\mathbf{b}(t), t))\, du$$

$$+ \int_0^t V(\mathbf{b}(u)) \int_u^t V(\mathbf{b}(s))f(\mathbf{b}(s), s)\, du\, ds + G$$

where G also has zero expectation by (14.9). Continuing in this way, one finds that

$$E(f(\mathbf{b}(0)), 0) = E\left(\left[\sum_{n=0}^{\infty}(-1)^n \int_{0<s_1<\cdots<s_n<t} \prod V(\mathbf{b}(s_i))\, ds_i\right] f(\mathbf{b}(t), t)\right)$$

If one recognizes the object in [...] as $\exp(-\int_0^t V(\mathbf{b}(s))\, ds)$ one has proven that

$$(e^{-tH}g)(\mathbf{0}) = E\left(\exp\left(-\int_0^t V(\mathbf{b}(s))\, ds\right) g(\mathbf{b}(t))\right)$$

In Section 16, we will use the general Itô lemma for an even slicker proof of the Feynman–Kac formula.

* * *

We next make a simple calculation:

Theorem 14.4 ($\nu = 1$)

$$E\left(\exp\left(\int f(s)\,db(s) + \int g(s)b(s)\,ds\right)\right) = \exp(\tfrac{1}{2}\alpha) \tag{14.10}$$

$$\alpha = \int |f(s)|^2\,ds + 2\int_{s<t} f(s)g(t)\,ds\,dt + \iint g(s)g(t)\min(s,t)\,ds\,dt \tag{14.11}$$

Proof The object in the exponential in (14.10) is clearly linear in b, so $\alpha = E([\int f(s)\,db + \int g(s)b(s)\,ds]^2)$. We recognize the first and third terms as just the expectation of $(\int f(s)\,db)^2$ and of $(\int g(s)b(s)\,ds)^2$. To compute the middle term pass to the Riemann sums defining $\int f(s)\,db$ and note that

$$E([b(s+\Delta s) - b(s)]b(t)) = \begin{cases} 0, & t < s \\ t - s, & s < t < s + \Delta s \\ \Delta s, & t > s + \Delta s \end{cases} \quad \blacksquare$$

We note that if f depends on some random variables independent of b, and E represents expectation with respect to b, then (14.10) still holds, e.g., for $\nu = 2$:

$$E\left(\exp\left(\int_0^1 f(b_2(s))\,db_1(s)\right)\right) = E\left(\exp\left(\frac{1}{2}\int_0^1 |f(b_2(s))|^2\,ds\right)\right)$$

which is $\int \exp(-(H_0 - \frac{1}{2}f^2))(x, 0)\,dx$.

* * *

Finally, we want to discuss the definition of stochastic integrals with respect to the Brownian bridge, α, and the oscillator process, q. The former is especially important since it will allow us to define the integral for almost every Wiener path with fixed endpoints. At first sight, this seems difficult, for the increments of α depend heavily on the present: since α must reach $\alpha(1) = 0$ at $t = 1$, it must know where it is at a given time to decide where to go next; more prosaically, one checks that for $s \leq t$

$$E([\alpha(t + \Delta t) - \alpha(t)]\alpha(s)) = -s\,\Delta t$$

and

$$E((q(t + \Delta t) - q(t))q(s)) = \tfrac{1}{2}e^{-(t-s)}[e^{-\Delta t} - 1]$$

The key to deciding how to overcome this problem is (5.6)

$$q(t) \doteq e^{-t}b(e^{2t})/\sqrt{2}$$

and

$$\alpha(t) \doteq (1 - t)b\left(\frac{t}{1 - t}\right) \tag{14.12}$$

These formulas would allow us to define $\int \cdot \, d\alpha$ and $\int \cdot \, dq$ in terms of db but, more simply, they suggest to us simple "increments" which are independent of the past; namely,

$$\gamma(t, \Delta t) = \alpha(t + \Delta t) - \alpha(t) + \frac{\Delta t}{1 - t}\alpha(t)$$

(since (14.12) says the increments of $\alpha(t)/(1 - t)$ are independent of the past) and

$$\lambda(t, \Delta t) = q(t + \Delta t) - e^{-\Delta t}q(t)$$

One trivially checks that for $s < t$:

$$E(\alpha(s)\gamma(t, \Delta t)) = 0 = E(q(s)\lambda(t, \Delta t))$$

and that

$$E(|\gamma(t, \Delta t)|^2) = (1 - t)^{-1}(1 - t - \Delta t)\,\Delta t$$
$$E(|\lambda(t, \Delta t)|^2) = \tfrac{1}{2}(1 - e^{-2\Delta t})$$

Both expectations go to Δt as $\Delta t \to 0$ and are bounded by Δt for all Δt. For nice enough f

$$\lim_{n\to\infty} \sum_{k=1}^{2^n} f\left(\alpha\left(\frac{k-1}{2^n}\right)\right)\gamma\left(\frac{k-1}{2^n}, \frac{1}{2^n}\right)$$

exists and has L^2-norm, $\int_0^1 E(f(\alpha(s))^2)\, ds$, for we need only mimic the construction of db. But clearly, by the continuity of paths:

$$\lim_{n\to\infty} \sum_{k=1}^{2^n} f\left(\alpha\left(\frac{k-1}{2^n}\right)\right)\alpha\left(\frac{k-1}{2^n}\right)\left[1 - \left(\frac{k-1}{2^n}\right)\right]^{-1}\frac{1}{2^n}$$

exists and equals $\int_0^1 f(\alpha(s))\alpha(s)\,(ds/(1 - s))$; there is no divergence at $s = 1$, since $\alpha(s) \to 0$ faster than $(1 - s)^{1/2-\varepsilon}$ by a law of iterated logarithm. As a result, for nice enough f's,

$$\lim_{n\to\infty} \sum_{k=1}^{n} f\left(\alpha\left(\frac{k-1}{2^n}\right)\right)\delta\alpha\left(\frac{k-1}{2^n}, \frac{1}{2^n}\right) \equiv \int_0^1 f(\alpha(s))\, d\alpha$$

exists and, since

$$\|a + b\|_2 \le \|a\|_2 + \|b\|_2 \qquad \text{and} \qquad \left\| \int a(s)\, ds \right\|_2 \le \int \|a(s)\|_2\, ds$$

we have

$$\left\| \int_0^1 f(\alpha(s))\, d\alpha \right\|_2 \le \left[\int_0^1 E(f(\alpha(s))^2)\, ds \right]^{1/2} + \int_0^1 [E((\alpha f(\alpha))^2)]^{1/2} \frac{ds}{1-s}$$

In this way, it is easy to see that one can define $\int_0^1 f(\alpha(s))\, d\alpha$ for almost every α, if f is in L^∞ near zero, and in L^p_{loc} for some $p > 2$. Similarly, one can define $\int_a^b f(q(s))\, dq$.

One can summarize the relevance of the above considerations for the full Wiener integral by:

Theorem 14.5 Let $g(\omega)$ be a function of the Wiener paths up to time t with the property that $\|g\|_\infty < \infty$ and

$$\mathbf{x}, \mathbf{y} \mapsto g\left(\boldsymbol{\omega}(s) + \left(\frac{s}{t}\right)\mathbf{y} + \left(1 - \frac{s}{t}\right)\mathbf{x}\right)$$

is continuous for almost every ω. Let F be a bounded continuous function on $\mathbb{R}^\nu$ and let f be a C^1-function on $\mathbb{R}^\nu$. Then

$$\int g(\omega) F\left(\int f(\boldsymbol{\omega}(s))\, d\boldsymbol{\omega} \right) d\mu_{0, \mathbf{x}, \mathbf{y}, t}$$

is continuous in $\mathbf{x}$, $\mathbf{y}$ for each fixed t.

Proof Rewrite the integral using (4.12) as

$$E_\alpha\left(\left(\left(\frac{s}{t}\right)\mathbf{y} + \left(1 - \frac{s}{t}\right)\mathbf{x} + \sqrt{t}\boldsymbol{\alpha}\left(\frac{s}{t}\right)\right) F\left(\sqrt{t} \int f(\ldots)\, d\alpha\right) (2\pi t)^{-\nu/2} e^{-|\mathbf{x}-\mathbf{y}|^2/2t}\right)$$

The integrand is uniformly bounded and thus continuity for almost every α will suffice. This is easy by hypothesis and construction of $d\alpha$. ∎

15. Schrödinger Operators with Magnetic Fields

Consider the quantum mechanical energy operator for a particle in a magnetic field $\mathbf{B}(x)$ with vector potential $\mathbf{a}(x)$ ($\mathbf{B} = \nabla \times \mathbf{a}$)

$$H(\mathbf{a}, V) = \tfrac{1}{2}(-i\nabla - \mathbf{a})^2 + V$$

where we will later discuss hypotheses on $\mathbf{a}$, V and the precise definition of H. We want to prove the following which we call the **Feynman–Kac–Itô** formula:

$$(f, e^{-tH(\mathbf{a}, V)} g) = \int e^{F(\boldsymbol{\omega}, t)} \overline{f(\boldsymbol{\omega}(0))} g(\boldsymbol{\omega}(t))\, d\mu_0 \tag{15.1}$$

where

$$F(\boldsymbol{\omega}, t) = -i\int_0^t \mathbf{a}(\boldsymbol{\omega}(s))\cdot d\boldsymbol{\omega} - \frac{i}{2}\int_0^t (\operatorname{div}\mathbf{a})(\boldsymbol{\omega}(s))\,ds - \int_0^t V(\boldsymbol{\omega}(s))\,ds \qquad (15.2)$$

In this section, we will give an analytic proof of (15.1) and in the next section a "stochastic integral" proof. We first learned of (15.1) from Nelson [195]; a probabilist would view it as a Cameron–Martin formula for "imaginary drifts" (see the discussion in Section 16) and so of considerable antiquity. There is even discussion of a's which are noncommuting matrices such as would arise in nonrelativistic couplings to Yang–Mills fields; see [6a, 40a, 168a].

Before turning to the proof we want to remark on several aspects of the formula itself. First, even though H has an a^2 term, no a^2 appears in F. The $\frac{1}{2}(\operatorname{div} a)$ term is a reflection of the fact that the kernel of e^{-tH} is Hermitian symmetric: Changing $\boldsymbol{\omega}(t)$ and $\boldsymbol{\omega}(0)$ by taking s into $t - s$ does not merely change the sign of $\int \mathbf{a}\cdot d\boldsymbol{\omega}$ because the differential $d\boldsymbol{\omega}$ has a preferred end. For this reason, the integral term in F should more naturally be the symmetric integral

$$S = \lim_{n\to\infty}\sum_{m=1}^{[2^n]}\frac{1}{2}\left[\mathbf{a}\left(\boldsymbol{\omega}\left(\frac{m}{2^n}\right)\right) + \mathbf{a}\left(\boldsymbol{\omega}\left(\frac{m-1}{2^n}\right)\right)\right]\cdot\left[\boldsymbol{\omega}\left(\frac{m}{2^n}\right) - \boldsymbol{\omega}\left(\frac{m-1}{2^n}\right)\right]$$

It is not hard to see that in terms of the Itô integral

$$S = \int_0^t \mathbf{a}(\boldsymbol{\omega}(s))\cdot d\boldsymbol{\omega} + \frac{1}{2}\int_0^t (\operatorname{div}\mathbf{a})\,ds$$

That is, the div a term restores the "time-reversal symmetry" destroyed by the Itô convention of having differentials point into the future.

One can also see the effect of this div a term if one considers gauge transformations. The change of $\mathbf{a}$ to $\tilde{\mathbf{a}} = \mathbf{a} + \operatorname{grad}\lambda$ should not change any physics since $\mathbf{B} = \nabla\times\mathbf{a} = \nabla\times\tilde{\mathbf{a}}$. In fact

$$e^{i\lambda}H(\mathbf{a}, V)e^{-i\lambda} = H(\tilde{\mathbf{a}}, V)$$

Thus, if $\mathbf{a}$ is replaced by $\tilde{\mathbf{a}}$ we must have

$$\tilde{F}(\boldsymbol{\omega}, t) = F(\boldsymbol{\omega}, t) + i\lambda(\boldsymbol{\omega}(0)) - i\lambda(\boldsymbol{\omega}(t)) \qquad (15.3)$$

To verify (15.3), we use Itô's lemma (Theorem (14.3):

$$\int_0^t (\nabla\lambda)(\boldsymbol{\omega}(s))\cdot d\boldsymbol{\omega} = \lambda(\boldsymbol{\omega}(t)) - \lambda(\boldsymbol{\omega}(0)) - \frac{1}{2}\int_0^t (\Delta\lambda)(\boldsymbol{\omega}(s))\,ds$$

Since $\operatorname{div}(\nabla\lambda) = \Delta\lambda$, we have that

$$\int_0^t \nabla\lambda \cdot d\boldsymbol{\omega} + \frac{1}{2}\int_0^t \operatorname{div}(\nabla\lambda)\, ds = \lambda(\boldsymbol{\omega}(t)) - \lambda(\boldsymbol{\omega}(0))$$

verifying (15.3). Note the div **a** term was essential for gauge invariance.

We now turn to the analytic proof of (15.1):

Lemma 15.1 Let $\mathbf{a} \in C_0^\infty(\mathbb{R}^\nu)$, real-valued. For $s > 0$, let

$$K_s(\mathbf{x}, \mathbf{y}) = (2\pi s)^{-\nu/2} \exp(-|\mathbf{x} - \mathbf{y}|^2/2s) \exp(\tfrac{1}{2}i[\mathbf{a}(\mathbf{x}) + \mathbf{a}(\mathbf{y})]\cdot(\mathbf{x} - \mathbf{y})) \tag{15.4}$$

Then

$$f \mapsto (Q_s f)(\mathbf{x}) = \int K_s(\mathbf{x}, \mathbf{y}) f(\mathbf{y})\, d^\nu y$$

defines a bounded operator on L^2 with $\|Q_s\| \le 1$. Moreover, for any $f \in D(H_0) = D(H_0(\mathbf{a})) = \{f \mid f, \nabla f, \Delta f \in L^2\}$:

$$\lim_{s \downarrow 0} \frac{d}{ds} Q_s f = -H_0(\mathbf{a}) f \tag{15.5}$$

where

$$H_0(\mathbf{a}) = \tfrac{1}{2}(-i\nabla - \mathbf{a})^2 \tag{15.6}$$

Proof As usual, let $P_s(\mathbf{x}, \mathbf{y}) = (2\pi s)^{-\nu/2} \exp(-|\mathbf{x} - \mathbf{y}|^2/2s)$ which is the integral kernel for e^{-sH_0}. Then $|K_s(\mathbf{x}, \mathbf{y})| \le P_s(\mathbf{x}, \mathbf{y})$ implies that

$$|(Q_s f)(\mathbf{x})| \le (e^{-sH_0}|f|)(\mathbf{x}) \tag{15.7}$$

so Q_s is as bounded operator with $\|Q_s\| \le \|e^{-sH_0}\| = 1$. Since $(d/ds)P_s = \frac{1}{2}\Delta_{\mathbf{y}} P_s$ and we can integrate by parts (since $\mathbf{a} \in C_0^\infty$ and $f \in D(H_0)$)

$$\frac{d}{ds}(Q_s f)(\mathbf{x}) = \frac{1}{2}\int P_s(\mathbf{x}, \mathbf{y}) \Delta_{\mathbf{y}}[e^{(1/2)i(\mathbf{a}(\mathbf{x}) + \mathbf{a}(\mathbf{y}))\cdot(\mathbf{x} - \mathbf{y})} f(\mathbf{y})]\, d^\nu y$$

Using

$$\Delta(e^{ih} f) = e^{ih}[\Delta f + 2i\nabla h \cdot \nabla f + (i\Delta h - (\nabla h)^2) f]$$

$$\nabla_{\mathbf{y}, i}[\tfrac{1}{2}(\mathbf{a}(\mathbf{x}) + \mathbf{a}(\mathbf{y}))\cdot(\mathbf{x} - \mathbf{y})] = -\tfrac{1}{2}(a_i(\mathbf{x}) + a_i(\mathbf{y})) + \frac{1}{2}\sum_j (\nabla_i a_j)(\mathbf{y})(x - y)_j$$

$$\Delta_y[\ldots] = -(\operatorname{div} \mathbf{a})(\mathbf{y}) + \tfrac{1}{2}(\Delta\mathbf{a})\cdot(\mathbf{x} - \mathbf{y})$$

and the fact that $P_s(\mathbf{x}, \mathbf{y}) \to \delta(\mathbf{x} - \mathbf{y})$ as $s \downarrow 0$, we obtain (15.5). ∎

Theorem 15.2 With $H_0(\mathbf{a})$, Q_t as above, we have that

$$\operatorname*{s-lim}_{n\to\infty}(Q_{t/n})^n = e^{-tH_0(\mathbf{a})} \tag{15.8}$$

Proof One need only mimic the proof [189] of Theorem 1.1. Let $S_t = e^{-tH_0(a)}$. For any $f \in D(H_0)$, we have that $\lim_{t\to 0}(d/dt)((S_t - Q_t)f) = 0$ by (15.5) and thus for $f \in D(H_0)$

$$\lim_{n\to\infty} \|n(S_{t/n} - Q_{t/n})f\| = 0$$

with convergence uniform on compact sets in the $D(H_0)$-norm. Since, as in (1.5),

$$\|(Q_{t/n}^n - S_t)f\| \leq n \sup_{0\leq s\leq t} \|(Q_{t/n} - S_{t/n})S_s f\|$$

the proof is complete. ∎

Theorem 15.3 For $\mathbf{a} \in C_0^\infty$, $V = 0$, (15.1–2) are valid.

Proof Clearly

$$(f, Q_{t/n}^n g) = \int e^{F_n(\boldsymbol{\omega}, t)} \overline{f(\boldsymbol{\omega}(0))} g(\boldsymbol{\omega}(t))\, d\mu_0$$

where

$$F_n(\boldsymbol{\omega}, t) = i \sum_{m=1}^{n} \frac{1}{2}\left[\mathbf{a}\left(\boldsymbol{\omega}\left(\frac{mt - t}{n}\right)\right) + \mathbf{a}\left(\boldsymbol{\omega}\left(\frac{mt}{n}\right)\right)\right] \cdot \left[\boldsymbol{\omega}\left(\frac{mt - t}{n}\right) - \boldsymbol{\omega}\left(\frac{mt}{n}\right)\right]$$

As $n \to \infty$, we have the L^2 convergence

$$F_{2^n}(\boldsymbol{\omega}, t) \to -i \int_0^t \mathbf{a}(\boldsymbol{\omega}(s)) \cdot d\boldsymbol{\omega} - \frac{i}{2}\int_0^t (\operatorname{div} \mathbf{a})(\boldsymbol{\omega}(s))\, ds$$

by the usual arguments. By passing to subsequences, we can be sure of convergence pointwise almost everywhere. Since

$$|e^{F_n}| \leq 1 \qquad \text{and} \qquad \bar{f}(\boldsymbol{\omega}(0))g(\boldsymbol{\omega}(t)) \in L^1(d\mu_0)$$

we see that

$$(f, e^{-tH_0(a)}g) = \lim_{n\to\infty}(f, Q_{t/n}^n g) \qquad \text{(by Theorem 15.2)}$$

$$= \int e^{F(\boldsymbol{\omega}, t)} \overline{f(\boldsymbol{\omega}(0))} g(\boldsymbol{\omega}(t))\, d\mu_0 \quad ∎$$

Now let $\mathbf{a} \in L^2_{\text{loc}}(\mathbb{R}^\nu)$. Then $-i\nabla_j - a_j$ is a symmetric and thus closable operator on C_0^∞. Let G_j be its closure. Let $H_0(\mathbf{a})$ be the unique positive self-adjoint operator with quadratic form domain $\bigcap_{j=1}^{\nu} D(G_j)$ with form

$$(f, H_0(\mathbf{a})g) = \sum_{j=1}^{\nu} (G_j f, G_j g)$$

The following result of Simon [260] which we state without proof is motivated in part by work of Kato [155].

Theorem 15.4 Let $\mathbf{a} \in L^2_{\text{loc}}(\mathbb{R}^\nu)$. Then C_0^∞ is a form core for $H_0(\mathbf{a})$. Moreover, if $\mathbf{a}_n \to \mathbf{a}$ in L^2_{loc} (i.e., $\|(\mathbf{a}_n - \mathbf{a})f\|_2 \to 0$ for all $f \in C_0^\infty$), then $(H_0(\mathbf{a}_n) + 1)^{-1} \to (H_0(\mathbf{a}) + 1)^{-1}$ strongly.

This result allows us to prove:

Theorem 15.5 Let $\mathbf{a} \in L^2_{\text{loc}}(\mathbb{R}^\nu)$ with div $\mathbf{a} = 0$ (distributional sense), and let V obey the hypotheses of Theorem 6.2. Then (15.1–2) are valid for $H(\mathbf{a}, V) = H_0(\mathbf{a}) + V$.

Proof First take $V = 0$ and let $\mathbf{a}_n \in C_0^\infty$ be chosen so that div $\mathbf{a}_n = 0$ and $\mathbf{a}_n \to \mathbf{a}$ in L^2_{loc}. By passing to a subsequence, we can assume that $\mathbf{a}_n \to \mathbf{a}$ pointwise as well. Then $\int \mathbf{a}_n(\boldsymbol{\omega}(t)) \cdot d\boldsymbol{\omega} \to \int \mathbf{a}(\boldsymbol{\omega}(t)) \cdot d\boldsymbol{\omega}$ almost everywhere in ω, and so, by the dominated convergence theorem, the right-hand side of (15.1) converges. By the strong resolvent convergence of Theorem 15.4 and the continuity of the functional calculus [214], the left-hand side converges. This establishes (15.1–2) for $\mathbf{a}$ in L^2_{loc} and $V = 0$. Bounded V's then follow by the Trotter formula and the first proof of Theorem 6.1. By following the proof of Theorem 6.2, one can accommodate general V's so long as one knows that V_- H_0-form bounded with relative bound $a < 1$ implies it is $H_0(\mathbf{a})$-form bounded with relative bound $a' < 1$. We will prove this below. ∎

One important consequence of (15.1–2) is the following.

Theorem 15.6 Under the hypotheses of Theorem 15.5,

$$|\exp(-tH(\mathbf{a}, V))\phi| \le \exp(-t(H_0 + V))|\phi| \tag{15.9}$$

pointwise.

Proof (15.9) is equivalent to

$$|(f, e^{-tH(\mathbf{a}, V)}g)| \le (|f|, e^{-t(H_0+V)}|g|)$$

which follows immediately from (15.1–2) if we note that $|e^F| = e^{\mathrm{Re}\,F}$ and Re F is independent of **a**. ∎

Remarks 1. (15.9) was originally conjectured by Simon [255] as a special case of a general conjecture on when $|e^{-tA}\phi| \leq e^{-tB}|\phi|$. Nelson (reported in [255]) remarked that this conjecture was an immediate consequence of the Feynman–Kac–Itô formula. The general conjecture was then proven independently by Hess *et al.* [127] and Simon [256] providing a purely functional analytic proof of (15.9) relating it to an inequality of Kato [153].

2. As is so often the case, one can avoid functional integrals and only use the "Trotter formula." For (15.9) follows easily form (15.7) and (15.8).

3. There is a circularity in the above in that the proof of Theorem 15.4 [260] begins with a *direct* proof of (15.9). In any event, (15.9) for smooth **a**'s can be proven as indicated (and this is essentially all that is proven in [255, 127, 256]). In some sense, the proof of (15.9) in [260] is the most elementary!

Simon's original motivation for (15.9) was concerned with problem (3) of Section 1:

Theorem 15.7 ([255]) If $H_0 + V$ has the property that $\mathrm{Tr}(e^{-t(H_0+V)}) < \infty$ for all t, then so does $H_0(\mathbf{a}) + V$ and

$$\mathrm{Tr}(e^{-t(H_0(\mathbf{a})+V)}) \leq \mathrm{Tr}(e^{-t(H_0+V)}) \tag{15.10}$$

Proof If $|A\phi| \leq B|\phi|$ pointwise and $\mathrm{Tr}(B^*B) < \infty$, it is easy to see that $\mathrm{Tr}(A^*A) \leq \mathrm{Tr}(B^*B)$. Thus (15.10) follows from (15.9). ∎

The real power of (15.9) in studying $H_0(\mathbf{a}) + V$ was realized by Avron *et al.* [4] and Combes *et al.* [37]. By combining (15.10) and (9.3) we recover the following theorem.

Theorem 15.8 ([37]) Let $\mathbf{a} \in L^2_{\mathrm{loc}}$ and let V obey the hypotheses of Theorem 9.2. Let

$$H(\hbar) = \frac{1}{2m}\left(\frac{\hbar}{i}\nabla - \mathbf{a}\right)^2 + V \tag{15.11}$$

Then

$$\mathrm{Tr}(e^{-\alpha H(\hbar)}) \leq \int \frac{d^\nu p\, d^\nu x}{(2\pi\hbar)^\nu} \exp\left(-\alpha\left[\frac{p^2}{2m} + V(\mathbf{x})\right]\right)$$

The proof in [37] does not use path integrals but rather the Golden–Thompson inequality.

By following the arguments in Section 10, we can recover the result of Combes *et al.* [37], originally obtained by different methods (see also [260b]).

Theorem 15.9 Let V be a continuous function on $\mathbb{R}^\nu$ with $\exp(-\beta V) \in L^1$. Let $\mathbf{a}$ be an $\mathbb{R}^\nu$-valued C^1-function on $\mathbb{R}^\nu$ with $\operatorname{div} \mathbf{a} = 0$ (for simplicity). Let $H(\hbar)$ be given by (15.11) and let

$$Z_Q(\hbar) = \operatorname{Tr}(\exp(-\beta H(\hbar)))$$

$$Z_C(\hbar) = \int \frac{d^\nu p\, d^\nu x}{(2\pi\hbar)^\nu} \exp\left[-\beta\left(\frac{p^2}{2m} + V(\mathbf{x})\right)\right]$$

Then $Z_Q/Z_C \to 1$ as $\hbar \to 0$.

Proof As in the proof of Theorem 10.1 by letting $W = \beta V$ and choosing a suitable multiple of $\mathbf{a}$, we need that

$$t^{\nu/2} \operatorname{Tr}(\exp[-tH(t^{-1/2}\mathbf{a}, t^{-1}W)]) \to (2\pi)^{-\nu/2} \int e^{-W}\, d^\nu x$$

By (15.1),

$$[\exp(-tH(t^{-1/2}\mathbf{a}, t^{-1}W))](\mathbf{x}, \mathbf{y}) = \int \exp\left[-t^{-1}\int_0^t W(\boldsymbol{\omega}(s))\, ds - i\int_0^t t^{-1/2}\mathbf{a}(\boldsymbol{\omega}(s)) \cdot d\boldsymbol{\omega}\right] d\mu_{0,\mathbf{x},\mathbf{y},t}$$

for almost all $\mathbf{x}$, $\mathbf{y}$. By Theorem 14.5, the right-hand side of this last expression is continuous in $\mathbf{x}$ and $\mathbf{y}$ so as usual:

$$\operatorname{Tr}(e^{-tH}) = \int\left[\int \cdots\, d\mu_{0,\mathbf{x},\mathbf{x};t}\right] dx$$

Rewriting everything in terms of the Brownian bridge, $\boldsymbol{\alpha}$, we see that we must show that

$$\int dx\, E_\alpha\left[\exp\left(-\int_0^1 W(\mathbf{x} + \sqrt{t}\boldsymbol{\alpha}(s))\, ds - i\int_0^1 \mathbf{a}(\mathbf{x} + \sqrt{t}\boldsymbol{\alpha}(s)) \cdot d\boldsymbol{\alpha}\right)\right]$$

$$\to \int e^{-W(\mathbf{x})}\, d^\nu x$$

as $t \downarrow 0$. Formally, this is true since $\int_0^1 \mathbf{a}(\mathbf{x}) \cdot d\boldsymbol{\alpha} = \mathbf{a}(\mathbf{x}) \cdot \int_0^1 d\boldsymbol{\alpha} = 0$ by $\int d\boldsymbol{\alpha} = \boldsymbol{\alpha}(1) - \boldsymbol{\alpha}(0) = 0$.

The actual proof is not much harder than this formal proof. Let

$$W_{\pm}(\mathbf{x}; \delta) = \begin{matrix}\max \\ \min\end{matrix}\{W(\mathbf{y}) \,|\, |\mathbf{x} - \mathbf{y}| \leq \delta\}$$

Let

$$F_0(\boldsymbol{\alpha}, t) = \exp\left(-\int_0^1 W(\mathbf{x} + \sqrt{t}\boldsymbol{\alpha}(s)) \, ds \right)$$

and

$$F(\boldsymbol{\alpha}, t) = F_0(\boldsymbol{\alpha}, t) \exp\left(-i \int_0^1 \mathbf{a}(\mathbf{x} + \sqrt{t}\boldsymbol{\alpha}(s)) \cdot d\boldsymbol{\alpha} \right)$$

Then

$$\begin{aligned} &\left| \int dx \, E_\alpha(F(\boldsymbol{\alpha}, t); \|\boldsymbol{\alpha}\|_\infty > R) \right| \\ &\qquad \leq \int dx \, E_\alpha(F_0(\boldsymbol{\alpha}, t); \|\boldsymbol{\alpha}\|_\infty > R) \\ &\qquad \leq \int dx \, E_\alpha(F_0(\boldsymbol{\alpha}, t)) - E(\|\boldsymbol{\alpha}\|_\infty \leq R) \int e^{-W_-(\mathbf{x}; R\sqrt{t})} \, dx \end{aligned}$$

By Theorem 10.1, the first term converges to $\int e^{-W(\mathbf{x})} \, dx$ as $t \to 0$ and as in that proof, $\int e^{-W_-(\mathbf{x}; R\sqrt{t})} \, dx \to \int e^{-W(\mathbf{x})} \, dx$ as $t \to 0$. Since $E(\|\boldsymbol{\alpha}\|_\infty \leq R) \to 1$, as $R \to \infty$ we can, given ε, find T and R so that for $t < T$:

$$\left| \int dx \, E_\alpha(F(\boldsymbol{\alpha}, t); \|\boldsymbol{\alpha}\|_\infty > R) \right| \leq \varepsilon$$

On the other hand, as $t \downarrow 0$,

$$E_\alpha(F(\boldsymbol{\alpha}, t); \|\boldsymbol{\alpha}\|_\infty \leq R) \to e^{-W(\mathbf{x})} E(\|\boldsymbol{\alpha}\|_\infty \leq R)$$

for each fixed $\mathbf{x}$ since $|F(\boldsymbol{\alpha}, t)| \leq \exp(-W_+(\mathbf{x}; R\sqrt{t}))$ which is dominated for $t < 1$ by a function in $L^1(D\alpha)$ and for each fixed $\boldsymbol{\alpha}$,

$$F(\boldsymbol{\alpha}, t) \to \exp\left(-\int_0^1 W(\mathbf{x}) \, ds - i \int_0^1 \mathbf{a}(\mathbf{x}) \cdot d\boldsymbol{\alpha} \right) = \exp(-W(\mathbf{x}))$$

For each fixed K, we conclude by the dominated convergence theorem that

$$\int_{|\mathbf{x}| < K} dx \, E_\alpha(F(\boldsymbol{\alpha}, t); \|\boldsymbol{\alpha}\|_\infty < R) \to \int_{|\mathbf{x}| < K} e^{-W(\mathbf{x})} E(\|\boldsymbol{\alpha}\|_\infty < R) \, dx$$

Finally, as in the first part,

$$\left| \int_{|\mathbf{x}|>K} E_\alpha(F(\boldsymbol{\alpha}, t); \|\boldsymbol{\alpha}\|_\infty < R)\, dx \right|$$

$$\leq \int_{|\mathbf{x}|>K} E_\alpha(F_0(\boldsymbol{\alpha}, t); \|\boldsymbol{\alpha}\|_\infty < R)\, dx$$

$$\leq \int E_\alpha(F_0)\, dx - E(\|\boldsymbol{\alpha}\|_\infty < R) \int_{|\mathbf{x}|<K} e^{-W_-(\mathbf{x}; R\sqrt{t})}\, dx$$

which can be made small if K is large, R is large, and t is small. ∎

A number of applications of (15.9) to spectral and scattering theory for $H(\mathbf{a}, V)$ have been made by Avron *et al.* [4]. Typical is the following which we used in the proof of Theorem 15.5.

Theorem 15.10 ([4]) Let $|V|$ be a multiplication operator which is H_0-form bounded with relative bound α. Then for $\mathbf{a} \in L^2_{\text{loc}}$, it is $H_0(\mathbf{a})$-form bounded with relative bound at most α.

Proof By (15.9) and $(A + \lambda)^{-1/2} = c \int e^{-\lambda t} e^{-At} t^{-1/2}\, dt$ (for suitable constant c), we have that

$$|(H_0(\mathbf{a}) + \lambda)^{-1/2}\phi| \leq (H_0 + \lambda)^{-1/2}|\phi|$$

so that

$$\| |V|^{1/2}(H_0(\mathbf{a}) + \lambda)^{-1/2}\phi\| \leq \| |V|^{1/2}(H_0 + \lambda)^{-1/2}|\phi| \|$$

which implies that

$$\| |V|^{1/2}(H_0(\mathbf{a}) + \lambda)^{-1/2}\| \leq \| |V|^{1/2}(H_0 + \lambda)^{-1/2}\|$$

Since B is A-form bounded ($B, A \geq 0$) with relative bound α if and only if $B^{1/2}(A + \lambda)^{-1/2}$ is bounded and $\lim_{\lambda\to\infty} \|B^{1/2}(A + \lambda)^{-1/2}\|^2 = \alpha$, the proof is complete. ∎

For fixed V and $\mathbf{a}$, let $K_\gamma(\mathbf{a}) = |V|^{1/2}(H_0(\mathbf{a}) + \gamma)^{-1}|V|^{1/2}$. By following the proof of Lieb's formula, one finds that for any F related to a positive f by (8.7),

$$|F(K_\gamma(\mathbf{a}))\phi| \leq F(K_\gamma(0))|\phi|$$

Since $F(K_\gamma(\mathbf{a}))$ is a positive operator, this implies that

$$\mathrm{Tr}(F(K_\gamma(\mathbf{a}))) \leq \mathrm{Tr}(F(K_\gamma(0))) \tag{15.12}$$

on account of the following.

Lemma 15.11 Let A and B be bounded *positive* operators on an L^2-space with $|A\phi| \leq B|\phi|$ pointwise. If B is trace class, then so is A and $\operatorname{Tr}(A) \leq \operatorname{Tr}(B)$.

Remark This is false if A is not positive.

Proof Let $\{S_i\}_{i=1}^n$ be a family of disjoint sets of finite measure and let P_S be the projection onto all functions of the form $\sum a_i \chi_{S_i}$. Then, $P_S A P_S = A_S$ and B_S are trace class and since $\operatorname{Tr}(A_S) = \sum_{i=1}^n \mu(S_i)^{-1}(\chi_{S_i}, A\chi_{S_i})$, the hypothesis implies that $\operatorname{Tr}(A_S) \leq \operatorname{Tr}(B_S) \leq \operatorname{Tr}(B)$. One can find a net of S's, S_α, so that $A_\alpha \to A$ strongly. That A is trace class then follows by a general result [259]. ∎

We can now follow Lieb's proof of the Cwickel–Lieb–Rosenbljum bound and then the proof of stability of matter to obtain the following theorem.

Theorem 15.12 ([4, 37]) Matter is stable in an arbitrary magnetic field; i.e., (9.22) continues to hold if $-\Delta_i$ is replaced by $-(i\partial_i - \mathbf{a}_i(\mathbf{x}))^2$. The constant c is independent of $\mathbf{a}$ and can be chosen to be the constant of the proof of Theorem 9.6.

Remark One might think that (15.12) implies that

$$N(V, \mathbf{a}) \equiv \dim(\text{spectral projection for } (-\infty, 0) \quad \text{for} \quad H(\mathbf{a}, V)) \leq N(V)$$

but this is not true; see [4].

* * *

Finally, we do a calculation of the explicit kernel of $e^{-tH_0(\mathbf{a})}$ when $a_1(x) = -\frac{1}{2}Bx_2$, $a_2(x) = \frac{1}{2}Bx_1$, $a_3(x) = 0$ [constant field $(0, 0, B)$]. The kernel is [84, 4]

$$\begin{aligned} e^{-tH_0(\mathbf{a})}(\mathbf{x}, \mathbf{y}) = {} & \left[\frac{B}{4\pi \sinh(\frac{1}{2}Bt)}\right]\left[\frac{1}{2\pi t}\right]^{1/2} \\ & \times \exp\left\{-\frac{1}{2t}(x_3 - y_3)^2 - \frac{B}{4}\coth\left(\frac{1}{2}Bt\right)[(x_2 - y_2)^2\right. \\ & \left. + (x_1 - y_1)^2] - \frac{1}{2}iB(x_1 y_2 - x_2 y_1)\right\} \end{aligned}$$

Since the $\partial^2/\partial z^2$ term commutes with the remainder, without loss, we restrict to two dimensions. For notational simplicity we consider the calculation with $t = 1$, $\mathbf{x} = (0, 0)$, $\mathbf{y} = (0, y)$. Actually using scaling covariance, translation invariance (not the usual realization of translations [5]), and rotation invariance, one can go from this special case to the general case. Thus, we need to compute

$$Z \equiv (2\pi t)^{-1} e^{-y^2/2t} \int e^{-iX} D\alpha_1 \, D\alpha_2$$

where α_1 and α_2 are two independent Brownian bridges and

$$X = \frac{1}{2} B\left[\int_0^1 \alpha_1 [d\alpha_2 + y\, ds] - \int_0^1 (\alpha_2 + sy)\, d\alpha_1\right]$$

Since X is quadratic in α, this integral is, in principle, explicitly calculable. First we note that

$$X = -B \int_0^1 (\alpha_2(s) + sy)\, d\alpha_1$$

The integration by parts is justified since

$$\begin{aligned} f(1)g(1) - f(0)g(0) &= \sum f\left(\frac{k-1}{2^n}\right)\left[g\left(\frac{k}{2^n}\right) - g\left(\frac{k-1}{2^n}\right)\right] \\ &\quad + \sum g\left(\frac{k-1}{2^n}\right)\left[f\left(\frac{k}{2^n}\right) - f\left(\frac{k-1}{2^n}\right)\right] \\ &\quad + \sum \left[g\left(\frac{k}{2^n}\right) - g\left(\frac{k-1}{2^n}\right)\right]\left[f\left(\frac{k}{2^n}\right) - f\left(\frac{k-1}{2^n}\right)\right] \end{aligned}$$

With $f = \alpha_2 + sy$ and $g = \alpha_1$ the left-hand side is zero, and the first two terms converge to Itô integrals and the last term to zero since α_1 and α_2 are independent. Next, we note that

$$E(d\alpha_1(s)\, d\alpha_1(t)) = \delta(s - t)\, ds - ds\, dt$$

which follows from $b(s) = \alpha(s) \dotplus sb(1)$ (orthogonal direct sum) and $db = d\alpha + (ds)b(1)$. Since X is linear in α_1,

$$\int e^{-iX} D\alpha_1 \equiv \exp\left(-\frac{1}{2}\int X^2 \, D\alpha_1\right)$$

so

$$Z = (2\pi t)^{-1} e^{-Y^2/2t} \int \exp(-\tfrac{1}{2}B^2 Y)\, D\alpha_2$$

with

$$Y = \int_0^1 (\alpha_2(s) + sy)^2 \, ds - \left(\int_0^1 (\alpha_2(s) + sy)\, ds\right)^2$$

The integral for Z can now be computed as we did in Section 6 in our second proof of Mehler's formula.

16. Introduction to Stochastic Calculus

In this section, we will present a "formal" treatment of some aspects of calculations with stochastic integrals. We will be formal in that we will use the general stochastic integral for nonanticipatory functions alluded to in Section 14, and that we will not give a proof or even precise hypotheses for the main result: the general Itô lemma. The reader can consult [86, 183] for more details. A **Stochastic integral** is a random function $c(\mathbf{b}, t)$ obeying

$$c(\mathbf{b}, t) = c(\mathbf{b}, 0) + \int_0^t f(\mathbf{b}, s) \cdot d\mathbf{b} + \int_0^t g(\mathbf{b}, s)\, ds \tag{16.1}$$

where f and g are nonanticipatory functionals with suitable L^p properties and $c(\mathbf{b}, 0)$ is independent of $\mathbf{b}$. One writes (16.1) in the shorthand:

$$dc = \mathbf{f} \cdot d\mathbf{b} + g\, ds \tag{16.2}$$

The following result is critical.

Theorem 16.1 (Itô's lemma) If $c_1, \ldots, c_m$ are stochastic integrals and u is a C^2-function on $\mathbb{R}^m \times [0, \infty)$ with some mild restrictions on growth at infinity, then

$$x = u(c_1, c_2, \ldots, c_m, t)$$

is a stochastic integral, and

$$dx = \sum_{i=1}^m \frac{\partial u}{\partial y_i} dc_i + \frac{\partial u}{\partial t} ds + \frac{1}{2} \sum_{i,j=1}^m \frac{\partial^2 u}{\partial y_i\, \partial y_j} dc_i\, dc_j \tag{16.3}$$

where $dc_i\, dc_j = \mathbf{f}_i \cdot \mathbf{f}_j\, ds (dc_i = \mathbf{f}_i \cdot d\mathbf{b} + g_i\, ds)$; i.e., $db_k\, db_l = \delta_{kl}\, ds$.

The pattern of proof is identical to that of the special case Theorem 14.3. Indeed, if the c_i's are approximated in the right way, it is the identical proof (see [183]).

Example 1 (product rule) If $dc_i = \mathbf{f}_i \cdot d\mathbf{b} + g_i\, ds$, then

$$d(c_1 c_2) = c_1\, dc_2 + c_2\, dc_1 + \mathbf{f}_1 \cdot \mathbf{f}_2\, ds \tag{16.4}$$

Example 2 (Feynman–Kac formula) Let us compute the differential of

$$c = \exp\left(-\int_0^t V(\mathbf{b}(s))\,ds\right) f(\mathbf{b}(t))$$

where $V, f \in C_0^\infty$. Since $z = \int_0^t V(\mathbf{b}(s))\,ds$ is a stochastic integral with $dz = V\,ds$, we have by Itô's lemma and $(ds)^2 = 0$ that $d(e^{-z}) = -e^{-z}V\,ds$. Again by Itô's lemma, $df = \nabla f \cdot d\mathbf{b} + \frac{1}{2}(\Delta f)\,ds$. Thus, using (16.4),

$$dc = \exp\left(-\int_0^t V(\mathbf{b}(s))\,ds\right)[\nabla f \cdot d\mathbf{b} - (Hf)\,ds]$$

where $H = -\frac{1}{2}\Delta f + Vf$. Thus, if we define

$$(Q_t f)(\mathbf{x}) = E\left(\exp\left(-\int_0^t V(\mathbf{x} + \mathbf{b}(s))\,ds\right) f(\mathbf{x} + \mathbf{b}(s))\right)$$

and we use (14.9) that $E(d\mathbf{b}) = 0$, we see that

$$(Q_t f)(\mathbf{x}) = f(\mathbf{x}) - \int_0^t ds\,(Q_s(Hf))(\mathbf{x})$$

which yields $Q_t = e^{-tH}$ and so a new proof of Feynman–Kac formula.

Example 3 (Feynman–Kac–Itô formula) Let us compute the differential of

$$c = \exp\left[-i\int_0^t \mathbf{a}(\mathbf{b}(s))\cdot d\mathbf{b} - \frac{i}{2}\int_0^t (\operatorname{div}\mathbf{a})(\mathbf{b}(s))\,ds\right] f(\mathbf{b}(t)) \equiv e^{-iz} f(\mathbf{b}(t))$$

with $\mathbf{a}, f \in C_0^\infty$. Then, by Itô's lemma

$$\begin{aligned} d(e^{-iz}) &= -ie^{-iz}\,dz - \tfrac{1}{2}e^{-iz}(dz)^2 \\ &= -ie^{-iz}[\mathbf{a}\cdot d\mathbf{b} + \tfrac{1}{2}(\nabla\cdot\mathbf{a})\,ds] - \tfrac{1}{2}e^{-iz}a^2\,ds \end{aligned}$$

so using (16.4)

$$dc = e^{-iz}[(\ldots)\,db + (\tfrac{1}{2}\Delta f - i\mathbf{a}\cdot\nabla f - \tfrac{i}{2}(\nabla\cdot\mathbf{a})f - \tfrac{1}{2}a^2 f)\,ds]$$

recognizing the occurrence of $-H_0(\mathbf{a})f$ [see (15.6)], and writing

$$(Q_t f)(x) = E\left(\exp\left(-i\int_0^t \mathbf{a}\cdot d\mathbf{b} + \int_0^t \frac{i}{2}\operatorname{div}\mathbf{a}\,ds\right) f(\mathbf{x} + \mathbf{b}(s))\right)$$

we find that

$$(Q_t f)(\mathbf{x}) = f(\mathbf{x}) - \int_0^t ds(Q_s(H_0(\mathbf{a})f))(\mathbf{x})$$

yielding the Feynman–Kac–Itô formula.

Example 4 (drift) Let $\mathbf{a} \in C_0^\infty$ and let

$$\mathbf{x}(t) = \mathbf{b}(t) + \int_0^t \mathbf{a}(\mathbf{b}(s))\, ds$$

which for obvious reasons is often called the **drift** generated by **a**. Then $d\mathbf{x} = d\mathbf{b} + \mathbf{a}\, ds$ so that

$$d(f(\mathbf{x}(t))) = \nabla f \cdot d\mathbf{b} + (\mathbf{a} \cdot \nabla f + \tfrac{1}{2}\Delta f)\, ds$$

Following our usual procedure

$$E(f(\mathbf{x}(t))) = (e^{-tX} f)(\mathbf{0})$$

where X is the differential operator

$$Xf = -\tfrac{1}{2}\Delta f - \mathbf{a} \cdot \nabla f \tag{16.5}$$

which is thus called the **generator of the drift**.

Example 5 (Cameron–Martin formula [30]) Let $dz = \mathbf{a} \cdot d\mathbf{b} - \frac{1}{2}a^2\, ds$ and $c = e^z f(\mathbf{b}(t))$. Then

$$dc = e^z[(\nabla f + \mathbf{a}f) \cdot d\mathbf{b} + (\mathbf{a} \cdot \nabla f + \tfrac{1}{2}\Delta f)\, ds]$$

so in terms of the generator X in (16.5):

$$E(e^z f(\mathbf{b}(t))) = (e^{-tX} f)(\mathbf{0})$$

This is usually summarized in the **Cameron–Martin formula**

$$\frac{Dx}{Db}(\text{Radon–Nikodym derivative}) = \exp\left(\int_0^t \mathbf{a}(\mathbf{b}(s))\, d\mathbf{b} - \frac{1}{2}\int_0^t a^2(\mathbf{b}(s))\, ds\right)$$

We can now explain how a classical probabilist would view the Feynman–Kac–Itô formula. If one writes $H_0(\mathbf{a}) = -\frac{1}{2}\Delta + i\mathbf{a} \cdot \nabla + \frac{1}{2}a^2 + \frac{1}{2}i(\operatorname{div} \mathbf{a})$, then one sees that $H_0(\mathbf{a})$ is the generator of a (complex) drift $-ia$ and a potential $\frac{1}{2}a^2 + \frac{1}{2}i(\operatorname{div} \mathbf{a})$. Then the combined Cameron–Martin formula for $\mathbf{a} \cdot \nabla$ and Feynman–Kac for $\frac{1}{2}a^2 + i \operatorname{div} \mathbf{a}$ yields the Feynman–Kac–Itô formula.

There is a connection between the Cameron–Martin formula and the Feynman–Kac formula which shows a connection between $P(\phi)_1$-process and drift processes—a connection which will reoccur in Section 19. Suppose that

$$H = -\frac{1}{2}\frac{d^2}{dx^2} + V$$

where we have added a constant to V which guarantees that $\inf \operatorname{spec}(H) = 0$. Let $H\Omega = 0$ and let L be the operator $Lf = \Omega^{-1}(H\Omega f)$. L is the generator of a contraction semigroup on $L^2(\mathbb{R}, \Omega^2\, dx)$ and $L1 = 0$. By the Feynman–Kac formula, e^{-tL} is given by

$$(e^{-tL}f)(x) = \Omega(x)^{-1}E\left(\exp\left(-\int_0^t V(x + b(s))\right)\Omega(x + b(t))f(x + b(t))\right) \tag{16.6}$$

But notice that

$$L = -\frac{1}{2}\frac{d^2}{dx^2} - a\frac{d}{dx}$$

where $a = \Omega^{-1}\Omega' = h'$ with $h = \ln \Omega$. Thus, by the Cameron–Martin formula:

$$(e^{-tL}f)(x) = E\left(\exp\left(\int_0^t a(x + b(s))\, db - \frac{1}{2}\int_0^t a^2(x + b(s))\, ds\right)f(x + b(t))\right) \tag{16.7}$$

Let us check that (16.6) and (16.7) agree. For simplicity take $x = 0$. By Itô's lemma (Theorem 14.3)

$$\int_0^t h'(b(s))\, db = h(b(t)) - h(b(0)) - \frac{1}{2}\int_0^t h''(b(s))\, ds$$

By explicit calculation

$$\begin{aligned} h'' + a^2 = a' + a^2 &= \Omega^{-2}[\Omega\Omega'' - (\Omega')^2 + (\Omega')^2] \\ &= \Omega^{-1}\Omega'' = 2V \end{aligned}$$

Thus

$$\int_0^t a(b(s))\, db - \frac{1}{2}\int_0^t a^2(b(s))\, ds = \ln \Omega(b(t)) - \ln \Omega(b(0)) - \int_0^t V(b(s))\, ds$$

establishing the equality of (16.6) and (16.7).

VI

Asymptotics

In this chapter, we discuss a variety of connections between Brownian motion and asymptotic behavior, either of other processes or functions of Brownian motion. This is a subject developed by Donsker and his collaborators. In Section 17, we show that normalized random walk approaches Brownian motion in a strong sense; in Section 18, we extend the formula

$$\lim_{n\to\infty}\left(\int_0^1 e^{-nf(x)}\,dx\right)^{1/n} = \exp[-\min(f)]$$

to various Gaussian processes and describe its relation to the asymptotic behavior of the Rayleigh–Schrödinger coefficients of certain perturbation problems; and in Section 19 we describe some deep results of Donsker and Varadhan and their relation to the Gibbs' variational principle of statistical mechanics.

17. Donsker's Theorem

At the beginning of Section 4, we introduced Brownian motion as an intuitive limit of random walks and we noted that

$$\lim_{n\to\infty} n^{-1/2}X_{[nt]} = b(t) \tag{17.1}$$

in some intuitive sense. Here we want to examine the sense in which (17.1) holds. The central limit theorem immediately tells us that if we consider fixed $t_1, \ldots, t_k$ then the joint distribution of $n^{-1/2}X_{[nt_i]}$ converges to that for $b(t_i)$. This simple result leaves open some important and natural questions.

For example, we found in Section 6 the probability distribution, P, for $|\{s \mid b(s) > 0; 0 \leq s \leq 1\}|$. One would guess that as $N \to \infty$,

$$N^{-1} \# \{n \leq N | X_n > 0\}$$

has a distribution approaching P but since this involves infinitely many X's, it is not clear that just knowing the convergence of the finite distributions is enough. In fact, we will see shortly by explicit example that it is certainly not enough for some reasonable functions. Thus, we need a stronger form of (17.1). The first hints of this strong form go back to Erdös and Kac [77] who proved an **invariance principle** namely if y_i are independent copies of some random variable with $E(y_i) = 0$, $E(y_i^2) = 1$, then the limiting distribution of a wide variety of functions of $(n^{-1/2} \sum_{i=1}^n y_i)$ is independent of which y_i is chosen. The reason for this is that a strong kind of convergence of (17.1) is involved—this was discovered by Donsker [53]. Kolmogorov–Prohorov [162] and then Prohorov [206, 207] introduced the key notions of weak convergence and tightness which yield an elegant framework for Donsker's result. For more information, the reader can consult Billingsley [14], whose treatment we follow closely in places, or Parthasarathy [198a].

Let $y_1, y_2, \ldots$ be independent, identically distributed random variables with $E(y_i) = 0$, $E(y_i^2) = 1$, and let $S_n = \sum_{i=1}^n y_i$. For fixed n, we want to define a process $X_n(t)$ with continuous sample paths by letting $X_n(k/n) = n^{-1/2} S_k$ for $0 \leq k \leq n$ and interpolating linearly in between; i.e. (with $[a]$ = integral part of a),

$$X_n(t) = n^{-1/2} S_{[nt]} + n^{-1/2}(nt - [nt]) y_{[nt]+1} \qquad (0 \leq t \leq 1) \tag{17.2}$$

It is trivial that the process $X_n(t)$ has continuous paths; indeed, with probability one, each y_j is finite and $X_n(t)$ is piecewise linear. Thus X_n induces a measure P_n on $C[0, 1]$. To state Donsker's theorem, we need the following notion.

Definition Let $\{P_n\}$, P_∞ be a family of Borel probability measures on $C[0, 1]$, the continuous functions on $[0, 1]$. We say P_n **converges weakly** to P_∞ if and only if

$$\int f \, dP_n \to \int f \, dP_\infty$$

for any bounded continuous function f from $C[0, 1]$ to $\mathbb{R}$.

WARNING "Weak" convergence is considerably *stronger* than convergence of the finite distributions. For example, if g_n is the function

$$g_n(t) = \begin{cases} nt, & 0 \leq t \leq 1/n \\ 2 - nt, & 1/n \leq t \leq 2/n \\ 0, & t \geq 2/n \end{cases}$$

and if P_n is the point mass at g_n, then the joint distribution of any $\omega(t_1), \ldots, \omega(t_m)$ converges to that associated to P_∞, the point mass at zero. But, there is no weak convergence since, e.g.,

$$E_n(e^{-\|\omega\|_\infty}) = e^{-1} \nrightarrow 1 = E_\infty(e^{-\|\omega\|_\infty})$$

Theorem 17.1 (Donsker's theorem [53]) Let P_n be the measure on $C[0, 1]$ induced by (17.2) (with y_i independent, identically distributed, and $E(y_i) = 0, E(y_i^2) = 1$). Let P_∞ be the measure on $C[0, 1]$ induced by Brownian motion. Then P_n converges to P_∞ weakly.

The remainder of the section is devoted to the proof and discussion of this theorem. First, we have some discussion of weak convergence. (See [14] for extensive discussion.)

Proposition 17.2 Let P_n converge to P weakly and A be a Borel set in $C[0, 1]$ with $P[\partial A] = 0$. Then $P_n(A) \to P(A)$.

Proof $C[0, 1]$ is a metric space, so every Borel measure is regular (see, e.g., Theorem 1.1 of [14]) and moreover Urysohn's lemma holds. It follows that for B open, any Borel probability measure, Q, and $\mathscr{C}$, **the family of continuous functions on** $C[0, 1]$:

$$Q(B) = \sup\left\{\int f\,dQ \mid f \in \mathscr{C}, 0 \le f \le 1, f = 0 \text{ on } B^c\right\}$$

By taking complements, we see that for any closed set D

$$Q(D) = \inf\left\{\int f\,dQ \mid f \in \mathscr{C}, 0 \le f \le 1, f = 1 \text{ on } D\right\}$$

Thus, since $P_n \to P$ weakly

$$\begin{aligned} P(A^{\text{int}}) &= \sup\left\{\int f\,dP \mid f \in \mathscr{C}, 0 \le f \le 1, f = 0 \text{ on } (A^{\text{int}})^c\right\} \\ &\le \underline{\lim}\, \sup\left\{\int f\,dP_n \mid f \in \mathscr{C}, \ldots\right\} \\ &= \underline{\lim}\, P_n(A^{\text{int}}) \le \underline{\lim}\, P_n(A) \end{aligned}$$

where we used $\int f\,dP = \lim_n \int f\,dP_n \le \underline{\lim}\, \sup(\int f\,dP_n \mid \cdots)$. Similarly

$$P(\bar{A}) \ge \overline{\lim}\, P_n(A)$$

If $P(A^{\text{int}}) = P(\bar{A})$, then the limit exists and equals $P(A)$. ▮

Typical of the consequences of Theorem 17.1 and Proposition 17.2 is the following corollary.

Corollary 17.3 Let X_n be the partial sums of the random walk of Section 4. Then

(a) $$\lim_{n\to\infty} E(n^{-1}\#\{j \le n \mid X_j > 0\} \le \alpha) = 2\pi^{-1} \text{ Arc sin } \sqrt{\alpha}$$

(b) $$\lim_{n\to\infty} E\left(\max_{j\le n} |X_j| \le \sqrt{n/t}\right) = \text{r.h.s. of (7.19)}$$

Proof If f is any function in $\mathscr{C}$, then Proposition 17.2 says that if $P_n \to P$ weakly, then $P_n(f \le \alpha) \to P(f \le \alpha)$ so long as $P(f = \alpha) = 0$. Thus (a) follows from Theorem 6.10 and (b) from Proposition 7.16. ▮

We now turn to the proof of Donsker's theorem.

Definition A sequence of Borel probability measures $\{P_n\}$ on $C[0, 1]$ is called **tight** if and only if for any $\varepsilon > 0$, there exists a compact $K \subset C[0, 1]$ with $P_n(K) > 1 - \varepsilon$ for all n.

Remark It is an interesting exercise to show that any single measure on $C[0, 1]$ is tight.

Theorem 17.4 (Prohorov's theorem) Let $\{P_n\}$ be a sequence of Borel probability measures on $C[0, 1]$ with the following two properties:

(a) The finite distributions converge: explicitly for any $0 \le t_1 < t_2 < \cdots < t_m \le 1$, there is a measure $d\mu_{t_i}$ on $\mathbb{R}^m$ so that as $n \to \infty$

$$\int F(\omega(t_1), \ldots, \omega(t_m))\, dP_n(\omega) \to \int F(x_1, \ldots, x_m)\, d\mu_{t_i}(x)$$

for all bounded continuous F on $\mathbb{R}^m$.

(b) The $\{P_n\}$ are tight.

Then, there is a measure P_∞ on $C[0, 1]$ so that $P_n \to P_\infty$ weakly and the finite distributions of P_∞ are the $d\mu_t$.

Proof For $r = 1, 2, \ldots$, pick K_r with $K_1 \subset K_2 \subset \cdots$ and $P_n(K_r) \ge 1 - r^{-1}$ for all n. Since the positive measures of mass at most one on a compact set are weakly compact (see, e.g., [214]), we can find a subsequence $P_{n(i)}$ so that $P_{n(i)} \upharpoonright K_r$ converge for each r to a measure $P_{\infty,r}$. Since the $P_{\infty,r}$

are easily seen to be consistent, we can define P_∞ on $\bigcup K_r$ and so on $C[0, 1]$. Given $f \in \mathscr{C}$, we have that

$$\int_{K_r} f\, dP_{n(i)} \to \int_{K_r} f\, dP_\infty$$

Since $P_\infty(C\backslash K_r) \leq 1/r$ by a simple argument and $P_n(C\backslash K_r) \leq 1/r$ by hypothesis, we see that for any r

$$\overline{\lim} \left| \int f\, dP_{n(i)} - \int f\, dP_\infty \right| \leq 2r^{-1} \| f \|_\infty$$

Since r is arbitrary $P_{n(i)} \to P_\infty$ weakly. Clearly, the $d\mu_t$ are the finite distributions of P_∞.

By the above argument, any subsequence has a weakly convergent subsubsequence and the limit must agree with P_∞ since Kolmogorov's theorem implies that there is at most one measure on $C[0, 1]$ with the $d\mu_t$ as finite distribution. By a general argument on sequential convergence, $P_n \to P_\infty$ weakly. ∎

Remark Prohorov's theorem is usually stated in the form: A family of Borel probability measures on a complete separable metric space has a compact weak closure if *and only if* the family is tight. We have isolated the half we need and added the extra argument.

The central limit theorem (Theorem 4.1) implies that under the hypotheses of Theorem 17.1, the finite distributions of the P_n converge to those of Brownian motion. Thus, Donsker's theorem is reduced to the proof of tightness. Notice that *by proving this we will*, at the same time, *have a proof that Brownian motion has a continuous version* independent of the arguments in Section 5. The compact sets needed in the proof of tightness will be uniformly bounded, uniformly equicontinuous families which are compact by the Arzelà–Ascoli theorem (see, e.g., [214]). The key estimate is the following whose proof is closely patterned on that of Doob's inequality (Theorem 3.5) and Lévy's inequality (Theorem 3.6.5).

Lemma 17.5 Under the hypotheses of Donsker's theorem:

$$\lim_{n\to\infty} E\left(\max_{i \leq n} |S_i| \geq \lambda\sqrt{n}\right) \leq c\lambda^{-3} \tag{17.3}$$

for each $\lambda > 0$, and a suitable $c < \infty$. The same result holds if, in the expectation, λ is replaced by a sequence, λ_n converging to λ as $n \to \infty$.

Proof Clearly, it suffices to consider the case $\lambda \geq 2\sqrt{2}$. Fix n. Let $A_i = \{\max_{j<i} |S_j| < \lambda\sqrt{n} \leq |S_i|\}$. Then

$$\begin{aligned} E\Big(\max_{i \leq n} |S_i| \geq \lambda\sqrt{n}\Big) &\leq E\Big(|S_n| \geq \tfrac{1}{2}\lambda\sqrt{n}\Big) + \sum_{i=1}^{n-1} E\Big(A_i \cap (|S_n| \leq \tfrac{1}{2}\lambda\sqrt{n})\Big) \\ &\leq E\Big(|S_n| \geq \tfrac{1}{2}\lambda\sqrt{n}\Big) + \sum_{i=1}^{n-1} E(A_i)E\Big(|S_n - S_i| \geq \tfrac{1}{2}\lambda\sqrt{n}\Big) \end{aligned} \tag{17.4}$$

since $|S_n - S_i|$ is independent of $S_1, \ldots, S_i$. But $\lambda \geq 2\sqrt{2}$, so

$$\begin{aligned} E(|S_n - S_i| \geq \tfrac{1}{2}\lambda\sqrt{n}) &\leq E(|S_n - S_i| \geq \sqrt{2n}) \\ &\leq (2n)^{-1}E(|S_n - S_i|^2) \\ &= \tfrac{1}{2}n^{-1}(n - i) \leq \tfrac{1}{2} \end{aligned}$$

Thus, since the A_i are disjoint and contained in $\{\max_{i \geq n} |S_i| \geq \lambda\sqrt{n}\}$, (17.4) implies that

$$\begin{aligned} \tfrac{1}{2}E\Big(\max_{i \leq n} |S_i| \geq \lambda\sqrt{n}\Big) &\leq E(|S_n| \geq \tfrac{1}{2}\lambda\sqrt{n}) \\ &\to (2\pi)^{-1/2} \int_{|x| \geq \lambda/2} e^{-x^2/2}\, dx \end{aligned} \tag{17.4'}$$

as $n \to \infty$, by the central limit theorem (Theorem 4.1). Using (3.4), we obtain (17.3) trivially. The extension to the sequence is easy if we note that for any $\varepsilon > 0$

$$\Big(\max_{i \geq n} |S_i| \geq \lambda_n\sqrt{n}\Big) \subset \Big(\max_{i \geq n} |S_i| \geq (1 - \varepsilon)\lambda\sqrt{n}\Big)$$

for n sufficiently large. ▮

Notice that if y_1 has an even distribution, one can replace the argument leading to (17.4′) by Lévy's inequality.

Proof of Theorem 17.1 To prove tightness, we claim it suffices to prove that for each positive integer k and each $\varepsilon > 0$, there is a $\delta(k;\varepsilon) > 0$ so that for all n

$$E\Big(\sup_{0 \leq u - t \leq \delta(k;\varepsilon)} |X_n(u) - X_n(t)| \leq k^{-1}\Big) \geq 1 - 2^{-k}\varepsilon \tag{17.5}$$

For assuming (17.5) holds, let

$$K_\varepsilon = \bigcap_{k=1}^{\infty} \Big\{f \;\Big|\; \sup_{|u-t| \leq \delta(k;\varepsilon)} |f(u) - f(t)| \leq k^{-1}\Big\} \cap \{f \mid f(0) = 0\}$$

Then K_ε is compact by the Arzelà–Ascoli theorem and by (17.5) and the fact that $X_n(0) = 0$, $P_n(K_\varepsilon) \geq 1 - \varepsilon$.

Now fix n. By the polygonal nature of the paths, the paths are piecewise C^1 with maximum derivative $\sup_{0 \leq i \leq n-1} |S_i - S_{i+1}| n^{+1/2}$ so

$$\begin{aligned} 1 - E&\left(\sup_{0 \leq u - t \leq \delta} |X_n(u) - X_n(t)| \leq k^{-1}\right) \\ &\leq 1 - E\left(\sup_{0 \leq i \leq n-1} |S_i - S_{i+1}| \leq n^{-1/2} k^{-1} \delta^{-1}\right) \\ &\leq \sum_{i=0}^{n-1} E(|S_i - S_{i+1}| \geq n^{-1/2} k^{-1} \delta^{-1}) \\ &\leq n[nk^2\delta^2 E(|S_i - S_{i+1}|^2)] = n^2 k^2 \delta^2 \end{aligned}$$

As a result for each fixed n, we can assure (17.5) by taking δ small. It follows that if (17.5) is proven for all $n \geq n_0(k, \varepsilon)$, we can conclude it for all n by shrinking $\delta(k, \varepsilon)$.

Now suppose that for each $n \geq n_0$ we can find $0 = t_0 < \cdots < t_l = 1$ with $\sup|t_i - t_{i-1}| \leq \delta(k, \varepsilon)$ (the t's may be n-dependent) so that

$$E\left(\sup_{0 \leq u - t_i \leq 2\delta(k, \varepsilon)} |X_n(u) - X_n(t_i)| \leq \tfrac{1}{2}k^{-1}\right) \geq 1 - 2^{-k}\varepsilon; \qquad i = 0, \ldots, l \tag{17.6}$$

Then (17.5) holds for $n \geq n_0$, since any u and t with $|u - t| \leq \delta$ are within 2δ of some common $t_i \leq \min(u, t)$.

Now given ε, k, fix δ so small that

$$c([\delta^{-1}] + 1)(2k)^3 \delta^{3/2} \leq \varepsilon/2^{k+1} \tag{17.7}$$

where c is the constant in (17.3).

For fixed n, let $0 \leq (u - t) < \delta$ and let $t = m/n$ for some m. Then, by the polygonal nature of the paths

$$|X_n(u) - X_n(t)| \leq \max_{0 \leq i \leq [n\delta] + 1} n^{-1/2} |S_{i+m} - S_m| \tag{17.8}$$

Now suppose that $k_0 \equiv [\delta n] \geq 1$.

Let $l = [n/[n\delta]] + 1$. Let $t_0 = 0$, $t_1 = k_0/n$, $\ldots$, $t_{l-1} = (l - 1)k_0/n$, $t_l = 1$. Then $|t_i - t_{i-1}| \leq k_0/n \leq \delta$, so bearing in mind that (17.6) just needs to hold for large n we see that it suffices that

$$\lim_{n \to \infty} \left[(l + 1)E\left(\max_{0 \leq i \leq [n\delta] + 1} |S_i| \geq \tfrac{1}{2}k^{-1}\sqrt{n}\right)\right] < \varepsilon/2^k$$

But, by (17.3), this limit is asymptotically bounded by

$$c\{[n/[n\delta]] + 1\}\{(2k)^{-1}\sqrt{n}([n\delta] + 1)^{-1/2}\}^{-3}$$

which approaches

$$c([\delta^{-1}] + 1)(2k)^3\delta^{3/2} < \varepsilon/2^k$$

by (17.7). ∎

18. Laplace's Method in Function Space

To understand the results we seek in this section, consider the following method for obtaining the leading behavior in Stirling's formula for the asymptotics of the gamma function

$$\Gamma(\alpha) = \int_0^\infty x^{\alpha-1}e^{-x}\,dx$$

If one changes variables to $x = \alpha y$ one finds that

$$\Gamma(\alpha) = \alpha^\alpha \int_0^\infty y^{-1}\exp[-\alpha(y - \ln y)]\,dy$$

To find the asymptotics, one locates the minimum of $y - \ln y$ which is one (occurring at $y = 1$). This suggests that the integral $\sim e^{-\alpha}$ and indeed one can prove easily that

$$\lim_{\alpha\to\infty}\left[\frac{\Gamma(\alpha)}{\alpha^\alpha}\right]^{1/\alpha} = e^{-1}$$

This method can be extended further by making a Gaussian approximation $(y - \ln y) \sim 1 + \frac{1}{2}(y-1)^2 + O((y-1)^3)$ and controlling the error. One obtains Stirling's formula $\Gamma(\alpha + 1) = \alpha^{\alpha+1/2}e^{-\alpha}(2\pi)^{1/2}(1 + O(\alpha^{-1}))$.

Here we want to develop an analog of this method for functional integrals. Suppose we have a Gaussian measure $d\mu_0$ on some function space with covariance $(\cdot, A\cdot)$. Formally

$$d\mu_0 = \exp(-\tfrac{1}{2}(x, A^{-1}x))\text{ “}dx\text{”} \tag{18.1}$$

where dx is a formal uncountably infinite product measure and includes an infinite normalization constant. The formal nature of (18.1) is indicated by the quotation marks about dx. In the papers quoted below, the formal nature is indicated by using a special symbol (“flat integral”) for integrals with respect to “dx.”

Suppose that we want to compute the asymptotics of

$$a(n) \equiv \int \exp(-nF(x/\sqrt{n}))\, d\mu_0(x) \tag{18.2}$$

Using (18.1) and changing variables $x/\sqrt{n} = y$, we see that (with change in normalization!)

$$a(n) = \int \exp(-n[\tfrac{1}{2}(x, A^{-1}x) + F(x)])\, \text{“}dx\text{”} \tag{18.3}$$

which suggests that

$$\lim_{n\to\infty} [-n^{-1} \ln a(n)] = \min_x [\tfrac{1}{2}(x, A^{-1}x) + F(x)] \equiv b \tag{18.4}$$

Of course, one must expect to have fairly strong continuity conditions on F *since the set of* x *with* $(x, A^{-1}x) < \infty$ *has* $d\mu_0$*-measure zero.* [This assertion that with respect to $\prod_{n=1}^{\infty} (2\pi)^{-1/2} \exp(-\frac{1}{2}\alpha_n^2)\, d\alpha_n$, the set of α's with $\sum \alpha_n^2 < \infty$ has measure zero is easily seen; for, as we showed in Section 3, $\overline{\lim}\, \alpha_n/\sqrt{2 \ln n} = 1$ with probability one, so $\alpha_n \to 0$ with probability zero.] Thus F must be “determined” by its values on this set of measure zero; i.e., it must be continuous in a topology in which $\{x \mid (x, A^{-1}x) < \infty\}$ is dense in a function space which has measure one.

The kind of problem we describe above was considered about ten years ago by two doctoral students of Donsker. Schilder [230] discussed Brownian motion and Pincus [204] more general Gaussian processes on [0, 1]. Schilder obtained (18.4) under suitable hypotheses on F. Moreover when the right-hand side has a unique minimum, he obtained an asymptotic series to all orders of $(\sqrt{n})^{-1}$ for $\exp(+nb) \int \exp(-nF(x/\sqrt{n}))\, d\mu_0(x)$. Pincus considered a problem related to but distinct from (18.4). Namely, he proved under suitable hypotheses on F and G that

$$\lim_{n\to\infty} \left[\int G(x/\sqrt{n}) \exp(-nF(x/\sqrt{n}))\, d\mu_0 \right] / a(n) = G(x^*)$$

whenever the minimum in (18.4) occurs at a unique point, x^*. If anything, the problem we want to discuss is easier than this, and our arguments are a modification of those of Pincus. Moreover, the methods are such that it should be possible, by combining the work of Pincus and Schilder, to develop rigorously Gaussian approximations and further results on $a(n)$ when F is smooth enough and $\frac{1}{2}(x, A^{-1}x) + F(x)$ has a unique minimum or a set of minima with a simple structure. (In the argument below, regions II and III are unimportant in any further terms in the asymptotic series.) We note that the general machinery of Donsker and Varadhan, discussed in Section 19, is applicable to these problems yielding (18.4); see, e.g., [57].

Recently, the kind of problems discussed above have produced considerable interest among particle physicists (who were unaware of the earlier work of Schilder and Pincus). The earliest work was done by Lipatov [164] whose ideas were developed by the Saclay school in a series beginning with Brézin *et al.* [24]. The ideas are well described by looking at the ground state of the quantum anharmonic oscillator; i.e., $E(\beta)$ indicates the lowest eigenvalue of $L_0 + \beta x^4$ where as usual $L_0 = -\frac{1}{2}\, d^2/dx^2 + \frac{1}{2}x^2 - \frac{1}{2}$. $E(\beta)$ has a formal perturbation series (see Section 20)

$$E(\beta) \sim \sum_{n=0}^{\infty} a_n \beta^n$$

This series is divergent but it can be summed [113] by a method known as Borel summability. Brézin *et al.* begin by writing [see (1.9)]

$$E(\beta) = -\lim_{T \to \infty} T^{-1} \ln \operatorname{Tr}(e^{-T(L_0 + \beta x^4)})$$

Consider first the asymptotics of

$$T^{-1} \ln \operatorname{Tr}(e^{-T(L_0 + \beta x^4)}) \sim \sum_{n=0}^{\infty} a_n(T)\beta^n$$

or more simply

$$\operatorname{Tr}(e^{-T(L_0 + \beta x^4)}) \sim \sum_{n=0}^{\infty} b_n(T)\beta^n \tag{18.5}$$

The point is that (18.5) can be written in terms of a suitable Gaussian integral,

$$\operatorname{Tr}(e^{-T(L_0 + \beta x^4)}) = \int \exp\left[-\beta \int_0^T x^4(s)\, ds\right] d\mu_{0,T}$$

by a Feynman–Kac formula. From this, one sees that

$$b_n(T) = \frac{(-1)^n}{n!} \int \left[\int_0^T x^4(s)\, ds\right]^n d\mu_{0,T}$$

or

$$b_n(T) = \frac{(-1)^n n^{2n}}{n!} \int \exp\left(-nF\left(\frac{x}{\sqrt{n}}\right)\right) d\mu_{0,T} \tag{18.6}$$

with

$$F(x) = -\ln\left[\int_0^T x^4(s)\, ds\right] \tag{18.7}$$

Our results below and Stirling's formula provide a *rigorous* proof that $[n^{-n}b_n(T)]^{1/n} \to \alpha(T)$ for an "explicit" $\alpha(T)$. By going several steps further, it is possible [24] to obtain the **Bender–Wu formula** ($\lim_{T\to\infty} \alpha(T) = 3$).

$$a_n \underset{n\to\infty}{\sim} (-1)^{n+1}\sqrt{6}\pi^{-3/2}3^n\Gamma(n + \tfrac{1}{2})(1 + O(1/n)) \tag{18.8}$$

but (18.8) has not yet been proven *by these methods* since it involves a formal interchange of n and T limits which is discussed further at the end of this section. In the case at hand, (18.8) is certainly correct since Bender and Wu [7] computed the first 75 a_n's and originally obtained (18.8) on a purely numerical basis. They also gave several "demonstrations" [8, 9, 10] of (18.8) using different ideas. Recently, a rigorous proof of (18.8) has been found [123a]. The importance of the work of Lipatov and the Saclay group is two-fold: first, they realized the relevance of (18.8) to doing numerical calculations with the Borel sums; secondly, their path integral method is *formally* applicable to field theories.

We now turn to a statement of the main result of this section: Below $(\cdot, \cdot)$ is the L^2 inner product; we use $\|\cdot\|_\infty$ and $\|\cdot\|_2$ for the L^p-norms. We will consider Gaussian processes $\{q(t)\}_{0\le t\le 1}$ with a covariance $\rho(s, t)$ obeying two conditions:

(a) For some $K < \infty$, $\theta \le \frac{1}{2}$:

$$|\rho(s, t)| \le K \tag{18.9}$$

$$|\rho(s, t) - \rho(s', t)| \le K|s - s'|^{2\theta} \tag{18.10}$$

(b) ρ is real and strictly positive definite; i.e., for $f \in C[0, 1]$, $f \not\equiv 0$,

$$\iint \rho(s, t)f(s)f(t)\, ds\, dt > 0$$

For the function F we will require [the F of (18.7) does not obey these conditions but see below].

(c) F is a Hölder continuous function on $C[0, 1]$, in the following sense.

$$|F(x) - F(y)| \le C_R\|x - y\|_2^{\theta_0} \tag{18.11}$$

for $x, y \in C[0, 1]$ with $\|x\|_\infty, \|y\|_\infty \le R$ where C_R may depend on R and $0 < \theta_0 \le 1$. Note that (18.11) is much stronger than ordinary Hölder continuity since it involves the L^2-norm and $\|\cdot\|_2 \le \|\cdot\|_\infty$.

(d) For all $x \in C[0, 1]$:

$$F(x) \ge -\tfrac{1}{2}c_1\|x\|_2^2 - c_2 \tag{18.12}$$

where $c_1\|A\| < 1$, and A is the operator on L^2 given by

$$(Ag)(s) = \int \rho(s, t)g(t)\,dt \tag{18.13}$$

Sometimes it is useful to replace (d) by

(d′) For some fixed c_1 and Δ and for any $\varepsilon > 0$ and all $x \in C[0, 1]$:

$$F(x) \geq -\tfrac{1}{2}c_1\|x\|_2^2 - c_2(\varepsilon) - \varepsilon\|A^{-\Delta}x\|_2^2 \tag{18.12′}$$

Moreover, $c_1\|A\| < 1$ and Δ is such that $\Delta < \frac{1}{2}$ and

$$\mathrm{Tr}(A^{1-2\Delta}) < \infty$$

Theorem 18.1 Suppose that (a)–(d) [or (d′)] hold. Let $d\mu_0$ be the Gaussian measure on $C[0, 1]$ with covariance ρ [supported on $C[0, 1]$ by (18.10) and Kolmogorov's lemma (Theorem 5.1)]. Then $D(A^{-1/2})$, the L^2-operator domain of $A^{-1/2}$ lies in $C[0, 1]$, the functional

$$H(x) = \tfrac{1}{2}(x, A^{-1}x) + F(x) \tag{18.14}$$

is bounded below on $D(A^{-1/2})$ and takes its minimum on at least one point x^* in $D(A^{-1/2})$. Moreover, (18.4) holds with $b = H(x^*)$.

Before turning to the proof of this theorem we note a slight extension:

Theorem 18.2 Suppose that (a), (b) hold and that F is a function on $C[0, 1]$ taking values in $(-\infty, \infty]$ so that

(e) $F(x) < \infty$ for some $x \in D(A^{-1/2})$.

(f) For each $m < \infty$, $F_m(x) = \min(F(x), m)$ obeys (c) and either (d) or (d′). Then all the conclusions of Theorem 18.1 remain true.

Proof Clearly F obeys (d′), so for ε small

$$H(x) \geq \tfrac{1}{2}(x, A^{-1}x) - \tfrac{1}{2}c_1(x, x) - c_2 - \varepsilon(x, A^{-2\Delta}x) \geq -c_2$$

since $c_1\|A\| < 1$. By (e), $H(x) < \infty$ for some $x \in D(A^{-1/2})$. Thus

$$M = 1 + \inf_x H(x) \equiv 1 + b < \infty$$

Let $a(n) = \int \exp(-nF(x/\sqrt{n}))\,d\mu_0$, and $a_m(n) = \int \exp(-nF_m(x/\sqrt{n}))\,d\mu_0$. Then, for $m > M$, $(a_m(n))^{1/n} \to e^{-b}$ by (f) and Theorem 18.1. But clearly

$$|a(n) - a_m(n)| \leq e^{-nM}$$

since $|e^{-nF} - e^{-nF_m}| \leq e^{-nM}$. ∎

Example Let us demonstrate the rigorous proof of the leading behavior of $b_n(T)$ for $T = 1$ (b_n given by (18.5); the general T case is similar). Define

$$\rho(s, t) = \mathrm{Tr}(xe^{-|s-t|L_0}xe^{-(1-|s-t|)L_0})[\mathrm{Tr}(e^{-L_0})]^{-1} \tag{18.15}$$

It is easy to check that ρ obeys (a), (b) with $\theta = \frac{1}{2}$ by obtaining an explicit formula for ρ (one can get the explicit formula by making an eigenfunction expansion for L_0 or more simply by noting that

$$\left(-\frac{d^2}{ds^2} + 1\right)\rho(s, t) = \delta(s - t)$$

with the condition that ρ be periodic at zero and one). Now let $Q(x) = \int_0^1 x^4(s)\, ds$. Then since $|a^4 - b^4| \leq 4[\max(a, b)]^3(a - b)$:

$$|Q(x) - Q(y)| \leq 4R^3\|x - y\|_2$$

if $R = \max(\|x\|_\infty, \|y\|_\infty)$. It follows that $F_m(x) = \min(m, -\ln Q)$ obeys (c) with $\theta_0 = 1$. Now, A^{-1} has eigenvalues $1 + (2\pi n)^2$ with eigenvectors $(2\pi)^{-1/2}e^{2\pi inx}$ so $\mathrm{Tr}(A^{1-2\Delta}) < \infty$ for $\Delta < \frac{1}{4}$. By a Hausdorff–Young and then a Hölder estimate, if f is in L^2 with (periodic) Fourier coefficients f_n and $2 < p < \infty$ [and $q = (1 - p^{-1})^{-1}$]:

$$\|f\|_p \leq C\left(\sum_n |f_n|^q\right)^{1/q} \leq D\left(\sum_n |n^\alpha f_n|^2\right)^{1/2}$$

so long as $\alpha(1/q - \frac{1}{2})^{-1} > 1$. Since $p < \infty$, we can take $\alpha < \frac{1}{2}$. But $\sum_n |n^\alpha f_n|^2 \leq \|A^{-(1/2)\alpha}f\|_2^2$. Thus, we have the simple Sobolev inequality:

$$Q(x) \equiv \|x\|_4^4 \leq c\|A^{-\alpha/2}x\|_2^4$$

so

$$\begin{aligned} F(x) = -\ln Q(x) &\geq -\ln c - 4\ln\|A^{-\alpha/2}x\|_2 \\ &\geq -c_2 - \varepsilon\|A^{-\alpha/2}x\|_2^2 \end{aligned}$$

verifying (d′). We summarize with the following.

Theorem 18.3 Let $b_n(T)$ be given by (18.5). Let

$$\alpha(T) \equiv \min\left\{\frac{1}{2}\int_0^T dt\left[x(t)^2 + \left(\frac{dx}{dt}\right)^2\right] - \ln \int_0^T x^4(t)\, dt\right\} \tag{18.16}$$

subject to the boundary condition that x be periodic. Then

$$\lim_{n\to\infty} [n!\,n^{-2n}(-1)^n b_n(T)]^{1/n} = e^{-\alpha(T)}$$

Remark It is easy to establish that the minimum in (18.16) obeys the nonlinear equation

$$-\frac{d^2x}{dt^2} + x - \frac{4x^3}{\int_0^T x^4(t)\,dt} = 0$$

so after scaling, it involves classical motion in a potential $x^4 - x^2$ which means that $\alpha(T)$ can be computed exactly up to quadratures which are given "explicitly" by elliptic functions; see, e.g., [24].

Now, we begin the proof of Theorem 18.1.

Lemma 18.4

(a) A is Hilbert–Schmidt and maps L^2 into $C[0, 1]$ and $A[x \mid \|x\|_2 \le 1]$ is compact in $C[0, 1]$.

(b) A is trace class, $A^{1/2}$ maps L^2 into $C[0, 1]$, and $A^{1/2}[x \mid \|x\|_2 \le 1]$ is compact in $C[0, 1]$.

(c) One has the estimates:

$$\|Ax\|_\infty \le C\|x\|_2; \qquad \|A^{1/2}x\|_\infty \le D\|x\|_2 \tag{18.17}$$
$$\|Ax\|_\infty \le D\|A^{1/2}x\|_2$$

Proof (a) By hypothesis (a)

$$\|Af\|_\infty \le K\int |f|\,dt \le K\left(\int |f|^2\,dt\right)^{1/2} \tag{18.18a}$$

$$|(Af)(s) - (Af)(s')| \le K|s - s'|^{2\theta}\|f\|_2 \tag{18.18b}$$

Thus A maps L^2 into $C[0, 1]$ and the image of the unit ball is a family of uniformly bounded, uniformly equicontinuous functions and so pre-compact by the Arzela–Ascoli theorem [214]. By the weak compactness of the ball, the image is closed. A is Hilbert–Schmidt since $\int \rho^2\,ds\,dt < \infty$.

(b) A is trace class since its kernel ρ is continuous, A is positive and $\int_0^1 \rho(s, s)\,ds < \infty$; see [259]. Define orthonormal eigenfunctions f_n and eigenvalues λ_n^2 by

$$(Af_n)(t) = \lambda_n^2 f_n(t) \tag{18.19}$$

We claim that for each s, t

$$\rho(s, t) = \sum_n \lambda_n^2 f_n(s) f_n(t) \tag{18.20}$$

For (18.20) clearly holds for almost all s and t. Let j_ε be an approximate identity in $C[0, 1]$, and let

$$f_{n,\varepsilon}(x) = f_n * j_\varepsilon(x)$$

$$\rho_\varepsilon(s, t) = \int \rho(s', t') j_\varepsilon(s - s') j_\varepsilon(t - t')\, ds'\, dt'$$

We have that

$$\rho_\varepsilon(s, s) = \sum \lambda_n^2 |f_{n,\varepsilon}(s)|^2$$

so since the f_n and ρ are continuous:

$$\begin{aligned}\sum_{n \le N} \lambda_n^2 |f_n(s)| &= \lim_\varepsilon \sum_{n \le N} \lambda_n^2 |f_{n,\varepsilon}(s)|^2 \\ &\le \lim_\varepsilon \rho_\varepsilon(s, s) = \rho(s, s)\end{aligned}$$

As a result,

$$\sum_n |\lambda_n f_n(s)|^2 \le K$$

for all s. Similarly,

$$\sum_n \lambda_n^2 |f_n(s) - f_n(s')|^2 \le 2K|s - s'|^{2\theta}$$

It follows that the right-hand side of (18.20) converges uniformly as $n \to \infty$ and so (18.20) holds.

Now, if $x(t) = \sum \beta_n f_n(t)$, then

$$|(A^{1/2}x)(t)|^2 \le |\sum \lambda_n \beta_n f_n(t)|^2 \le \|x\|_2^2 \sum |\lambda_n f_n(t)|^2 = \rho(t, t)\|x\|_2^2$$

Similarly,

$$\begin{aligned}|(A^{1/2}x)(s) - (A^{1/2}x)(s')|^2 &\le \|x\|_2^2[\rho(s, s) + \rho(s', s') - \rho(s, s') - \rho(s', s)] \\ &\le 2K|s - s'|^{2\theta}\|x\|_2^2\end{aligned}$$

Thus $A^{1/2}$ maps L^2 into $C[0, 1]$ and the image of the unit ball is compact as in (a).

(c) The first two estimates are just the assertions that A and $A^{1/2}$ are bounded from L^2 to C. (18.17) is the second estimate for $A^{1/2}x$. ∎

Lemma 18.5 $H(x) = \frac{1}{2}(x, A^{-1}x) + F(x)$ is bounded below on $D(A^{-1/2})$ and $H(x)$ takes its minimum $H(x^*)$ at one point, x^*, at least.

Proof By (18.12′) [hypothesis (d′)],

$$\begin{aligned} H(x) &\geq \tfrac{1}{2}(x, A^{-1}x) - \tfrac{1}{2}c_1\|x\|_2^2 - c_2 - \varepsilon(x, A^{-1/2}x) \\ &\geq -c_2 + \tfrac{1}{2}\delta(x, A^{-1}x) \end{aligned}$$

if $1 - c_1\|A\| - 2\varepsilon\|A\|^{1/2} \geq \delta$. Thus H is bounded below and on the set where $H(x) \leq \inf_y H(y) + 1$, $(x, A^{-1}x)$ is bounded from above. Thus $H(x)$ takes its infimum on a compact subset of $C[0, 1]$ (by Lemma 18.4(b)). Since $H(x)$ is easily seen to be lower semicontinuous, we conclude that H takes its minimum value. ∎

The key to the proof of Theorem 18.1 is the following:

Lemma 18.6 Let $\gamma > 0$. Then $d\mu_0(x + \gamma Ax)$ is $d\mu_0$ absolutely continuous, and

$$d\mu_0(x + \gamma Ax) = D(\gamma)\exp[-\tfrac{1}{2}\gamma^2(Ax, x) - \gamma(x, x)]\, d\mu_0(x) \quad (18.21)$$

and $D(\gamma) = \det(1 + \gamma A) \equiv \prod_{n=1}^{\infty}(1 + \gamma\lambda_n^2)$.

Proof Introduce the realization of the Gaussian process $d\mu_0$:

$$x(t) = \sum_{n=1}^{\infty} \alpha_n \lambda_n f_n(t) \quad (18.22)$$

where the α_n are random variables distributed by

$$\bigtimes_{n=1}^{\infty} (2\pi)^{-1/2}\exp(-\tfrac{1}{2}\alpha_n^2)\, d\alpha_n$$

Then $x \to (1 + \gamma A)x$ corresponds to $\alpha_n \to (1 + \gamma\lambda_n^2)\alpha_n$. Thus if $d\mu_0^N = \bigtimes_{n=1}^{N}(2\pi)^{-1/2}\exp(-\tfrac{1}{2}\alpha_n^2)\, d\alpha_n$:

$$d\mu_0^N(x + \gamma Ax) = \prod_{n=1}^{N}\{(1 + \gamma\lambda_n^2)\exp[-\tfrac{1}{2}\gamma^2\lambda_n^4\alpha_n^2 - \gamma\lambda_n^2\alpha_n^2]\}\, d\mu_0^N$$

Now as $N \to \infty$, $\sum \lambda_n^2\alpha_n^2 \to (x, x)$, $\sum \lambda_n^4\alpha_n^2 \to (x, Ax)$ so the exponential converges in $L^1(d\mu_0)$ by the monotone convergence theorem to

$$\exp(-\tfrac{1}{2}\gamma^2(Ax, x) - \gamma(x, x))$$

Since $\sum \lambda_n^2 < \infty$, $\prod_{n=1}^{N}(1 + \gamma\lambda_n^2) \to D(\gamma) < \infty$. ∎

Remark Lemma 18.6 is essentially an explicit form of one-half of Theorem 2.5.

Therefore, we can write $a(n)$, given by (18.2),

$$a(n) = \det(1 + \sqrt{n}\, A) E[\exp(-nQ_n(x))] \tag{18.23}$$

$$Q_n(x) = \tfrac{1}{2}(x, Ax) + F(Ax + n^{-1/2}x) + n^{-1/2}(x, x) \tag{18.24}$$

since both sides of (18.23) are

$$\int \exp(-nF[(1/\sqrt{n})(x + \sqrt{n}\, Ax)])\, d\mu_0(x + \sqrt{n}\, Ax)$$

Now since $A \geq 0$:

$$1 \leq \det(1 + \sqrt{n}\, A) = \prod_{j=1}^{\infty} (1 + \sqrt{n}\, \lambda_j^2) \leq \exp\left[\sqrt{n}\left(\sum_{j=1}^{\infty} \lambda_j^2\right)\right]$$

so that (18.23) says that to conclude the proof of Theorem 18.1, it suffices to show that

$$[E(\exp(-nQ_n(x)))]^{1/n} \to \exp(-H(x^*))$$

Since $Q_n(x)$ is almost $H(Ax)$, this is beginning to look controllable. We need three more preparatory lemmas:

Lemma 18.7 For some γ and C:

$$E(\|x\|_\infty \geq a) \leq Ce^{-\gamma a^2} \tag{18.25}$$

Remark Marcus and Shepp [180] and Fernique [82a] have shown that for any Gaussian process with paths in $C[0, 1]$ and covariance ρ:

$$\lim_{a \to \infty} a^{-2} \log E(\|x\|_\infty \geq a) = -\frac{1}{2 \sup_s[\rho(s, s)]}$$

See also [57].

Proof This proof is a simple extension of the idea behind Kolmogorov's lemma (Theorem 5.1). By hypothesis (a):

$$E(|x(s) - x(s')|^2) \leq 2K|s - s'|^{2\theta}$$

so, since the $x(s)$ are Gaussian, for $C_0 \geq 1$, $n \geq 0$

$$E\left(\left|x\left(\frac{k}{2^n}\right) - x\left(\frac{k-1}{2^n}\right)\right| \geq C_0 2^{-n\theta/2}\right) \leq 2\int_y^\infty (2\pi)^{-1/2} e^{-x^2/2}\, dx$$

$$\leq D_1 \exp(-D_2 C_0^2 2^{n\theta})$$

where $y = C_0(2K)^{-1/2}2^{n\theta/2}$ and D_1, D_2 are suitable constants. Thus,

$$E\left(\left|x\left(\frac{k}{2^n}\right) - x\left(\frac{k-1}{2^n}\right)\right| \geq C_0 2^{-n\theta/2}, \text{ some } 0 \leq k \leq 2^n, \text{ some } n\right)$$

$$\leq D_1 \sum_{n=0}^{\infty} 2^n \exp(-D_2 C_0^2 2^{n\theta})$$

$$\leq D_3 \exp(-D_2 C_0^2)$$

If $|x(k/2^n) - x((k-1)/2^n)| \leq C_0 2^{-n\theta/2}$ for all k and n, then writing $t = \sum a_j/2^j$, $a_j = 0$ or 1 we see that

$$|x(t) - x(0)| \leq \sum a_j C_0 2^{-j\theta/2} = C_0 B$$

for some constant B. Thus,

$$E(|x(t) - x(0)| \geq C, \text{ some } t) \leq D_3 \exp(-D_2 B^{-2} C^2)$$

If we note that

$$E(\|x\|_\infty \geq a) \leq E(|x(0)| \geq a/2) + E(|x(t) - x(0)| \geq a/2, \text{ some } t)$$

we obtain (18.25). ▮

Lemma 18.8 Suppose $\mathrm{Tr}(A^{1-2\Delta}) < \infty$. For ε sufficiently small:

$$E(\exp[\varepsilon(x, A^{-2\Delta}x)]) < \infty$$

Proof In terms of the realization (18.22)

$$(x, A^{-2\Delta}x) = \sum_{n=1}^{\infty} \lambda_n^{2-4\Delta}\alpha_n^2$$

If one looks at Theorem 3.11 and its claimed extension to nonpositive matrices, one sees that

$$\int \exp\left(\sum_{n=1}^{\infty} \beta_n \alpha_n^2\right)\left[\bigtimes_{n=1}^{\infty} (2\pi)^{-1/2} \exp(-\tfrac{1}{2}\alpha_n^2)\, d\alpha_n\right] < \infty$$

if (and only if) $\beta_n < \frac{1}{2}$ for all n and $\sum \beta_n < \infty$. ▮

Lemma 18.9 For any $y \in L^2$ and $\beta > 0$, there is a c with

$$E(x \mid \|A^{1/2}x - y\|_2 \leq \beta;\ \|x\|_\infty \leq c) > 0 \qquad (18.26)$$

Proof Since $\|x\|_\infty < \infty$, almost everywhere in x, we need only prove (18.26) with the condition $\|x\|_\infty \le c$ dropped. In terms of the realization (18.22)

$$\begin{aligned}\|A^{1/2}x - y\|_2^2 &= \sum_{n=1}^{\infty} (\alpha_n \lambda_n^2 - y_n)^2 \\ &\le \sum_{n=1}^{N} (\alpha_n \lambda_n^2 - y_n)^2 + 2\sum_{N+1}^{\infty} (\alpha_n \lambda_n^2)^2 + 2\sum_{N+1}^{\infty} |y_n|^2\end{aligned}$$

Choosing first N with $2\sum_{N+1}^{\infty} |y_n|^2 \le \beta^2/4$, we see that it suffices to show that

$$E\left(\sum_{N+1}^{\infty} (\alpha_n \lambda_n^2)^2 \le \frac{\beta^2}{8}\right) > 0$$

If $|\alpha_n| \le (\sqrt{2}/4)\beta\lambda_n^{-1}(\sum \lambda_m^2)^{-1/2}$ then $\sum (\alpha_n \lambda_n^2)^2 \le \beta^2/8$, so we only need

$$E(|\alpha_n| \le \gamma\lambda_n^{-1}, \text{ all } n) > 0 \tag{18.27}$$

for any $\gamma > 0$. But

$$E(|\alpha_n| \le \gamma\lambda_n^{-1}, \text{ all } n) = \prod_n \left[1 - E\left(|\alpha_n| \ge \frac{\gamma}{\lambda_n}\right)\right]$$

so (18.27) follows from $\sum E(|\alpha_n| \ge \gamma/\lambda_n) < \infty$. This is easy from

$$E(|\alpha_n| \ge a) \le C \exp(-\tfrac{1}{2}a^2) \le 2Ca^{-2} \qquad \text{and} \qquad \sum \lambda_n^2 < \infty \quad \blacksquare$$

Proof of Theorem 18.1 *Lower bound* Given ε, we claim that we can find β so that $\|A^{1/2}x - A^{-1/2}x^*\|_2 \le \beta$ implies

$$H(Ax) - H(x^*) \le \varepsilon \tag{18.28}$$

For

$$\begin{aligned}H(Ax) - H(x^*) \le{}& \tfrac{1}{2}|(A^{1/2}x, A^{1/2}x) - (A^{-1/2}x^*, A^{-1/2}x^*)| \\ &+ |F(Ax) - F(x^*)|\end{aligned}$$

The first factor is small if $\|A^{1/2}x - A^{-1/2}x^*\|_2$ is small by the continuity of the norm. The second factor is small by the continuity of F in $\|\cdot\|_\infty$ and (18.17) which implies that

$$\|Ax - x^*\|_\infty \le D\|A^{1/2}x - A^{-1/2}x^*\|_2 \tag{18.29}$$

This verifies (18.28). Pick c so that (18.26) holds. Then on the set

$$S = \{x \mid \|A^{1/2}x - A^{-1/2}x^*\|_2 \le \beta, \|x\|_\infty \le c\}$$

we have that

$$\begin{aligned}Q_n(x) &= H(Ax) + F(Ax + x/\sqrt{n}) - F(Ax) + \tfrac{1}{2}n^{-1/2}(x, x) \\ &\le \varepsilon + H(x^*) + C_R c n^{-(1/2)\theta_0} + \tfrac{1}{2}c^2 n^{-1/2}\end{aligned}$$

where C_R is the constant in (18.11), we have used $\|x\|_2 \leq \|x\|_\infty$ and R is first chosen so that $\|Ax\|_\infty \leq R/2$ for $x \in S$ [use (18.29)] and then so that $cn^{-1/2} < R/2$. Thus for n large:

$$E(e^{-nQ_n}) \geq \exp(-n[\varepsilon + H(x^*) + O(n^{-\theta_0/2})])E(S)$$

Since $E(S) > 0$, we have that $\underline{\lim}(1/n)\ln E(e^{-nQ_n}) \geq -\varepsilon - H(x^*)$. Since ε is arbitrary we have the required lower bound.

Upper bound: General strategy Consider the three regions:

$$S_{\text{I}} = \{x \mid \|Ax\|_\infty \leq R,\ \|x\|_\infty \leq R\sqrt{n}\}$$
$$S_{\text{II}} = \{x \mid \|x\|_\infty \geq R\sqrt{n}\}$$
$$S_{\text{III}} = \{x \mid \|Ax\|_\infty \geq R\}$$

We will control the contributions of S_{II} and S_{III} for R sufficiently large and S_{I} for any R. We suppose that (d′) holds. The reader can check that the extra condition on ρ, $\text{Tr}(A^{1-2\Delta}) < \infty$ is only used (via Lemma 18.8) to control the extra $\varepsilon\|A^{-\Delta}x\|_2^2$ terms present in case (d′). Thus case (d) goes through with minimal changes.

Upper bound: Region I As in the lower bound, we have for $x \in S_{\text{I}}$:

$$\begin{aligned} Q_n(x) &= H(Ax) + F(Ax + x/\sqrt{n}) - F(Ax) + \tfrac{1}{2}n^{-1/2}(x, x) \\ &\geq H(x^*) - C_{2R}\|x\|_2/\sqrt{n} + \tfrac{1}{2}n^{-1/2}\|x\|_2^2 \\ &\geq H(x^*) - \tfrac{1}{2}n^{-1/2}C_{2R}^2 \end{aligned}$$

by using $-xy + \frac{1}{2}y^2 \geq -\frac{1}{2}x^2$. It follows that

$$E(e^{-nQ_n} \mid x \in S_{\text{I}}) \leq e^{-nH(x^*)}\exp(\tfrac{1}{2}n^{1/2}C_{2R}^2)$$

yielding the required upper bound.

Upper bound: Region II We begin with the estimate from (18.12′):

$$\begin{aligned} Q_n(x) \geq{} & \tfrac{1}{2}(Ax, x) + \tfrac{1}{2}n^{-1/2}(x, x) - \tfrac{1}{2}c_1\|Ax + (x/\sqrt{n})\|_2^2 \\ & - c_2(\varepsilon) - \varepsilon\|A^{-\Delta}(Ax + (x/\sqrt{n}))\|_2^2 \end{aligned}$$

Now

$$\|Ax + (x/\sqrt{n})\|_2^2 \leq (\|A\| + (2/\sqrt{n}))(Ax, x) - (1/n)(x, x)$$

and

$$\begin{aligned} \|A^{-\Delta}(Ax + (x/\sqrt{n}))\|^2 \leq{} & \|A^{1-2\Delta}\|[(Ax, x) + 2n^{-1/2}(x, x)] \\ & + n^{-1}(x, A^{-2\Delta}x) \end{aligned}$$

so

$$Q_n(x) \geq (a_n/2)(Ax, x) + b_n(x, x) - c_2(\varepsilon) - \varepsilon n^{-1}(x, A^{-2\Delta}x) \quad (18.30)$$

$$a_n = (1 - c_1\|A\| - 2c_1 n^{-1/2} - 2\varepsilon\|A\|^{1-2\Delta}) \quad (18.30a)$$

$$b_n = \tfrac{1}{2}n^{-1/2}[1 - \tfrac{1}{2}c_1 n^{-1/2} - 4\varepsilon\|A\|^{1-2\Delta}] \quad (18.30b)$$

One can choose ε small and then n so large that $a_n, b_n > 0$. It follows that

$$\begin{aligned} E(e^{-nQ_n} \mid S_{\mathrm{II}}) &\leq e^{nc_2}E(e^{\varepsilon(x, A^{-2\Delta}x)} \mid S_{\mathrm{II}}) \\ &\leq e^{nc_2}E(e^{2\varepsilon(x, A^{-2\Delta}x)})^{1/2}E(S_{\mathrm{II}})^{1/2} \\ &\leq c\exp(nc_2 - \tfrac{1}{2}n\gamma R^2) \end{aligned}$$

where we used the Schwarz inequality in the second step and Lemmas 18.7 and 18.8 in the last inequality (ε may have to be decreased). So long as we choose R so large that $c_2 - \frac{1}{2}\gamma R^2 \leq -H(x^*)$, we have the requisite bound on S_{II}.

Upper bound: Region III Use (18.30) and choose ε so small and n so large that $b_n > 0$ and $a_n \geq \delta > 0$ for some fixed δ. Notice that in region III, $\|Ax\|_2 \geq D^{-1}\|Ax\|_\infty \geq D^{-1}R$ by (18.17). Thus,

$$Q_n(x) \geq (\delta/2)D^{-2}R^2 - c_2(\varepsilon) - \varepsilon n^{-1}(x, A^{-2\Delta}x)$$

so that

$$E(e^{-nQ_n} \mid S_{\mathrm{III}}) \leq E(e^{\varepsilon(x, A^{-2\Delta}x)})\exp(nc_2 - (\delta/2)nD^{-2}R^2)$$

So long as $c_2 - (\delta/2)D^{-2}R^2 \leq -H(x^*)$, we have the required bound on S_{III}. ∎

At the present moment, a complete proof of the Bender–Wu formula (18.8) has only been obtained by very different methods [123a]. However, one can imagine a proof along the lines above; only one crucial step is missing. The passage from the asymptotics of $b_n(T)$ to those of $a_n(T)$ is not hard. One only needs to use the Taylor series for $\ln(1 + x)$. The net result is that for T fixed and n large $a_n(T) \sim n^n\alpha(T)^n n^{b(T)}(\gamma(T) + O(1/n))$ (this assumes that one can push Theorem 18.1 to higher order for some very special degenerate x^*), where $\alpha(T) \to \alpha$, etc., as $T \to \infty$. The missing point in the proof is justifying the interchange of n and T limits. The subtlety of the missing step is indicated by the following example.

Example (Donsker and Simon, unpublished) Let

$$f(T, \beta) = -(1/T)\ln \operatorname{Tr}(\exp(-T(L_0 + \tfrac{1}{2}\beta x^2)))$$

Then $f(\infty, \beta) \equiv \lim_{T\to\infty} f(T, \beta) = \frac{1}{2}[(\beta + 1)^{1/2} - 1]$. Thus, for n large and $T = \infty$:

$$a_n(\infty) \sim \tfrac{1}{2}\pi^{-1/2}(-1)^{n+1}n^{-3/2}(1 + O(1/n)) \tag{18.31}$$

On the other hand,

$$\begin{aligned}\mathrm{Tr}(\exp(-T(L_0 + \tfrac{1}{2}\beta x^2))) &= \sum_{n=0}^{\infty} \exp(-[(n + \tfrac{1}{2})\sqrt{1 + \beta} - \tfrac{1}{2}]T) \\ &= e^{(1/2)T}\tfrac{1}{2}[\sinh((T/2)\sqrt{1 + \beta})]^{-1}\end{aligned}$$

Using $\sinh(x) = x \prod_{n=1}^{\infty} (1 + x^2/(n\pi)^2)$ we obtain

$$f(T, \beta) = \frac{1}{2T} \ln(1 + \beta) + \frac{1}{T} \sum_{n=1}^{\infty} \ln\left(1 + \frac{\beta}{\gamma_n(T)}\right) - \frac{1}{2} + \frac{1}{T} \ln\left(2 \sinh \frac{T}{2}\right)$$

with $\gamma_n(T) = 1 + T^{-2}(2n\pi)^2$. Thus for *fixed* T as $n \to \infty$,

$$a_n(T) \sim (2T)^{-1}(-1)^{n+1}n^{-1}(1 + O(1/n)) \tag{18.32}$$

Comparing (18.31) and (18.32), we see that the leading $(-1)^n$ terms agree but the n^b and constant terms are wrong. Analytically, it is clear why this happens. The square root singularity for $n = \infty$ is the limit of an infinite number of logarithmic singularities at $-\gamma_n(T)$ which coalesce as $T \to \infty$. For fixed T, only one log contributes as $n \to \infty$.

Since (18.8) has been proven by other means [123a], the interchange is permissible for the anharmonic oscillator; the issue is to understand why.

* * *

The ideas of this section are useful for understanding the imaginary time analog of the classical limit for solutions of the time dependent Schrödinger equation. Let

$$H_\hbar = (-\hbar^2/2m)\Delta + V(y) \tag{18.33}$$

The classical limit question involves studying $\exp(-itH_\hbar/\hbar)\psi$ as $\hbar \to 0$. We will instead consider

$$a(t, y, \hbar; \psi) \equiv [\exp(-tH_\hbar/\hbar)\psi](y) \tag{18.34}$$

in the limit as $\hbar \to 0$. The quantity when $\psi \equiv 1$ will be denoted by $a(t, y, \hbar)$ (see Section 25 for the definition of $\exp(-sH_\hbar)$ on L^∞). We will state one result and then discuss its significance and the kinds of extensions possible.

Theorem 18.10 Let V be bounded from below and C^1 on $\mathbb{R}$. Let $H_\hbar$ be given by (18.33) and $a(t, y, \hbar)$ by (18.34) with $\psi \equiv 1$. Then

$$\lim_{\hbar \to 0} -\hbar \ln[a(t, y_1, \hbar)] = \alpha(t, y_1) \equiv \min_{\gamma \in \Gamma_{y_1}} \left\{ \int_0^t \left[\frac{1}{2m} \gamma'(s)^2 + V(\gamma(s)) \right] ds \right\} \tag{18.35}$$

where Γ_{y_1} is the set of C^1-paths with $\gamma(0) = y_1$.

Proof By scaling y and/or V one can (by changing V) suppose that $t = m = 1$. By translation covariance, we can suppose also that $y_1 = 0$. The operator A^{-1} with quadratic form $\frac{1}{2} \int_0^1 \gamma'(s)^2\, ds$, with domain the closure of the form domain Γ_0, is precisely $A^{-1} = -d^2/dt^2$ with boundary conditions $\gamma(0) = 0$, $\gamma'(1) = 0$ (see [217]). The integral kernel of A, the inverse of this A^{-1}, is just $\rho(s, t) \equiv \min(s, t)$. Noticing that $F(b) = \int_0^t V(b(s))\, ds$ obeys (18.11, 12) on account of the hypothesis on V, we see that Theorem 18.1 is applicable so that

$$\alpha(1, 0) = \lim_{\hbar \to 0} -\hbar \ln[\tilde{a}(1, 0, \hbar)]$$

where

$$\tilde{a}(1, 0, \hbar) = \int \exp\left[-\hbar^{-1} \int_0^1 V(\hbar^{1/2} b(s))\, ds \right] Db$$

with Db the Brownian motion measure. Thus, by the Feynman–Kac formula,

$$\tilde{a}(1, 0, \hbar) = \left[\exp\left(-\left[\frac{-1}{2} \frac{d^2}{dy^2} + \hbar^{-1} V_\hbar(y) \right] \right) \psi \right](0)$$

where ψ is the function which is identically one and $V_\hbar(y) = V(\hbar^{1/2} y)$. Let $(W\eta)(y) = \eta(\hbar^{1/2} y)$. Then using

$$W_\hbar H_\hbar W_\hbar^{-1} = \left[\frac{-1}{2} \frac{d^2}{dy^2} + \hbar^{-1} V_\hbar(y) \right] \hbar$$

we see that

$$\begin{aligned} \tilde{a}(1, 0, \hbar) &= [\exp(-H_\hbar \hbar^{-1}) W_\hbar \psi](\hbar^{-1/2} 0) \\ &= a(1, 0, \hbar) \end{aligned}$$

since $W_\hbar \psi = \psi$ and $\hbar^{-1/2} 0 = 0$. ∎

A closely related theorem but with different methods and hypotheses has been proven by Truman [280c]. The applicability of the Pincus–Schilder methods to this context seems to be new.

We examine the connection of the above result with "the classical limit" in a series of remarks.

Remarks 1. Standard calculus of variations arguments show that the minimum of

$$E(\gamma) = \int_0^t [(2m)^{-1}\gamma'(s)^2 + V(\gamma(s))]\, ds$$

over γ's with $\gamma(t) = y_1$ [by replacing $\gamma(\cdot)$ by $\gamma(t - \cdot)$ this minimum is $\alpha(t, y_1)$] obeys

$$\gamma''(s) = m(\nabla V)(\gamma(s)) \text{ (Note: } \textit{not } -\nabla V) \tag{18.36}$$

with "boundary conditions,"

$$\gamma'(0) = 0, \qquad \gamma(t) = y_1 \tag{18.37}$$

That is, if we let $Y_1(y_0, s)$ denote the solution of (18.36) with initial condition $Y_1(y_0, 0) = y_0$, $\partial Y_1/\partial s(y_0, 0)$, then the existence of the minimizing γ guaranteed by Theorem 18.1 assures us that $Y_1(y_0, t) = y_1$ always has a solution $Y_0(y_1, t)$; it must be that the minimizing γ is just

$$\gamma_0(s) = Y_1(Y_0(y_1, t), s) \tag{18.38}$$

If there is more than one solution Y_0, we must pick one that minimizes E.

We note that the study of "mixed boundary conditions" like (18.37) is characteristic of the Hamiltonian–Jacobi theory. We also emphasize that the corresponding classical solution obeys Newton's equation with $F = \nabla V$; i.e., in going from the solution of the Schrödinger equation, $e^{-itH}\psi$, where the $\hbar \to 0$ limit is *formally* given by ordinary classical mechanics to the solution of the heat equation $e^{-tH}\psi$, the sign of V changes in the corresponding classical mechanics. This is a well-known phenomena.

2. It is easy to extend the Pincus–Schilder theory to allow independent Brownian motions $\gamma_i(s)$ and thereby to extend Theorem 18.10 to ν-dimensional y's.

3. If we replace $\psi = 1$ by $\psi(y) = e^{-tS_0(y)/\hbar}$, then the formal minimum problem is to minimize

$$\int_0^t \left[\frac{1}{2m}(\gamma'(s))^2 + V(\gamma(s))\right] ds + S_0(\gamma(0))$$

subject to $\gamma(t) = y_1$. The solution to this problem obeys (18.36) but the boundary conditions (18.37) are now

$$\gamma'(0) = (\nabla S_0)(\gamma(0)); \qquad \gamma(t) = y_1$$

which arise in the full Hamiltonian–Jacobi theory. The only problem with this kind of extension of Theorem 18.10 is that

$$F(b) = \int_0^1 V(b(s))\, ds + S_0(b(1))$$

does not obey condition (18.11); i.e., F is Hölder continuous in $\|\cdot\|_\infty$ but *not* in $\|\cdot\|_2$. It may be possible to extend Theorem 18.1 to handle this case. Alternately, it may be possible to treat certain S_0's (certainly $S_0(y) = -ay^2$) by writing $\exp(-\hbar^{-1}S_0)$ as a conditional expectation.

$$\exp(-\hbar^{-1}S_0(b(1))) = E\left(\exp\left(-\hbar^{-1}\int_0^{1+\delta} S_\delta(b(s), s, \hbar)\, ds\right)\middle| b(1)\right)$$

for suitable S_δ.

4. If there is a unique γ in Γ_{y_0} minimizing E and V has extra smoothness, then Schilder's evaluation of the full asymptotic series for $a(t, y, \hbar)$ is valid and one obtains the series in terms of the Gaussian approximation of $E(\gamma)$ near the minimizing $\gamma(s)$ given by (18.38). The covariance of the corresponding Gaussian process will enter. This covariance is just the integral kernel of the inverse of $-(d^2/dt^2) + (d^2V/dy^2)(\gamma_0(t))$ with suitable boundary conditions (the "Feynman Green's function" of DeWitt–Morette). In particular, one finds that

$$e^{+\hbar\alpha}a \to (\partial Y_1/\partial y_0)^{-1/2} \tag{18.39}$$

5. If the minimum is unique, the method of Pincus shows that for suitable g's,

$$\int e^{-nF(b/\sqrt{n})}g(b/\sqrt{n})\, Db \Big/ \int e^{-nF(b/\sqrt{n})}\, Db \to g(b^*)$$

b^* being the unique minimum. Doing the scaling in Theorem 18.10 and using 18.31, one finds that for suitable ψ's,

$$a(t, y_1, \hbar; \psi)e^{+\alpha\hbar} \to (\partial Y_1/\partial y_0)^{-1/2}\psi(Y_0(y_1, t)) \tag{18.40}$$

6. The celebrated formula of Maslov is just the analog of (18.40) for $e^{-itH_\hbar/\hbar}\psi$. His formula follows from *formal* application of stationary phase ideas to the *formal* Feynman integral. This is notoriously difficult to make rigorous; for some partial results, see Truman [280c], Yajima [288], and the papers of Fujiwara [97a,b] and Hagedorn [291].

19. Introduction to the Donsker–Varadhan Theory

In this section, we wish to give a brief introduction to an important method of Donsker and Varadhan. Since these authors have seven papers [54–60] covering roughly 220 journal pages on the subject, we cannot hope to give a comprehensive overview; see especially [54] for more details. We intend to concentrate here on one aspect of the theory, namely the connection with the Gibbs' variational principle of statistical mechanics. To get the flavor of the results, we begin with a theorem of Kac [138] which motivated the theory:

Theorem 19.1 Let $V \geq 0$ be in $L^1_{\text{loc}}(\mathbb{R})$ and let

$$L_0 = -\tfrac{1}{2}(d^2/dx^2 - x^2 + 1)$$

as usual. Then

$$-\lim_{t\to\infty} \frac{1}{t} \ln\left[\int Dq \exp\left(-\int_0^t V(q(s))\, ds\right)\right]$$
$$= \inf\left\{\sigma(f) + i(f) \mid f \in C_0^\infty(\mathbb{R});\ \int f^2\, dx = 1\right\} \tag{19.1}$$

with

$$\sigma(f) = (f, L_0 f) \qquad \text{and} \qquad i(f) = \int V(x) |f(x)|^2\, dx$$

Proof The right-hand side of (19.1) is the bottom of the spectrum of $L \equiv L_0 + V$ by the Rayleigh–Ritz principle. The left-hand side is

$$-\lim_{t\to\infty} t^{-1} \ln(\Omega_0, e^{-tL}\Omega_0)$$

by the Feynman–Kac formula, so (19.1) follows from (1.19). ∎

The Donsker–Varadhan theory gives suitable limits as $t \to \infty$ of path integrals as an infinimum of a variational object. An example of the theory is the following.

Theorem 19.2 Let W be a bounded positive continuous symmetric function on $\mathbb{R}^2$. Then

$$-\lim_{t\to\infty} t^{-1} \ln\left[\int Dq \exp\left(-t^{-1} \int_0^t ds \int_0^t dv\, W(q(s), q(v))\right)\right]$$
$$= \inf\left\{\sigma(f) + i_2(f) \mid f \in C_0^\infty(\mathbb{R});\ \int f^2\, dx = 1\right\} \tag{19.2}$$

where $i_2(f) = \int W(x, y) |f(x)|^2 |f(y)|^2\, dx\, dy$.

The proof of (19.2) is not nearly so easy as that of (19.1) because there is no Feynman–Kac formula for rewriting the left-hand side of (19.2). Below, we will settle for proving one-half of (19.2); namely,

$$\varliminf_{t\to\infty} t^{-1} \ln\left[\int Dq \exp\left(-t^{-1}\int_0^t ds \int_0^t dv\, W(q(s), q(v))\right)\right] \geq -[\sigma(f) + i_2(f)] \tag{19.3}$$

for every $f \in C_0^\infty$ with $\int f^2\, dx = 1$. We want to describe an analogy between (19.2) and the Gibbs' variational principle of statistical mechanics. The connection of (19.1) and this principle has already been noted by Guerra *et. al.* [120] who consider extensions to $P(\phi)_2$ Euclidean field theories (see [120, 121, 96]). This work is partial motivation for our presentation here.

We begin by describing the Gibbs' principle for classical lattice gases following Ruelle [225] (see also [226, 227, 133, 119]). Let $\mathbb{Z}^\nu$ denote the lattice of points in $\mathbb{R}^\nu$ with integral coefficients. Let Λ be a "box" $\mathbb{Z}^\nu \cap [0, L)^\nu$ and let $|\Lambda| = L^\nu$ be the number of points in Λ. Let J be a function on $\mathbb{Z}^\nu$ which has bounded support (we restrict ourselves to pair interactions with finite range; were we to consider connections with tangent functions to the pressure [225, 133], this would be a serious restriction). Define

$$Z_\Lambda(J) = \sum_{\sigma_i = \pm 1,\, i \in \Lambda} e^{-U_\Lambda(J)(\sigma)}$$

$$U_\Lambda(J)(\sigma) = \sum_{i,\, j \in \Lambda} \sigma_i \sigma_j J(i - j)$$

$$p_\Lambda(J) = |\Lambda|^{-1} \ln Z_\Lambda(J)$$

Then, we will show that

(a) $$\lim_{\Lambda\to\infty} p_\Lambda(J) \equiv p(J)$$

exists for $\Lambda \to \infty$ in the sense of taking boxes all of whose sides go to infinity. Let ρ be a probability measure on $\{-1, 1\}^{\mathbb{Z}^\nu}$. If we restrict ρ to the variables $\{\sigma_i\}_{i\in\Lambda}$, we obtain a measure ρ_Λ on $\{-1, 1\}^\Lambda$. ρ is called **translation invariant** if ρ_Λ and $\rho_{\Lambda+a}$ are identical under the natural identification of $\{-1, 1\}^\Lambda$ and $\{-1, 1\}^{\Lambda+a}$. One defines

$$S_\Lambda(\rho) = -\sum_{\sigma_i = \pm 1,\, i\in\Lambda} \rho_\Lambda(\sigma)[\ln \rho_\Lambda(\sigma)]$$

where $\rho_\Lambda(\sigma)$ is the ρ_Λ measure of $\{\sigma\}$. We will also show that

(b) $$\lim_{\Lambda\to\infty} |\Lambda|^{-1} S_\Lambda(\rho) \equiv s(\rho)$$

exists. Finally, we will show that

(c) $$\lim_{\Lambda \to \infty} |\Lambda|^{-1} \int U_\Lambda \, d\rho \equiv U(J, \rho)$$

exists. The **Gibbs' variational inequality** asserts that

(d) $$p(J) \geq s(\rho) - U(J, \rho)$$

for any translation invariant ρ. The full **Gibbs' principle** asserts that

(e) $$p(J) = \sup_\rho (s(\rho) - U(J, \rho))$$

The above describes the interaction of a family of spins, p is the free energy (the letter p comes from the fact that in a lattice gas language, it is a pressure) and s an entropy per unit volume.

Proof of (*a*) For simplicity take Λ to run through cubes of side l. Fix l_0 and let $l = nl_0 + a$ with $0 \leq a < l_0$. Decompose the l^ν cube into n^ν cubes of volume l_0^ν and a leftover "strip" of volume $l^\nu - (nl_0)^\nu$. Write

$$U_{[l^\nu]} = \sum_{i=1}^{n^\nu} U_{[l_0^\nu]}(i) + R \tag{19.4}$$

where $U_{[l^\nu]}$ is the interaction within the l^ν cube and $U_{[l_0^\nu]}(i)$ is the interaction within the ith $[l_0^\nu]$ cube. R is what is left over, i.e., interaction between the cubes and within the strip. Now, suppose that J has range b; i.e., $J(n) = 0$ if $|n| \geq b$. Then R only involves spins in the strip and in regions within b of the boundary of a l_0^ν cube. Thus,

$$|R(\sigma)| \leq \{[l^\nu - (nl_0)^\nu] + n^\nu[l_0^\nu - (l_0 - b)^\nu]\}|J|_1$$

with $|J|_1 = \sum_n |J(n)|$, so

$$|R(\sigma)| \leq n^\nu |J|_1 \{[l_0^\nu - (l_0 - b)^\nu] + l_0^\nu[(1 + (1/n))^\nu - 1]\}$$

From (19.4) we have that:

$$\exp(|l|^\nu p_l) \leq \exp(|R|_\infty)\exp(n^\nu |l_0|^\nu p_{l_0})$$

so that

$$\left[\frac{l}{nl_0}\right]^\nu p_l \leq p_{l_0} + |J|_1 \left\{\left[1 - \left(1 - \frac{b}{l_0}\right)^\nu\right] + \left[\left(1 + \frac{1}{n}\right)^\nu - 1\right]\right\}$$

Taking $l \to \infty$, we see that

$$\overline{\lim}\, p_l \leq p_{l_0} + |J|_1 \left[1 - \left(1 - \frac{b}{l_0}\right)^\nu\right]$$

Now taking $l_0 \to \infty$,

$$\overline{\lim}\, p_l \le \underline{\lim}\, p_{l_0}$$

i.e., the limit exists. Finiteness of the limit follows from $|p_\Lambda| \le |J|_1$. ∎

Remark The above proof has the advantage of extending to rather general interactions. For the case at hand, there is a simpler proof: Write $U_{\Lambda_1 \cup \Lambda_2} = U_{\Lambda_1} + U_{\Lambda_2} + I_{\Lambda_1;\Lambda_2}$. Then $\sum \exp(-U_{\Lambda_1} - U_{\Lambda_2}) I_{\Lambda_1;\Lambda_2} = 0$, since $I_{\Lambda_1;\Lambda_2}$ is linear in the spins in Λ_1 and we can separately invert the spins in Λ_1. Thus, by Jensen's inequality:

$$Z_{\Lambda_1 \cup \Lambda_2} \ge Z_{\Lambda_1} Z_{\Lambda_2} e^{-0} = Z_{\Lambda_1} Z_{\Lambda_2}$$

so $\ln Z_\Lambda$ is superadditive implying existence of the limit.

Proof of (*b*) We first claim that for disjoint Λ_1 and Λ_2:

$$S_{\Lambda_1 \cup \Lambda_2}(\rho) \le S_{\Lambda_1}(\rho) + S_{\Lambda_2}(\rho) \tag{19.5}$$

On $\{-1, 1\}^{\Lambda_1 \cup \Lambda_2}$ define functions ρ_{12}, ρ_1, and ρ_2 by $\rho_{12} = \rho_{\Lambda_1 \cup \Lambda_2}$ and $\rho_1(\sigma_1, \sigma_2) = \rho_{\Lambda_1}(\sigma_1)$, $\rho_2(\sigma_1, \sigma_2) = \rho_{\Lambda_2}(\sigma_2)$. Then, since

$$\rho_1(\sigma_1) = \sum_{\sigma_2} \rho_{12}(\sigma_1, \sigma_2)$$

we have that

$$\begin{aligned} S_{\Lambda_1 \cup \Lambda_2} - S_{\Lambda_1} - S_{\Lambda_2} &= \sum_\sigma \rho_{12}[\ln \rho_1 + \ln \rho_2 - \ln \rho_{12}] \\ &= \sum_\sigma \rho_{12} \ln\left[\frac{\rho_1 \rho_2}{\rho_{12}}\right] \\ &\le \ln\left(\sum_\sigma \rho_{12}\left[\frac{\rho_1 \rho_2}{\rho_{12}}\right]\right) \\ &= \ln 1 = 0 \end{aligned}$$

proving (19.5). In the above, the inequality comes from Jensen's inequality if we note that ρ_{12} is a probability measure and that $-\ln$ is convex. We have also used the fact that $\rho_1 \rho_2$ is a probability density on $\Lambda_1 \cup \Lambda_2$.

From (19.5) and a standard argument $\lim |\Lambda|^{-1} S_\Lambda(\rho)$ exists and equals $\inf |\Lambda|^{-1} S_\Lambda(\rho)$. The finiteness of this infimum comes from the fact that $S_\Lambda \ge 0$. ∎

Proof of (*c*) Since ρ is translation invariant,

$$\int U_\Lambda \, d\rho = \sum_n J(n) N(n, \Lambda) \int \sigma_0 \sigma_n \, d\rho$$

where $N(n, \Lambda) = \#\{(i, j) \mid i - j = n;\ i, j \in \Lambda\}$. Since it is easy to see that $|\Lambda|^{-1}N(n, \Lambda) \to 1$ as $\Lambda \to \infty$ for each n,

$$|\Lambda|^{-1} \int U_\Lambda \, d\rho \to \sum_n J(n) \int \sigma_0 \sigma_n \, d\rho \quad \blacksquare$$

Proof of (d) By Jensen's inequality and the fact that $\sum \rho_\Lambda(\sigma) = 1$:

$$\sum_{\substack{\sigma_i = \pm 1 \\ i \in \Lambda}} e^{-U_\Lambda(\sigma)} = \sum e^{-U_\Lambda} e^{-\ln \rho_\Lambda} \rho_\Lambda$$

$$\geq \exp\left(S_\Lambda(\rho) - \int U_\Lambda \, d\rho\right)$$

Taking logs, dividing by $|\Lambda|^{-1}$, and taking $\Lambda \to \infty$, (d) results. $\blacksquare$

Proof of (e) In order to understand both the proof and the content of the Gibbs' principle, we note that to get equality in the argument in (d), we must have

$$\ln \rho_\Lambda = -U_\Lambda + \text{const}$$

since exp is strictly convex. Thus, the finite volume analog of the Gibbs' principle picks out the finite volume Gibbs' distribution

$$e^{-U_\Lambda(\sigma)}/Z_\Lambda$$

We will therefore try to take a measure, ρ, as much like this Gibbs measure as possible. Fix Λ_0 and cover $\mathbb{Z}^\nu$ with nonoverlapping copies of $\Lambda_0 : \Lambda_\alpha$ with $\alpha \in \mathbb{Z}^\nu$. Let

$$\rho_0 = \prod_{\alpha \in \mathbb{Z}^\nu} [\exp(-U_{\Lambda_\alpha})/Z_{\Lambda_0}]$$

ρ_0 is not translation invariant, but it is periodic. If τ_i is translation by i units, then

$$\rho_1 = |\Lambda_0|^{-1} \sum_{i \in \Lambda_0} \tau_i \rho_0$$

is translation invariant. We must compute $s(\rho_1)$ and $U(J, \rho_1)$. Since $-x \ln x$ is concave,

$$S_\Lambda(\rho_1) \geq |\Lambda_0|^{-1} \sum_{i \in \Lambda_0} S_\Lambda(\tau_i \rho_0)$$

Since $\ln x$ is monotone:

$$S_\Lambda(\rho_1) \leq |\Lambda_0|^{-1} \sum_{i \in \Lambda_0} S_\Lambda(\tau_i \rho_0) + \ln |\Lambda_0|$$

Moreover, it is easy to see that

$$\lim_{\Lambda\to\infty} |\Lambda|^{-1} S_\Lambda(\tau_i \rho_0) = |\Lambda_0|^{-1} S_{\Lambda_0}(\rho_0)$$

The last three relations imply that

$$\begin{aligned} s(\rho_1) &= |\Lambda_0|^{-1} S_{\Lambda_0}(\rho_0) \\ &= |\Lambda|_0^{-1} \ln Z_{\Lambda_0} + \sum_{\substack{\sigma_i = \pm 1 \\ i \in \Lambda_0}} U_{\Lambda_0}(\sigma) \exp(-U_{\Lambda_0}(\sigma)) Z_{\Lambda_0}^{-1} \end{aligned}$$

Using the fact that ρ_0 is invariant under reversal of the σ's in any Λ_α, we see that this last term is $U(J, \rho_0)$ and so, by linearity, it equals $U(J, \rho_1)$. Thus,

$$s(\rho_1) - U(J, \rho_1) = |\Lambda_0|^{-1} \ln Z_{\Lambda_0}$$

It follows that

$$\sup_\rho [s(\rho) - U(J, \rho)] \geq p(J)$$

which, given (d), proves (e). ▮

Before turning to the path integral case, it is useful to consider changes of s, p, U under change of "reference measure," i.e., instead of summing over $\sigma_i = \pm 1, i \in \Lambda$, suppose we pick a fixed probability measure $\rho^{(0)}$ on $\{-1, 1\}^{\mathbb{Z}^\nu}$ and let

$$\tilde{p}(J) = \lim_{\Lambda\to\infty} |\Lambda|^{-1} \int e^{-U_\Lambda(J)} \, d\rho^{(0)}$$

$$\tilde{s}(J) = -\lim_{\Lambda\to\infty} |\Lambda|^{-1} \int \ln[\rho_\Lambda / \rho_\Lambda^{(0)}] \, d\rho_\Lambda$$

We have in mind that $\rho^{(0)}$ is some kind of limit of $e^{-U_\Lambda(J_0)}/Z_\Lambda(J_0)$. *Formally* one expects that in this case:

$$\tilde{p}(J) = p(J + J_0) - p(J_0)$$

$$\tilde{s}(\rho) = s(\rho) - U(J_0, \rho) - p(J_0)$$

and it should be possible to prove this using the methods of [96]. If these formulas hold, then the Gibbs' principle is equivalent to

$$\tilde{p}(J) = \sup_\rho [\tilde{s}(\rho) - U(J, \rho)]$$

The point of this reformulation is that the reference measure in the path integral case is Dq which, in lattice approximation, is not a product measure, but a nontrivial $\rho^{(0)}$.

All of the above with the exception of the change of reference measure is fairly standard and is further discussed in [226, 227, 133]. One aspect that we have not considered is the connection with the tangent structure of the pressure [133, 226, 227], a subject which could well be relevant to the Donsker–Varadhan theory.

Let us begin by studying (following [120]) the relation between (19.1) and the Gibbs' principle. Fix V and consider the $P(\phi)_1$-process associated to $L_0 + V$ as a measure $d\rho_V$ on $C(\mathbb{R})$. Then (6.10) asserts that $d\rho_V \upharpoonright \sum_{[a,b]}$ (which indicates the σ-algebra generated by $\{q(s) | a \le s \le b\}$) is Dq absolutely continuous with Radon–Nikodym derivative

$$\rho_{[a,b]} = g(q(a))g(q(b))e^{-U_{(a,b)}}e^{+E(V)(b-a)} \tag{19.6a}$$

where $h_V = g\Omega_0$ is the ground state of $L_0 + V$, $E(V) = \inf \operatorname{spec}(L_0 + V) = -\lim_{b\to\infty} b^{-1} \ln(\int e^{-U_{(0,b)}}\, Dq)$, and $U_{(a,b)} = \int_a^b V(q(s))\, ds$. Clearly $-E(V)$ is the analog of the pressure. Let us compute the analog of the entropy

$$s(\rho_V) = -\lim_{b\to\infty} b^{-1} \int (\ln \rho_{[0,b]})\rho_{[0,b]}\, Dq$$

Using the above formula for $\rho_{[a,b]}$, the integral in question is easily seen to be

$$bE(V) - b(h_V, Vh_V) + 2(h_V, (\ln g)h_V)$$

where the inner products are in $L^2(\mathbb{R})$. Thus,

$$s(\rho_V) = -E(V) + (h_V, Vh_V) = -(h_V, L_0 h_V)$$

so the entropy per unit volume is the "free Hamiltonian expectation." We claim that

$$-E(V) = \sup_W (s(\rho_W) - U(V, \rho_W)) \tag{19.6b}$$

with

$$U(V, \rho_W) = \lim_{b\to\infty} b^{-1} \int U_{[0,b]}(V)\, d\rho_W = (h_W, Vh_W)$$

For (19.6b) is just the assertion that

$$E(V) = \inf_W (h_W, L_0 h_W) + (h_W, Vh_W)$$

i.e., (19.1)!

One can extend (19.6b) by allowing more general measures ρ on $C(\mathbb{R})$ than the ρ_W. All that is required [120] is translation invariance, absolute continuity of $d\rho \upharpoonright \Sigma_{[0,b]}$, and a weak growth restriction on the L^p-norm of

$d\rho \restriction \Sigma_{[0,b]}/Dq$ ("weak-temperedness"). In that case, one can prove the existence of $s(\rho)$, $U(V, \rho)$ and the inequality in (19.6b). The proofs closely follow those of (a)–(d) above.

[Parenthetically, we note that one direction for extending (19.6b) is to the $P(\phi)_2$ field theory of Section 24: the Gibbs' inequality is proven in [120], the Gibbs' principle in [121], and in [96] it is shown that the supremum is actually realized for the ρ's connected with the field theory.]

The formal connection of Theorem 19.2 and the variational principle should be clear by this point. That this connection extends to some proofs is illustrated by the following, patterned after the proof of (d) above.

Proof of (19.3) We prove the result for f's in $\mathscr{S}(\mathbb{R})$ which are strictly positive. By a limiting argument, the result extends to f's which are positive, continuous, and piecewise C^1. Since $i_2(f) = i_2(|f|)$ and $\sigma(f) \geq \sigma(|f|)$ (by "Kato's inequality"; see [153] and (1.1)); (19.3) then follows. Let $V = \frac{1}{2}[f''f^{-1} - x^2 + 1]$. The $P(\phi)_1$-process, ρ, for $L_0 + V$ has Radon–Nikodym derivative (19.6a) with $g = f\Omega_0^{-1}$. Fix t and let

$$U_t = t^{-1} \int_0^t ds \int_0^t dv\, W(q(s), q(v))$$

Then, by Jensen's inequality

$$\begin{aligned} \int Dq\, e^{-U_t} &= \int \rho_{[0,t]} e^{-U_t} e^{-\ln \rho}\, Dq \\ &\geq \exp\left(S_t(\rho) - \int U_t\, d\rho\right) \end{aligned} \tag{19.7}$$

As above,

$$t^{-1} S_t(\rho) \to -\sigma(f)$$

Let $\mu_t(x, y)$ be the joint distribution of $q(s)$, $q(s + t)$ with respect to $d\rho$. Then

$$t^{-1} \int U_t\, d\rho = t^{-2} \int_0^t ds \int_0^t dv \left[\int W(x, y)\, d\mu_{|s-v|}(x, y)\right]$$

By the ergodicity of $d\rho$ (see Example 2 of Section 3), one has that

$$\lim_{s \to \infty} \int W(x, y)\, d\mu_s(x, y) = \int W(x, y) |f(x)|^2 |f(y)|^2\, dx\, dy$$

so that

$$t^{-1} \int U_t\, d\rho \to i_2(f)$$

Thus (19.7) implies (19.3).

Remarks 1. Although it appears different, the above proof is (approximately) identical to that in [54]. In [54], Donsker–Varadhan do not talk about "$P(\phi)_1$-processes" but rather of "drift-processes." By the discussion at the end of Section 16, the two notions are identical.

2. (suggested by Varadhan) There are two special aspects of the Donsker–Varadhan theory that are not present in the general Gibbs' variational framework: First, in the Donsker–Varadhan theory, the σ-functional is a functional of a density (namely $|f|^2$) on $\mathbb{R}$, i.e., of the state at a single time, while in the Gibbs' framework s is a functional of the state for all times. Secondly, the minimizing state in the case of the Donsker–Varadhan theory above is a $P(\phi)_2$-process, i.e., a Markov process. If we consider the special family of one-dimensional lattice systems with interaction $J + J_0$ where J_0 is a nearest neighbor coupling and J is "mean field interaction," we have a lattice analog of the kind of interactions treated in Theorem 19.2. We notice that for ergodic states ρ, one expects that $U(J, \rho)$ is only a function i_J of $\rho_{\{0\}}$, the restriction of ρ to functions of σ_0. Thus

$$\begin{aligned}\sup_\rho[\tilde{s}(\rho) - U(J, \rho)] &= \sup_\alpha\left(\left[\sup_{\{\rho \mid \rho_{\{0\}} = \alpha\}} \tilde{s}(\alpha)\right] - i_J(\alpha)\right)\\ &= \sup_\alpha(-I(\alpha) - i_J(\alpha))\end{aligned}$$

where α denotes a probability distribution on ± 1 and

$$I(\alpha) = \inf_{\{\rho \mid \rho_{\{0\}} = \alpha\}} (-\tilde{s}(\alpha)).$$

This calculation "explains" the first special feature of the Donsker–Varadhan theory noted above. Next suppose that ρ is some state which gives equality in $\tilde{p}(J) \geq \tilde{s}(\rho) - U(J, \rho)$. Let β be $\rho_{\{0,1\}}$, the measure on $\{-1, 1\}^2$ obtained by restricting to functions of σ_0 and σ_1. Let us suppose that $\beta(s_0, s_1) > 0$ for all four values of s_0 and s_1. By an argument below, there is a Markov process ρ' with $\rho'_{\{0,1\}} = \beta$ and a "nearest neighbor interaction", J' so that $p(J') = s(\rho') - U(J', \rho')$. Since $U(J, \rho)$ is only a function of $\rho_{\{0\}}$, it is only a function of $\rho_{\{0,1\}}$ and so $U(J, \rho) = U(J, \rho')$. $U(J', \rho)$ is only a function of $\rho_{\{0,1\}}$ since J' consists only of "nearest neighbor interactions"; similarly $U(J_0, \rho)$ is only a function of $\rho_{\{0,1\}}$. Thus from the Gibbs' inequality

$$s(\rho') - U(J', \rho') \geq s(\rho) - U(J', \rho)$$

and the equalities (definition of $\tilde{s}$)

$$\tilde{s}(\rho) = s(\rho) - U(J_0, \rho) - p(J_0); \qquad \tilde{s}(\rho') = s(\rho') - U(J_0, \rho') - p(J_0)$$

we conclude that

$$\tilde{s}(\rho') \geq \tilde{s}(\rho)$$

and thus ρ' also gives equality in the Gibbs' principle. This explains the second special feature of the Donsker–Varadhan theory noted above.

In the above we wanted to construct an interaction J' and Markovian state ρ' once we are given β. This can be done as follows: Given J', we consider a matrix $J(x, y) = e^{-J'(x, y)}$. The basic transfer matrix formalism of statistical mechanics [120] shows that a state ρ' giving equality in the Gibbs' principle has

$$\rho'_{\{0, \dots, n\}} = \phi(s_0) \prod_{i=1}^{n} T(s_{i-1}, s_i)\phi(s_n) \,/\, \alpha^n \sum \phi(s_0)^2$$

where ϕ is the (necessarily unique up to constant) eigenvector of T with positive components and α is the corresponding eigenvalue. This ρ' is Markovian. Thus given β, we seek J' with

$$\beta(x, y) = \phi(x)\phi(y)e^{-J'(x, y)}/\alpha \sum \phi(x)^2$$

This is easy; we let $\phi(x) \equiv (\sum_y \beta(x, y))^{1/2}$ and $-J'(x, y) = \ln \beta(x, y) - \ln \phi(x) - \ln \phi(y)$. Then ϕ is an eigenvector of T with eigenvalue one and $\sum \phi(x)^2 = 1$ and β is the associated $\rho'_{\{0, 1\}}$.

We close this section with a somewhat imprecise description of the general framework in which Donsker and Varadhan imbed Theorems 19.1 and 19.2.

The oscillator **local time** is the probability measure $L_t(q, \cdot)$ defined for each path q by

$$L_t(q, A) = t^{-1} |\{s \mid q(s) \in A\}|$$

where, as usual, $|\cdots|$ indicates Lebesgue measure. Notice that

$$\int_0^t V(q(s))\, ds = t \int V(x) L_t(q, dx)$$

$$\frac{1}{t} \int_0^t \int_0^t W(q(s), q(v))\, ds\, dv = t \int W(x, y) L_t(q, dx) L_t(q, dy)$$

Thus, Donsker–Varadhan consider general maps Φ from $\mathcal{M}$, the probability measures on $\mathbb{R}$ to $\mathbb{R}$. The general form of their variational principle is

$$\lim_{t \to \infty} \frac{1}{t} \ln \int Dq\, e^{-t\Phi(L_t(q, \cdot))} = -\inf_{\mu \in \mathcal{M}} [I(\mu) + \Phi(\mu)] \tag{19.8}$$

where

$$\begin{aligned} I(\mu) &= \infty, && \text{if } \mu \text{ is not absolutely continuous with respect to } dx \\ &= (f, L_0 f), && \text{if } \mu = f^2\, dx \end{aligned}$$

Of course, there are technical hypotheses on Φ for (19.8) to hold, and more general "reference" processes than Dq are considered. Donsker–Varadhan express a related result which is almost equivalent to (19.8); namely, for suitable sets C in $\mathscr{M}$:

$$E(L_t(q, \cdot) \in C) \sim \exp\left(-t \inf_{\mu \in C} I(\mu)\right) \tag{19.9}$$

Let us mention two explicit applications of the machinery, especially (19.9) extended to much more general Markov processes than Dq.

(a) If one takes independent copies of a fixed Gaussian process and uses the fact that sums of independent Gaussian variables are Gaussian, (19.9) gives information on the distribution of the Gaussian process in suitable asymptotic regions. In this way, Donsker–Varadhan [57] recover the result of Fernique–Marcus–Shepp quoted in the remark following Lemma 18.7 and also some versions of Theorem 18.1.

(b) In Section 22, we will discuss the Wiener sausage $W_\delta(t)$ which is a set-valued random function of the Brownian path, $\mathbf{b}(s)$, defined by

$$W_\delta(t) = \{\mathbf{x} \mid \text{dist}(\mathbf{x}, \mathbf{b}(s)) \le \delta, \text{ some } s \in [0, t]\}.$$

There we will prove that the volume of $W_1(t)$ obeys

$$t^{-1} |W_1(t)| \to 2\pi \tag{19.10}$$

(in $\nu = 3$ dimensions) as $t \to \infty$ for almost every path $\mathbf{b}$. This suggests that $E(e^{-\alpha|W_1(t)|}) \sim e^{-2\pi\alpha t}$ as $t \to \infty$. In fact this is false! Donsker–Varadhan prove that [58]

$$\lim_{t \to \infty} t^{-3/5} \ln E(e^{-\alpha|W_1(t)|}) = \beta \tag{19.11}$$

for an explicit constant $\beta \in (0, \infty)$. (Without the explicit value of β, this result had been conjectured by Kac and Luttinger [144] on physical grounds.) (19.11) says that $E(e^{-\alpha|W_1(t)|})$ goes to zero more slowly than one would expect on the basis of (19.10). The reason is that $|W_1(t)|$ is large on sets of measure going to zero as $t \to \infty$ but not as fast as would be necessary for the $e^{-2\pi\alpha t}$ behavior. Let us sketch the first step in the argument of [58] which shows the relation to Theorems 19.1 and 19.2. As we will show in Section 22, the scaling covariance of $b(s)$ implies that in ν-dimensions

$$|W_1(t)| \doteq \lambda^{\nu/2} |W_{1/\sqrt{\lambda}}(t\lambda^{-1})|$$

Choosing $\lambda = t^{2/\nu+2}$ (chosen so that $\lambda^{\nu/2} = t\lambda^{-1}$) and letting $s = t^{\nu/\nu+2}$, we see that

$$\lim_{t\to\infty} t^{-\nu/\nu+2} \log E(\exp(-\alpha|W_1(t)|)) = \lim_{s\to\infty} s^{-1} \log E(\exp(-\alpha s|W_{s^{-1/\nu}}(s)|))$$

Except for the "mild" $s^{-1/\nu}$ dependence of δ, we are precisely in the situation of (19.8) since $|W|$ is only a function of the image of b and thus of the support of $L_t(b, \cdot)$. See [58] for the remainder of the argument.

VII

Other Topics

20. Perturbation Theory for the Ground State Energy

Perturbation theory for discrete eigenvalues goes back to Rayleigh's classic *The Theory of Sound* [213]; it was rediscovered by Schrödinger in his famous series [232] written at the dawn of the "new" quantum theory. These Rayleigh–Schrödinger series were placed on a firm mathematical footing by Rellich [218] with further developments by Kato [149] and Sz-Nagy [271]. All these authors dealt primarily with the regular case where the series are convergent. A typical case where the series are not convergent is $L(\beta) = L_0 + \beta x^4$ where as usual $L_0 = -\frac{1}{2}(d^2/dx^2) + \frac{1}{2}x^2 - \frac{1}{2}$. Indeed, the Bender–Wu formula (18.8) says that the series is divergent. One can also prove this divergence directly (see [7] and below). In this case, Kato [149] and Titchmarsh [277] proved that the Rayleigh–Schrödinger series were asymptotic. See [88, 152, 217, 219] for reviews of these results.

In the forties and early fifties, Tomonaga, Schwinger, Feynman, and Dyson (see [233] for a collection of relevant papers) developed some systematic series for certain objects in quantum field theory. One of special interest is the "energy per unit volume." If one specializes to one space–time dimension (i.e., *zero* space dimensions), the Hamiltonian of the quantum field is just $L(\beta)$ and its ground state energy is just this energy per unit volume. While the coefficients of this series can be shown to be equal to those of the Rayleigh–Schrödinger series for this case, they appear quite different and more compact. One goal in this section is to present the Feynman series for $E(\beta) \equiv \inf \operatorname{spec}(L(\beta))$ and prove that it is asymptotic. Our arguments below are essentially a specialization of those of Dimock [52] from two space–time dimensions to one.

Even though the perturbation series for $E(\beta)$ is divergent, the function can be recovered from the series by a summability method known as Borel summability. Using the Rayleigh–Schrödinger theory, this was first proven by Graffi *et al.* [113]. By working a little harder on the estimates below, one could prove Borel summability using the path integral realizations. Indeed, this is just the specialization from two space–time dimensions to one of some work of Eckmann *et al.* [72].

Throughout this section $E(\beta)$ denotes the lowest eigenvalue of $L(\beta) = L_0 + \beta x^4$. Our goal is first to establish that $E(\beta)$ has an asymptotic series $\sum a_n \beta^n$ as $\beta \downarrow 0$ and then to obtain explicit expressions for the a_n. Recall first the following definition.

Definition A function $f(\beta)$ on $(0, a)$ $(a > 0)$ is said to have $\sum b_n \beta^n$ as **asymptotic series, as** $\beta \downarrow 0$, written

$$f(\beta) \sim \sum b_n \beta^n \qquad (\beta \downarrow 0)$$

if and only if, for each N,

$$\lim_{\beta \downarrow 0} \left[f(\beta) - \sum_{n=0}^{N} b_n \beta^n \right] \Big/ \beta^N = 0$$

A function f has at most one asymptotic series but the existence of functions like $\exp(-\beta^{-1})$ with zero asymptotic series implies that two distinct functions may have the same asymptotic series.

Lemma 20.1

(a) If $f(\beta)$ is C^∞ on $(0, a)$ and for each n, $b_n = (n!)^{-1} \lim_{\beta \downarrow 0} d^n f/d\beta^n$ exists, then $\sum b_n \beta^n$ is an asymptotic series for f. In this case we say that f is C^∞ on $[0, a)$.

(b) If $f_N(\beta)$ is a sequence of C^∞-functions on $[0, a)$ and $f(\beta) = \lim_{N\to\infty} f_N(\beta)$ exists and for each n, there is a C_n with

$$\left| \frac{d^n f_N}{d\beta^n}(\beta) \right| \le C_n, \qquad 0 < \beta < a, \quad \text{all } N$$

then f is C^∞ on $[0, a)$ and

$$\frac{d^n f}{d\beta^n}(\beta) = \lim_{N\to\infty} \frac{d^n f_N}{d\beta^n}(\beta), \qquad 0 \le \beta < a \tag{20.1}$$

(including $\beta = 0$*).*

Proof (a) Taking $\beta_0 \downarrow 0$ in Taylor's theorem with remainder, one sees that

$$f(\beta) = \sum_{n=0}^{N} b_n \beta^n + \int_0^{\beta} dt_1 \int_0^{t_1} dt_2 \cdots \int_0^{t_N} dt_{N+1} \frac{d^{N+1}f}{d\beta^{N+1}}(t_{N+1})$$

Since $d^{N+1}f/d\beta^{N+1}$ has a limit as $\beta \downarrow 0$, the last multiple integral is easily seen to be bounded by $C_N \beta^{N+1}$.

(b) Suppose that we know the limit on the right-hand side of (20.1) exists. Then by using the dominated convergence theorem in Taylor's theorem with remainder, one easily sees that f is C^∞ on $[0, a)$ and that (20.1) holds. By induction, we only need to show that if $|d^2 f_N/d\beta^2| \le C_2$ and $f_N(\beta)$ converges, then $df_N/d\beta$ converges. Fix $\beta > \beta_0$ and let $A_N = df_N/d\beta\,(\beta_0)$. Then

$$\begin{aligned} |[f_N(\beta) - f_M(\beta)] - [f_N(\beta_0) - f_M(\beta_0)] \\ - (\beta - \beta_0)[A_N - A_M]| \le C_2(\beta - \beta_0)^2 \end{aligned}$$

by Taylor's theorem. Taking $N, M \to \infty$ we see that

$$|\overline{\lim}\, A_N - \underline{\lim}\, A_N| \le C_2 |\beta - \beta_0|$$

so taking $\beta \to \beta_0$ we see that A_N converges. ∎

The basic strategy of proof that $E(\beta)$ has an asymptotic series is the following: By (1.9) and the Feynman–Kac formula:

$$E(\beta) = \lim_{N \to \infty} E_N(\beta) \tag{20.2}$$

$$E_N(\beta) = -(2N)^{-1} \ln \int \exp\left(-\beta \int_{-N}^{N} q^4(s)\, ds\right) Dq \tag{20.3}$$

Each $E_N(\beta)$ is C^∞ on $[0, \infty)$ so we will prove a bound

$$\left| \frac{d^n E_N}{d\beta^n}(\beta) \right| \le C_n, \qquad 0 \le \beta \le 1 \tag{20.4}$$

and thus conclude that $E(\beta)$ is C^∞ on $[0, 1)$ by Lemma 20.1(b) and therefore that $E(\beta)$ has an asymptotic series by Lemma 20.1(a).

To prove (20.4), we use a compact expression for $d^n E_N/d\beta^n$ in terms of the Ursell functions of (12.14). Let $\langle X_1, \ldots, X_n \rangle_{T, \beta, N}$ denote the Ursell function $u_n(X_1, \ldots, X_n)$ with respect to the measure $\exp(-\beta \int_{-N}^{N} q^4(s)\, ds)\, Dq$/Normalization. Then, by (12.14) and the linearity of u_n:

$$\begin{aligned} \frac{d^n E_N}{d\beta^n} &= -(2N)^{-1}(-1)^n \int_{-N}^{N} ds_1 \cdots \int_{-N}^{N} ds_n \langle q^4(s_1), \ldots, q^4(s_n) \rangle_{T, \beta, N} \\ &= (-1)^{n+1} n! (2N)^{-1} \int_{-N < s_1 < \cdots < s_n < N} \langle q^4(s_1), \ldots, q^4(s_n) \rangle_{T, \beta, N}\, d^n s \end{aligned} \tag{20.5}$$

Now we claim that for $s_1 < \cdots < s_n$ and $0 \le \beta \le 1$,

$$\langle q^4(s_1), \ldots, q^4(s_n)\rangle_{T,\beta,N} \le Q_n \exp(-D_n|s_n - s_1|) \tag{20.6}$$

From this, one easily sees that the integral in (20.5) is bounded as N varies so that (20.4) holds. Clearly (20.6) follows from

$$\langle q^4(s_1), \ldots, q^4(s_n)\rangle_{T,\beta,N} \le Q_n \exp(-D'|s_j - s_{j-1}|) \tag{20.7}$$

all j, for take $D_n = D'/n$ and $|s_j - s_{j-1}|$ maximal. [At this point we are being very crude and losing all hope of a careful estimate on the growth of C_n with n; more careful analysis ([72]) shows that one can take D_n rather than D' independent of n. Our n^{-1} dependence of D_n leads to an extra factor of n^n in estimating (20.5).]

To prove (20.7), we use Cartier's formula, (12.11), which says that

$$\langle q^4(s_1), \ldots, q^4(s_n)\rangle_{T,\beta,N} = \langle \tilde{X}_1(s_1) \cdots \tilde{X}_n(s_n)\rangle_{\beta,N,\sim}$$

where $\langle\cdot\rangle_{\beta,N,\sim}$ is an expectation with respect to n independent copies of $\langle\cdot\rangle_{\beta,N}$ and $[\omega = \exp(2\pi i/n)]$

$$\tilde{X}(s_i) = \sum_{j=0}^{n-1} \omega^j q_j(s_i)^4$$

Now, let $\tilde{L}(\beta) = L_1(\beta) + \cdots + L_n(\beta)$ on $L^2(\mathbb{R}^n)$ where L_i denotes L as an operator on the ith variable and let $\hat{L}(\beta) = \tilde{L}(\beta) - nE(\beta)$. Let

$$\tilde{\Omega}_0 = \prod_{i=1}^{n} \Omega_0(x_i)$$

Finally, let $\tilde{X} = \sum_{j=0}^{n-1} \omega^j x_j^4$. Then, by a Feynman–Kac formula:

$$\langle \tilde{X}(s_1) \cdots \tilde{X}_n(s_n)\rangle = (\tilde{\Omega}_0, e^{-t_0\hat{L}}\tilde{X} \cdots \tilde{X}e^{-t_n\hat{L}}\tilde{\Omega}_0)Z^{-1} \tag{20.8}$$

with $Z = (\tilde{\Omega}_0, e^{-2N\hat{L}}\tilde{\Omega}_0)$, $t_0 = s_1 + N$, $t_1 = s_2 - s_1, \ldots, t_{n-1} = s_n - s_{n-1}$, $t_n = N - s_n$. Let $\tilde{\Omega}$ be the ground state for $\hat{L}$. Then, as in the proof of (12.16), for any $i = 2, \ldots, n$,

$$(\tilde{\Omega}_0, e^{-t_0\hat{L}}\,\tilde{X} \cdots e^{-t_{i-1}\hat{L}}\tilde{\Omega}) = 0$$

and thus in (20.8), we can replace $e^{-t_{i-1}\hat{L}}$ by $e^{-t_{i-1}\hat{L}}(1 - P)$ where $P = (\tilde{\Omega}, \cdot)\tilde{\Omega}$. Let $E_2(\beta)$ be the second lowest eigenvalue of $L(\beta)$. Then $E_2 - E$ is nonvanishing and continuous for all β in $[0, 1]$ so $\varepsilon \equiv \inf_{0\le\beta\le 1}(E_2 - E) > 0$ and

$$\|e^{-t_{i-1}\hat{L}}(1 - P)\| \le e^{-\varepsilon t_{i-1}}$$

It follows that

$$\langle \tilde{X}(s_1) \cdots \tilde{X}(s_n)\rangle \le Z^{-1}e^{-\varepsilon t_{i-1}}\|\tilde{X}e^{-t_{i-2}\hat{L}} \ldots \tilde{\Omega}_0\|\,\|\tilde{X}e^{-t_i\hat{L}} \cdots \tilde{\Omega}_0\|$$

Each $\|\cdots\|$ can be rewritten using the Feynman–Kac formulas as a Dq integral for which Hölder's inequality can be applied to bound

$$\int e^{-U}\tilde{X}_1 \cdots \tilde{X}_m \, Dq$$

by $\prod (\int e^{-U}|\tilde{X}_i|^m \, Dq)^{1/m}$. Thus, using the fact that $Z \geq (\tilde{\Omega}_0, \hat{\Omega})^2$ is uniformly bounded for $\beta \in [0, 1]$, we see that:

$$\langle \tilde{X}(s_1) \cdots \tilde{X}(s_n) \rangle \leq Ce^{-\varepsilon t_{i-1}} \sup_{\substack{s,t,\, 0 \leq \beta \leq 1 \\ 1 \leq m \leq n-1}} [1 + (\tilde{\Omega}_0, e^{-s\hat{L}}|\tilde{X}|^m e^{-t\hat{L}}\tilde{\Omega}_0)]$$

so (20.7) follows if we show this last supremum is finite. This can be accomplished by considering three cases, $s, t \leq 1, t \geq 1, s \geq 1$.

Case 1 $s, t \leq 1$.

$$(\tilde{\Omega}_0, e^{-s\hat{L}}|\tilde{X}|^m e^{-t\hat{L}}\tilde{\Omega}_0) = \int |\tilde{X}(q(0))|^m e^{(s+t)E(\beta)} \exp\left(-\beta \int_{-s}^{t} \sum q_j^4(u) \, du\right) Dq$$

which can be controlled since $q^4 \geq 0$ and $\int |\tilde{X}(q(0))|^m \, Dq < \infty$.

Case 2 $t \geq 1$. It suffices to show that $|\tilde{X}|^m e^{-\hat{L}}$ is bounded. By the Feynman–Kac formula and $q^4 \geq 0$,

$$|(e^{-\hat{L}}f)(q)| \leq |(e^{-\hat{L}_0}f)(q)| e^{+E(\beta)}$$

so $|\tilde{X}|^m e^{-\hat{L}}$ bounded follows from the fact that $|\tilde{X}|^m e^{-\hat{L}_0}$ is bounded.

Case 3 $s \geq 1$. By Case 2, $e^{-\hat{L}}|\tilde{X}|^m = (|\tilde{X}|^m e^{-\hat{L}})^*$ is bounded.

We have thus proven (20.7) and therefore we have the following.

Theorem 20.2 $E(\beta)$ has an asymptotic series $\sum_{n=0}^{\infty} a_n \beta^n$ as $\beta \downarrow 0$.

Now we want to identify a_n. Looking at (20.5), the following is the obvious guess.

Theorem 20.3 The coefficients a_n of Theorem 20.2 are given by

$$a_n = \frac{(-1)^{n+1}}{n!} \int_{\substack{-\infty < s_2 < \infty \\ \vdots \\ -\infty < s_n < \infty}} \langle q^4(0), q^4(s_2), \ldots, q^4(s_n) \rangle_T \, d^{n-1}s \tag{20.9}$$

where $\langle \cdot, \ldots, \cdot \rangle_T$ is an Ursell function with respect to Dq.

Remark The same proof as below shows that the analog of (20.9) with the left-hand side replaced by $(n!)^{-1}\, d^nE/d\beta^n$ and the right-hand side by $\langle\cdots\rangle_{T,\beta}$ holds for all β if $\langle\cdots\rangle_{T,\beta}$ denotes Ursell functions with respect to the $P(\phi)_1$-process associated to $L(\beta)$.

Proof By the proof of Theorem 20.2, $a_n = \lim_{N\to\infty} a_n(N)$ where

$$a_n(N) = \frac{(-1)^{n+1}}{n!}(2N)^{-1}\int_{-N}^{N} ds_1 \cdots \int_{-N}^{N} ds_n\, f(s_1, \ldots, s_n; N)$$

$$f(s_1, \ldots, s_n; N) = \langle q^4(s_1), \ldots, q^4(s_n)\rangle_{T, N, \beta=0}$$

(for $\beta = 0$, f is independent of N but we write the proof so as to be applicable to $\beta \neq 0$). As in the proof of Theorem 20.2, in the region $s_1 < \cdots < s_n$;

$$|f(s_1, \ldots, s_n; N) - f(s_1, \ldots, s_n; \infty)| \leq C[e^{-\varepsilon(N+s_1)} + e^{-\varepsilon(N-s_n)}]e^{-D_n|s_n - s_1|}$$

so that

$$a_n(N) = \left[(-1)^{n+1}(2N)^{-1}\int_{-N<s_1<\cdots<s_n<N} ds_1 \cdots ds_N\, f(s_1, \ldots, s_n; \infty)\right] + O(N^{-1})$$

Now, the right-hand side of (20.9) is equal to

$$\begin{aligned}
&(-1)^{n+1}\int_{0<s_2<\cdots<s_n} ds_2 \cdots ds_n\, f(0, s_2, \ldots, s_n; \infty)\\
&\quad =(-1)^{n+1}(2N)^{-1}\int_{\substack{-N<s_1<N\\ s_1<s_2<\cdots<s_n}} ds_1 \cdots ds_N\, f(s_1, \ldots, s_n; \infty)\\
&\quad \equiv \tilde{a}_n
\end{aligned}$$

since $f(\ldots; \infty)$ only depends on the successive time differences after reordering. Thus,

$$|\tilde{a}_n - a_n(N)| = O(N^{-1}) + (2N)^{-1}\int_{\substack{-N<s_1<N<s_n\\ s_1<\cdots<s_n}} ds_1 \cdots ds_N\, f(s_1, \ldots, s_n; \infty)$$

But $|f(s_i)| \leq Ce^{-D_n|s_n - s_1|}$, so the integral is dominated by

$$\begin{aligned}
&C(2N)^{-1}\int_{-N}^{N} ds_1 \int_{N}^{\infty} |s_n - s_1|^{n-2}e^{-D_n|s_n - s_1|}\, ds_n\\
&\quad \leq C_n(2N)^{-1}\int_{-N}^{N} ds_1\, e^{-D_n|N-s_1|/2}\\
&\quad \leq C_n(2N)^{-1}\int_{0}^{\infty} dy\, e^{-D_n y/2} = O(N^{-1})
\end{aligned}$$

Thus $a_n(N) \to \tilde{a}_n$. ∎

Ursell functions are very complicated sums so the expression (20.9) is of limited value. To go beyond it we must use the fact that Dq is a Gaussian measure. As a preliminary, we note two facts:

Lemma 20.4 Let $X_1, \ldots, X_{2k}$ be jointly Gaussian random variables. Then

$$\langle X_1 \cdots X_{2k} \rangle = \sum_{\text{pairings}} \langle X_{i_1} X_{j_1} \rangle \cdots \langle X_{i_k} X_{j_k} \rangle$$

where $\sum_{\text{pairings}}$ denotes the sum over all $(2k)!/2^k k!$ ways of breaking $(1, \ldots, 2k)$ into k pairs.

Proof Both sides are multilinear and symmetric in the X's so it suffices to check the case $X_1 = \cdots = X_{2k}$. That is, we need to show that for Gaussian X:

$$\langle X^{2k} \rangle = \frac{2k!}{2^k k!} \langle X^2 \rangle^k$$

But this follows from

$$\langle \exp(\alpha X) \rangle = \exp(\tfrac{1}{2}\alpha^2 \langle X^2 \rangle) \quad \blacksquare$$

Lemma 20.5 Let $X_1, \ldots, X_{2k}$ be Gaussian with respect to some expectation, $\langle \cdot \rangle$. Let $Y_1 = X_1 \cdots X_{l_1}$, $Y_2 = X_{l_1+1} \cdots X_{l_2}, \ldots, Y_m = X_{l_{m-1}+1} \cdots X_{l_m}$ with $l_m = 2k$. Then the Ursell function with respect to $\langle \cdot \rangle$ is given by:

$$u_m(Y_1, \ldots, Y_m) = \sum_{\substack{\text{connected}\\ \text{pairings}}} \langle X_{i_1} X_{j_1} \rangle \cdots \langle X_{i_k} X_{j_k} \rangle \tag{20.10}$$

where a pairing is connected if and only if $\{1, \ldots, 2k\}$ is connected after one joins $(1, \ldots, l_1)$ together, $\ldots$, $(l_{m-1} + 1, \ldots, l_m)$ together, and then i_1 to j_1, i_2 to j_2, etc.

Example Take $k = 3$; $Y_1 = X_1 X_2 X_3$, $Y_2 = X_4$, $Y_3 = X_5 X_6$. Of the 15 pairings, only the three with 5 paired to 6 are not connected.

Proof We first claim that for *any* random variables

$$\langle Y_1 \cdots Y_m \rangle = \sum_{P \in \mathscr{P}} \prod_{\pi_i \in P} u(\pi_i) \tag{20.11}$$

where $\mathscr{P}$ is the family of all partitions $P = \{\pi_1, \ldots, \pi_l\}$ of $\{1, \ldots, m\}$ and $u(\pi_i) \equiv u_k(Y_{j_1}, \ldots, Y_{j_k})$ with $\pi_i = \{j_1, \ldots, j_k\}$. (20.11) is equivalent to

$$u_m(Y_1, \ldots, Y_m) = \langle Y_1 \cdots Y_m \rangle - \sum_{\substack{P \in \mathscr{P}\\ \#(P) \geq 2}} \prod u(\pi_i) \tag{20.11'}$$

This may be proven inductively by checking Percus' axioms (see Proposition 12.11) for u_m *defined* by (20.11′) assuming that (20.11) holds for fewer than m Y's.

On account of (20.11′), if we are given $\langle Y_{i_1} Y_{i_2} \cdots Y_{i_l} \rangle$ for any subset $\{i_1, \ldots, i_l\}$ of $\{1, \ldots, m\}$ and candidates $a(\pi)$ for the $u(\pi)$ and if the $a(\pi)$ satisfy (20.11) for any $\langle Y_{i_1} \cdots Y_{i_l} \rangle$, then $a(\pi) = u(\pi)$. In this way, we can verify (20.10). For given Gaussian X's, $\langle Y_{i_1} \cdots Y_{i_l} \rangle$ is a sum over pairings by Lemma 20.4. Any pairing induces a partition of $\{i_1, \ldots, i_l\}$ by looking at the connected subsets induced by joining $\{1, \ldots, l_i\}$, etc., and then the pairs. Clearly

$$\langle Y_{i_1} \cdots Y_{i_l} \rangle = \sum_{P \in \mathscr{P}(i_1, \ldots, i_l)} \prod_{\pi_i \in P} \sum_{\substack{\text{connected} \\ \text{pairings} \\ \text{of } \pi_i}} \langle X_{i_1} X_{j_1} \rangle \cdots$$

so that the candidates for $u(\pi)$ given by (20.10) obey (20.11). ∎

We are now ready to give the **Feynman rules** for the coefficients a_n of (20.9). By a **labeled *n*-graph**, Γ, for q^4 we mean n labeled points $1, \ldots, n$ and some lines joining the points with the property that exactly four lines come out of each point. Moreover, we label from 1 to 4 those lines associated to each point. A typical labeled three-graph is shown in Figure 2. Clearly there is a one–one correspondence between labeled n-graphs and pairings that enter in $\langle q^4(s_1), \ldots, q^4(s_n) \rangle_T$. We define the **value**, $v(\Gamma)$, **of a graph** as follows: For each line l joining i and j let $g(l) = \frac{1}{2} \exp(-|s_i - s_j|)$. Then

$$v(\Gamma) = \frac{(-1)^{n+1}}{n!} \int_{\substack{s_1 = 0 \\ -\infty < s_2 < \infty, \ldots, -\infty < s_n < \infty}} \prod_{l \in \Gamma} g(l) \, ds_2 \cdots ds_n \qquad (20.12)$$

In general, the integral in (20.12) will be convergent if and only if Γ is connected. In colloquial usage, one does not distinguish between Γ and $v(\Gamma)$ and calls them both, **Feynman graphs**, **Feynman diagrams**, or **Feynman integrals**. The last term is unfortunate since the integral in (20.12) is very different from the formal object of (1.8) which is also called a Feynman integral. (We note that the Feynman integral of *Homology and Feynman Integrals* [131] is of the type of (20.12)—more precisely, it is of the more

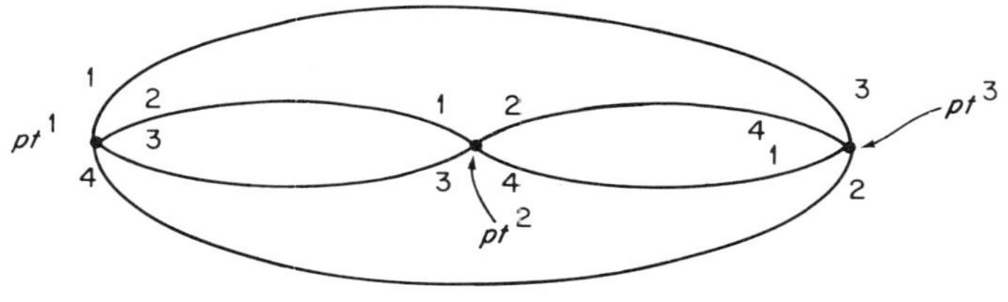

Figure 2. A typical labeled three-graph.

general type associated below to Green's functions—the object of [131] is to study the analyticity properties of $v(\Gamma)$ as the external parameters defining the Green's function are varied.) Clearly (20.9) and Lemma 20.5 imply that the a_n of Theorem 20.2 is given by

$$a_n = \sum \{v(\Gamma) \mid \text{connected labeled } n\text{-graphs } \Gamma\} \tag{20.13}$$

Notice that the value of a graph is independent of any of the labeling. For this reason, it is useful sometimes to consider **unlabeled graphs** γ. We set $v(\gamma) = v(\Gamma)$ for any $\Gamma \in \gamma$, and $s(\gamma)$ is the number of labeled Γ's which become γ after removing all labels. We leave it to the reader to show that the **symmetry numbers**, $s(\gamma)$, are given by:

$$s(\gamma) = n!(4!)^n \left[M(\gamma) \prod_{i<j} m_{ij}(\gamma)! \prod_i 2^{m_{ii}(\gamma)} \right]^{-1}$$

where $m_{ij}(\gamma)$ is the number of lines joining points i and j in some $\Gamma \in \gamma$ and $M(\gamma)$ is the number of ways of labeling the points leading to identical point-labeled graphs. For example, if $n = 2$, there are two connected unlabeled graphs, shown in Figure 3. Moreover, $s(\gamma_1) = 4! = 24$ [$M(\gamma) = n!$ here;

Figure 3. Unlabeled connected two-graphs.

a typical γ with $n = 3$ and with $M(\gamma) = 2$, not $3!$ is shown in Figure 4] and $s(\gamma_2) = 4!4!/2!2 \cdot 2 = 72$. Moreover,

$$v(\gamma_1) = -\frac{1}{2}\int_{-\infty}^{\infty} \left(\frac{1}{16}\right) e^{-4|t|}\, dt = -\frac{1}{64}$$

$$v(\gamma_2) = -\frac{1}{2}\int_{-\infty}^{\infty} \left(\frac{1}{16}\right) e^{-2|t|}\, dt = -\frac{1}{32}$$

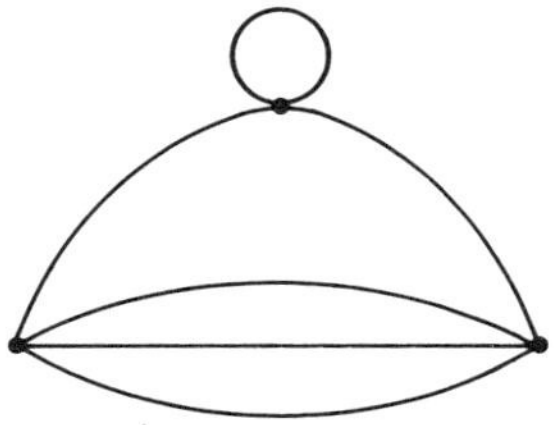

Figure 4. A graph with $M(\gamma) \neq n!$.

so using

$$a_n = \sum \{s(\gamma)v(\gamma) \mid \gamma \text{ an unlabeled connected graph}\} \tag{20.14}$$

we see that

$$a_2 = -21/8$$

Let us use these graphs to show that

$$n!AB^n \le (-1)^{n+1}a_n \le n!CD^n \tag{20.15}$$

(see also [7, 241]).

Lemma 20.6 Let $N(n)$ be the number of connected labeled n-graphs. Then,

$$(n-1)!(2n)!/(n!2^n) \le N(n) \le (4n)!/[(2n)!2^{2n}]$$

Proof The upper bound is just the number of ways of pairing $4n$ objects which is the total number of labeled diagrams, connected or not. To get the lower bound, pick one of the $(n-1)!$ orderings of $2, \ldots, n$, say, $i_1, \ldots, i_{n-1}$. Let line 2 from point 1 be joined to line 1 from point i_1, line 2 from i_1 to line 1 from i_2, etc., and line 2 from i_n to line 1 from point 1. Pair the remaining $2n$ lines in an arbitrary way. Each of these $(n-1)!(2n)!/n!2^n$ graphs is connected. ∎

The point is that the lemma implies that

$$(n!)^2ab^n \le N(n) \le (n!)^2c\, d^n \tag{20.16}$$

Lemma 20.7 For some fixed β, δ and all Γ:

$$\beta^n \le (-1)^{n+1}v(\Gamma)n! \le \delta^n \tag{20.17}$$

Proof To get the lower bound we consider points s_i with $0 \le s_i \le \frac{1}{2}$ in the integral in (20.12). Then $g(l) \ge (4e)^{-1/2}$ in that region so

$$(-1)^{n+1}n!v(\Gamma) \ge [(4e)^{-1}]^n$$

To get the upper bound, we first claim that one can pick $(n-1)$ lines $l_1, \ldots, l_{n-1} \in \Gamma$ which connect the graph. Assuming this fact, we let $x_k = s_i - s_j$ for the line l_k. The fact that $l_1, \ldots, l_{n-1}$ connect the points implies that any s_i is a linear combination of x_k's so the coordinate transformation

$s_i \to x_k$ is nonsingular. Moreover, it is easy to see that the s_i are combinations of x's with integral coefficients and vice versa. Thus the transformation has Jacobian 1. Since $|g(l)| \le \frac{1}{2}$,

$$(-1)^{n+1} n! v(\Gamma) \le \int \left[\prod_{k=1}^{n-1} g(l_k)\, dx_k \right] (1/2)^{n+1}$$

$$= \left[\int_{-\infty}^{\infty} \frac{1}{2} e^{-|t|}\, dt \right]^n (1/2)^{n+1} = (1/2)^{n+1}$$

This leaves the proof of the claim about $l_1, \ldots, l_{n-1}$. For $n = 2$, the result is obvious for an arbitrary graph, even without the four lines per point condition. Suppose we know the result for $n \le n_0 - 1$ for such an arbitrary graph. Given an n point connected graph pick some line, l_0, joining i and j. Consider the graph obtained by fusing i and j. It can be connected by $n_0 - 2$ lines by the induction hypothesis. Adding the line l_0, we connect the original graph with $n_0 - 1$ lines. ∎

(20.16) and (20.17) clearly imply the following theorem.

Theorem 20.8 (20.15) holds.

Actually, one can estimate the quantity $N(n)$ of Lemma 20.6 for n large rather exactly. Let

$$G(n) = (4n)!/[(2n)!2^{2n}] = (4n-1)(4n-3)\cdots(1)$$

be the number of graphs connected or not. Then

$$\lim_{n\to\infty} N(n)/G(n) = 1 \tag{20.18}$$

Indeed, for $n \ge 2$

$$|G(n) - N(n)| \le CG(n)n^{-1} \tag{20.19}$$

which is even stronger than (20.18). Our proof of (20.19) is motivated by ideas of Cvitanovic, Lautrup, and Pearson [39a]. We begin by considering the functions:

$$a(t) = (2\pi)^{-1/2} \int_{-\infty}^{\infty} \exp(-\tfrac{1}{2}x^2 - tx^4)\, dx$$

$$c(t) = \ln a(t)$$

Then $a(t)$ has a diagramatic perturbation expansion identical to that of the q^4 oscillator except that now every graph has value 1. Thus (with $G(0) = 1$, $N(0) = 0$),

$$a(t) \sim \sum_{n=0}^{\infty} A(n)(-t)^n$$

$$c(t) \sim \sum_{n=0}^{\infty} C(n)(-t)^n$$

with $A(n) = G(n)/n!; C(n) = N(n)/n!$. Equating coefficients of t^n in $a(t) = \exp(c(t))$, we see that

$$A(n) = C(n) + \sum_{k=2}^{n} (k!)^{-1} \sum_{j_1 + \cdots + j_k = n} C(j_1) \cdots C(j_k) \qquad j_i \geq 1 \qquad (20.20)$$

(20.20) can also be obtained by realizing that an arbitrary labeled graph breaks into k connected pieces. The number of graphs in k pieces is just

$$\sum_j (k!)^{-1}[n!/j_1! \cdots j_k!] N(j_1) \cdots N(j_k)$$

Now let

$$B(k, n) = \sum_{j_1 + \cdots + j_k = n} A(j_1) \cdots A(j_k) \qquad j_i \geq 1$$

Noting that $A(n-j)A(j)$ is decreasing in j until $j = [n/2]$ and then increasing, one sees that for $n \geq 2$:

$$B(2, n) \leq A(n)\left[\frac{2n}{(4n-1)(4n-3)} + n\left(\frac{n(n-1)}{(4n-1)(4n-3)(4n-5)(4n-7)}\right)\right]$$

$$\leq Dn^{-1}A(n) \qquad (20.21)$$

for a suitable constant D. In the above, the first contribution comes from the $j = 1$ and $j = n - 1$ terms in the sums and the second by bounding the remaining sum by $nA(2)A(n-2)$. Now

$$B(k, n) = \sum_{j_1=1}^{n-k+1} A(j_1)B(k-1; n-j_1)$$

From this formula and (20.21) one easily proves inductively in k that

$$B(k, n) \leq D^k n^{-1} A(n) \qquad (20.22)$$

Using (20.20), (20.22), and the trivial bound $C(j) \leq A(j)$, we obtain (20.19) with $C = e^D$. By working with these same ideas, one can obtain fairly easily explicit terms in an asymptotic expansion

$$N(n)/G(n) \sim 1 + b_1 n^{-1} + b_2 n^{-2} + \cdots$$

* * *

We close this section with a discussion of two further aspects of the Feynman graphical analysis: "graphs in p-space," and graphs for Green's functions. In (20.12) introduce a variable p_l for each l and write

$$g(l) = \frac{1}{2\pi} \int \frac{e^{ip_l(s_i - s_j)}}{p_l^2 + 1}\, dp_l$$

Doing this involves choosing a direction (from i to j) on each line. Proceeding formally, we do each s_i integration as follows: The s_i-dependence is

$$\int \exp\left[is_i\left(\sum_{\text{in}} p_l - \sum_{\text{out}} p_l\right)\right] dp_i = 2\pi\delta\left(\sum_{\text{in}} p_l - \sum_{\text{out}} p_l\right)$$

where $\sum_{\text{out}} p_l$ is over those lines pointing from i to some j and $\sum_{\text{in}} p_l$ is over those lines pointing from some j to i. Then

$$v(\Gamma) = \frac{(-1)^{n+1}}{n!} \int \left\{\prod_{l \in \Gamma} (p_l^2 + 1)^{-1}(2\pi)^{-1}\, dp_l\right\} \prod_{i=2}^{n} 2\pi\delta\left(\sum_{\text{in}} p_l - \sum_{\text{out}} p_l\right)$$

While the above manipulations are formal, it is not hard to justify the final result. In essence, the last integral is a generalized convolution and its equality to $v(\Gamma)$ is just a generalization of the usual Fourier transform formula relating products and convolutions. Graphically, the above can be described as follows: Each line has a momentum p_l with conservation of momentum enforced at each vertex $i = 2, \ldots, n$ (it then follows automatically at vertex 1). The number of free momenta to integrate over (usually called **loop momenta** since they can be chosen to "flow" in loops) is thus not $2n$ but only $n + 1$. This "momentum space" analysis is common for two reasons: In quantum field theory, it is most usual to discuss renormalization theory in "p-space" (e.g., Weinberg [284]) although more recently x-space analysis has become more common (e.g., Speer [262]); see Hepp [125] for a review of renormalization theory. In discussing scattering theory in quantum field theory, the Fourier transforms of the Green's functions we are about to discuss are natural and these have a p-space analysis.

Finally, one often considers the (truncated) **Euclidean Green's functions** defined by

$$G_n(s_1, \ldots, s_n) = \langle q(s_1), \ldots, q(s_n) \rangle_{T, N=\infty, \beta}$$

It is not hard to show by mimicking the proof of Theorem 20.2, that

$$\frac{\partial^k}{\partial \beta^k} G_n(s_1, \ldots, s_n)$$

$$= \int_{-\infty < t_i < \infty} (-1)^k \langle q(s_1), \ldots, q(s_n), q^4(t_1), \ldots, q^4(t_k) \rangle_{T, \beta} \, d^k t$$

Thus G_n has an asymptotic series given by a sum of a suitable kind of connected Feynman graph.

21. Dirichlet Boundaries and Decoupling Singularities in Scattering Theory

We have already seen in Section 7 (see Lemma 7.10) that Dirichlet boundary conditions are closely connected to the exclusion of Wiener paths from certain regions. However, because we dealt with Wiener paths with a fixed initial point, i.e., with Brownian motion, we required a rather strong regularity condition to establish the connection. In this section, we begin by establishing a weaker kind of connection under the most general circumstances; we will then describe some ideas of Deift–Simon [45] on the connection of those exclusion formulas and a problem in scattering theory.

The following involves the general operator $H_{\mathrm{D};S}$ defined in Section 7; i.e., the operator on $L^2(\mathbb{R}^\nu \backslash S)$ which is the Friedrich's extension of $-\Delta$ on $C_0^\infty(\mathbb{R}^\nu \backslash S)$.

Theorem 21.1 Let S be an *arbitrary* closed set in $\mathbb{R}^\nu$. Then for any $f, g \in L^2(\mathbb{R}^\nu \backslash S)$:

$$(f, e^{-tH_{\mathrm{D};S}} g) = \int_Q \overline{f(\boldsymbol{\omega}(0))} g(\boldsymbol{\omega}(t)) \, d\mu_0(\omega) \tag{21.1}$$

where Q is the set of paths:

$$Q = \{\boldsymbol{\omega} \mid \boldsymbol{\omega}(s) \notin S, 0 \le s \le t\}$$

Proof ([252]) Extend the form of $H_{\mathrm{D};S}$ to $L^2(\mathbb{R}^\nu)$ by viewing it as a non-densely defined quadratic form.

Write $\mathbb{R}^\nu \backslash S = \bigcup_{n=1}^{\infty} K_n$ where each $\bar{K}_n$ is compact, K_n is open, and $\bar{K}_n \subset K_{n+1}$. Choose $\phi_n \in C_0^\infty$ so that $\phi_n = 1$ on $\bar{K}_n$ and $\operatorname{supp} \phi_n \subset K_{n+1}$. Define

$$V(\mathbf{x}) = \begin{cases} \sum_n |\nabla\phi_n(\mathbf{x})|^2 + \operatorname{dist}(\mathbf{x}, S)^{-3}, & \mathbf{x} \notin S \\ \infty, & \mathbf{x} \in S \end{cases}$$

Let $Q(V) = \{f \in L^2 \mid \int V(\mathbf{x})|f(\mathbf{x})|^2\, dx < \infty\}$. We first claim that $Q(V) \cap Q(-\Delta) \subset Q(H_{\mathrm{D};S})$, for $f \in Q(V)$ implies that $\sum \|(\nabla\phi_n)f\|^2 < \infty$ so that if $f \in Q(V) \cap Q(-\Delta)$, then $(\nabla\phi_n)f \to 0$ in L^2. Thus $\phi_n f \to f$, $\nabla(\phi_n f) \to \nabla f$; i.e., $f \in Q(H_{\mathrm{D};S})$. But clearly, $C_0^\infty(\mathbb{R}^\nu \backslash S) \subset Q(V) \cap Q(-\Delta)$. Thus, by the monotone convergence theorems for forms [152, 253, 254],

$$e^{-tH_{\mathrm{D};S}} = \operatorname*{s-lim}_{\lambda \downarrow 0} \exp[-t(-\tfrac{1}{2}\Delta + \lambda V)] \tag{21.2}$$

By approximating V with $V_m = \min(m, V)$, one sees that

$$(f, \exp[-t(-\tfrac{1}{2}\Delta + \lambda V)]g) = \int \overline{f(\boldsymbol{\omega}(0))} g(\boldsymbol{\omega}(t)) \exp\left(-\lambda \int_0^t V(\boldsymbol{\omega}(s))\, ds\right) d\mu_0(\omega) \tag{21.3}$$

But since V has a $\operatorname{dist}(\mathbf{x}, S)^{-3}$ term and Wiener paths are Hölder continuous of order $\frac{1}{3}$, $\int_0^t V(\boldsymbol{\omega}(s))\, ds = \infty$ if $\boldsymbol{\omega}(s) \in S$ for some s. Conversely $\int_0^t V(\boldsymbol{\omega}(s))\, ds < \infty$ if ω never hits S since $\boldsymbol{\omega}$ then lies in some K_n and moreover, $\operatorname{dist}(\boldsymbol{\omega}(s), S)^{-3}$ is bounded. Thus

$$Q = \left\{\boldsymbol{\omega} \,\middle|\, \int_0^t V(\boldsymbol{\omega}(s))\, ds < \infty\right\}$$

so that (21.2, 3) imply (21.1). ∎

Before turning to our main application of (21.1), we remark upon some other connections:

(1) (21.1) gives an expression for $P_{\mathrm{D};S}(\mathbf{x}, \mathbf{y}; t)$ holding almost everywhere in $\mathbf{x}, \mathbf{y}$. It is an ideal tool for studying rather subtle properties of $P_{\mathrm{D};S}$ such as whether it approaches zero as $\mathbf{x} \to S$.

(2) Intuition connected with (21.1) is useful in deciding what kinds of singularities in V can destroy simplicity of the ground state for $-\Delta + V$ (see [80]).

(3) Motivated by (21.1), Klauder [159] has noted an interesting phenomenon in singular perturbation theory. Let $L_0 = -\frac{1}{2}\, d^2/dx^2 + \frac{1}{2}x^2 - \frac{1}{2}$, as usual, and let $V = x^{-\alpha}$ ($\alpha > 2$). Then (e.g., [215]) $L(\beta) = L_0 + \beta V$ is

essentially self-adjoint on $C_0^\infty(\mathbb{R}\backslash\{0\})$ for $\beta > 0$. Let $E(\beta)$ denote the ground state energy of $L(\beta)$. Then

$$\lim_{\beta \downarrow 0} E(\beta) = 1 \neq \inf \operatorname{spec}(L_0) = 0$$

For, if one writes a Feynman–Kac formula for $L(\beta)$, $\int_0^t V(\omega(s))\, ds = \infty$ for any (Hölder continuous) path crossing zero. Thus, taking $\beta \downarrow 0$, we recover the Feynman–Kac formulas for "L_0" with a Dirichlet condition at zero, call it L_D and not L_0. We have $\inf \operatorname{spec}(L_D) = 1$. In fact, our proof of (21.1) is motivated by [159]. See [43, 246] for a discussion of the Klauder phenomena using purely quadratic form ideas; see [160, 114, 123] for more detailed information on $E(\beta)$ for β small; see [79] for a more thorough analysis of the Wiener paths and the Feynman–Kac formula for $L(\beta)$.

We now turn to the application to scattering theory. Given an operator, H, which is somehow like $H_0 = -\frac{1}{2}\Delta$ at spatial infinity, the two fundamental "foundational" questions of scattering theory are the **existence of the wave operators**

$$\Omega^{\pm}(H, H_0) = \operatorname*{s-lim}_{t \to \mp\infty} e^{itH} e^{-itH_0}$$

and their **completeness**

$$\operatorname{Ran} \Omega^{\pm}(H, H_0) = \operatorname{Ran} P_{ac}(H)$$

where $P_{ac}(H)$ is the projection onto all vectors whose H-spectral measures are absolutely continuous. An extensive literature has developed to study the case $H = H_0 + V$ mainly based on perturbation formulas like

$$(H - z)^{-1} - (H_0 - z)^{-1} = -(H_0 - z)^{-1} V (H - z)^{-1}$$

which do not distinguish between positive and negative singularities of V. (See [216] for a review.) Motivated by an example of Pearson [200], described further below, Deift and Simon [45] tried to see what local singularities of V had no effect on the existence and completeness questions; i.e., they sought theorems which asserted that $\Omega^{\pm}(H_0 + V, H_0)$ existed and were complete if $\Omega^{\pm}(H_0 + W, H_0)$ existed and were complete and if $W - V$ had compact support (earlier Kupsch–Sandhas [163] had studied the analogous existence question). For simplicity, we suppose that $W = 0$; i.e., we consider here V's of compact support; it is easy to extend the results to the more general setting. Deift–Simon exploited the semigroup e^{-tH}, which is more sensitive to the sign of V, Feynman–Kac formulas, and some rather elaborate machinery. Some small simplifications of the case $V \geq 0$ occur in [252]. Using different machinery (and in particular, neither path integrals nor Dirichlet boundaries) Combescure–Ginibre [38] recovered the Deift–Simon results with fewer technicalities (see also [216] for an exposition of their work); below we use some of the ideas of [38] to simplify the details of the Deift–Simon approach. Independently of Deift–Simon, related results were obtained by Pearson [201] and Semenov [236].

Suppose first that $V \geq 0$ has compact support. We will suppose that $V \in L^1(\mathbb{R}^\nu \backslash G)$ where G is a closed set of measure zero. In that case, $H_0 + V$ is a densely defined closed form on $Q(H_0) \cap Q(V)$ and so defines an operator H. By Theorem 6.2, e^{-tH} is given by a Feynman–Kac formula. Let S be a sphere which surrounds supp(V). Let H_0' be the Dirichlet Laplacian $H_{D;S}$ and let $H' = H_0' + V$. The strategy is to prove that $e^{-H} - e^{-H_0}$ is trace class by proving that $e^{-H'} - e^{-H}$, $e^{-H_0} - e^{-H_0'}$, and $e^{-H'} - e^{-H_0'}$ are trace class and then use the following (for a proof, see, e.g., [216]).

Theorem 21.2 Let A, B be self-adjoint operators which are bounded from below. Suppose that $e^{-A} - e^{-B}$ is in $\mathcal{I}_1$, the trace class. Then

$$\Omega^{\pm}(A, B) = \operatorname*{s-lim}_{t \to \mp\infty} e^{itA} e^{-itB} P_{\text{ac}}(B)$$

exist and have the same range as $P_{\text{ac}}(A)$.

The following illustrates the use of (21.1).

Lemma 21.3 Let H_0' be $H_{D;S}$ where S is the boundary of any compact convex set. Then, for all $\mathbf{x}, \mathbf{y} \in \Omega$, the unbounded component of $\mathbb{R}^\nu \backslash S$:

$$0 \leq (e^{-tH_0} - e^{-tH_0'})(\mathbf{x}, \mathbf{y}) \leq (2\pi t)^{-\nu/2} \exp(-[\text{dist}(\mathbf{x}, S)^2 + \text{dist}(\mathbf{y}, S)^2]/4t) \tag{21.4}$$

In particular,

$$(1 + X^2)^m (e^{-tH_0} - e^{-tH_0'})(1 + X^2)^k \tag{21.5}$$

is Hilbert–Schmidt for any integers, k and m.

Proof By symmetry, it suffices to prove (21.4) with $4t$ replaced by $2t$ and with the dist($\mathbf{y}$, S) term replaced by zero. Given $\mathbf{x} \in \Omega$, let $\mathbf{z}_0$ be the point in S closest to x and let π be the plane through $\mathbf{z}_0$ perpendicular to $\mathbf{x} - \mathbf{z}_0$. By some simple geometry, π separates S and $\mathbf{x}$. Thus, any Wiener path joining $\mathbf{x}$ and $\mathbf{y}$ which passes through S must pass through π; i.e., since (21.1) implies for almost every $\mathbf{x}$, $\mathbf{y}$

$$(e^{-tH_0} - e^{-tH_{D;S}})(\mathbf{x}, \mathbf{y}) = \int_{\boldsymbol{\omega}(s) \in S, \text{ some } 0 \leq s \leq t} d\mu_{0,x,y;t}(\boldsymbol{\omega}) \tag{21.6}$$

we have

$$0 \le (e^{-tH_0} - e^{-tH_0'})(\mathbf{x}, \mathbf{y}) \le (e^{-tH_0} - e^{-tH_{D;\pi}})(\mathbf{x}, \mathbf{y})$$
$$= e^{-tH_0}(\mathbf{x}, \mathbf{y}') \tag{21.7}$$

where $\mathbf{y}'$ is the image of $\mathbf{y}$ under reflection about π. In obtaining (21.7), we use the method of images formula for $e^{-tH_{D;\pi}}$. (21.7) and the trivial

$$|(e^{-tH_0} - e^{-tH_0'})(\mathbf{x}, \mathbf{y})| \le e^{-tH_0}(\mathbf{x}, \mathbf{y})$$

imply that

$$|(e^{-tH_0} - e^{-tH_0'}))(\mathbf{x}, \mathbf{y})| \le \min[e^{-tH_0}(\mathbf{x}, \mathbf{y}), e^{-tH_0}(\mathbf{x}, \mathbf{y}')]$$

Since either

$$\operatorname{dist}(\mathbf{x}, \mathbf{y}) \ge \operatorname{dist}(\mathbf{x}, \mathbf{z}_0) \qquad \text{or} \qquad \operatorname{dist}(\mathbf{x}, \mathbf{y}') \ge \operatorname{dist}(\mathbf{x}, \mathbf{z}_0)$$

we have proven (21.4). The formula (21.4) implies the kernel of (21.5) is square integrable for $\mathbf{x}, \mathbf{y} \in \Omega$. A similar proof yields a

$$(2\pi t)^{-\nu/2} \exp(-\operatorname{dist}(\mathbf{x}, \Omega)^2/2t)$$

bound for $\mathbf{x} \in \Omega$, $\mathbf{y} \notin \Omega$ and a $(2\pi t)^{-\nu/2}$ bound for $\mathbf{x}, \mathbf{y} \notin \Omega$. ∎

Theorem 21.4 If $V \ge 0$ has compact support $[V \in L^1(\mathbb{R}^\nu \backslash G)$, H as above$]$ then $e^{-H} - e^{-H_0}$ is trace class and, in particular, $\Omega^{\pm}(H, H_0)$ exist and are complete.

Proof Let $A = e^{-H_0} - e^{-H_0'}$, $B = e^{-H} - e^{-H'}$, $C = e^{-H_0'} - e^{-H'}$. We prove A, B, and C are trace class. Write $A = PQ + Q'P'$, where

$$P = e^{-H_0/2}(1 + X^2)^{-k}$$
$$Q = (1 + X^2)^k(e^{-H_0/2} - e^{-H_0'/2})$$
$$Q' = (e^{-H_0/2} - e^{-H_0'/2})(1 + X^2)^k$$
$$P' = (1 + X^2)^{-k}e^{-H_0'/2}$$

By the lemma, Q, Q' are Hilbert–Schmidt for any k and it is easy to see that P, P' are Hilbert–Schmidt for k large. Thus A is a trace class. Since the Feynman–Kac formula [including (21.4)] and $V \ge 0$ show that the integral kernels of P, Q, P', Q' dominate the analogs with H_0, H_0' replaced by H, H', we see that B is trace class. Finally if Ω, Ω' are the unbounded and bounded components of $\mathbb{R}^\nu \backslash S$, then

$$e^{-H_0'} = e^{-K} \oplus e^{-L} \qquad \text{on} \quad L^2(\Omega') \oplus L^2(\Omega)$$
$$e^{-H'} = e^{-(K+V)} \oplus e^{-L}$$

Now e^{-Kt} has a bounded kernel and thus, since Ω' is bounded, e^{-tK} is Hilbert–Schmidt and so e^{-2tK} is trace class. Again, by the Feynman–Kac formula, $e^{-t(K+V)}$ is also Hilbert–Schmidt and so trace class. Thus $C = e^{-K} \oplus 0 - e^{-(K+V)} \oplus 0$ is also trace class. ∎

Remark By very different methods, Davies and Simon [41] have proven that H's of the above type also have empty singular continuous spectrum.

We now want to extend this result to allow V to have a negative part. The idea, also important in Section 25, will be to exploit Hölder's inequality in the form:

$$\int_Q e^{-W}\, d\mu \le \left(\int_Q d\mu\right)^{1/p} \left(\int e^{-qW}\, d\mu\right)^{1/q}$$

which will lead to a bound on an integral kernel $K(\mathbf{x}, \mathbf{y})$ by $L(\mathbf{x}, \mathbf{y})^a M(\mathbf{x}, \mathbf{y})^{1-a}$. Thus, we will need the following lemma.

Lemma 21.5 Let $p = 2^n$ $(n \ge 1)$. Suppose that $A, B \in \mathscr{I}_p$ $(=\{A \,|\, |A|^p \in \mathscr{I}_1\})$ with positive integral kernels $L(x, y)$, $M(x, y)$ [on some measure space $L^2(X, d\nu)$]. Let C be an integral operator with kernel K obeying

$$|K(x, y)| \le L(x, y)^a M(x, y)^{1-a} \tag{21.8}$$

for $0 \le a \le 1$. Then $C \in \mathscr{I}_p$.

Proof (by induction on n) $n = 1$ is easy, since $A \in \mathscr{I}_2$ if and only if its integral kernel is in $L^2(X \times X, d\nu \otimes d\nu)$ and $L, M \in L^2$ implies $L^a M^{1-a} \in L^2$ by Hölder's inequality. Suppose the result is known for $n < n_0$. (21.8) implies that (by Hölder's inequality)

$$\begin{aligned} |K^*K(x, y)| &\le \int |K(z, x)|\,|K(z, y)|\, d\nu(z) \\ &\le (L^*L)(x, y)^a (M^*M)(x, y)^{1-a} \end{aligned}$$

so by the induction hypotheses $K^*K \in \mathscr{I}_{p/2}$. ∎

Lemma 21.6 Let $H(\alpha) = H_0 + \alpha V$. Let $p = 2^n (n \ge 1)$. If

$$(1 + X^2)^{-k} \exp(-tH(1 + \delta))$$

is in $\mathscr{I}_p$ for all large k (both $\delta > 0$ and t fixed), then

$$(1 + X^2)^l [\exp(-tH) - \exp(-tH')](1 + X^2)^m$$

is in $\mathscr{I}_p$ for all l, m. Here $H = H(1)$ and $H' = H_{\mathrm{D};S} + V$ for S the boundary of a convex compact set.

Proof Let P be the family of paths hitting S. Then

$$|(e^{-tH} - e^{-tH'})(\mathbf{x}, \mathbf{y})| = \left| \int_P \exp\left(- \int_0^t V(\omega(s))\, ds \right) d\mu_{0,\mathbf{x},\mathbf{y};t} \right|$$

$$\leq \left(\int_P d\mu_{0,\mathbf{x},\mathbf{y};t} \right)^{1/\gamma} \left(\int \exp\left(-(1+\delta) \int V \right) d\mu_0 \right)^{1/1+\delta}$$

where $\gamma^{-1} + (1+\delta)^{-1} = 1$. Let

$$K(x, y) = (1 + x^2)^l (e^{-tH} - e^{-tH'})(\mathbf{x}, \mathbf{y})(1 + y^2)^m$$

$$L(x, y) = (1 + x^2)^\alpha (e^{-tH_0} - e^{-tH_0'})(\mathbf{x}, \mathbf{y})(1 + y^2)^\beta$$

$$M(x, y) = (1 + x^2)^{-k} e^{-tH(1+\delta)}(\mathbf{x}, \mathbf{y})$$

with $\alpha = \gamma[l + (1+\delta)^{-1}k]$; $\beta = \gamma m$. Then $|K| \leq L^{1/\gamma} M^{1/(1+\delta)}$. By hypothesis, M is an $\mathscr{I}_p$ kernel, and by Lemma 21.3, L is $\mathscr{I}_p$, so K is $\mathscr{I}_p$ by the last lemma. ∎

Lemma 21.7 Let $V_- \leq 0$. If $H_0 + (1+\delta)V_-$ is bounded from below for some $\delta > 0$, then $H' = H_0' + V_-$ obeys $(1 + X^2)^{-k} e^{-tH'}$ is in $\mathscr{I}_2$ for all large k.

Proof As in the proof of Theorem 21.4:

$$e^{-tH'} - e^{-tH_0'} = (e^{-t(K+V_-)} - e^{-tK}) \oplus 0$$

Moreover, $(1 + X^2)^{-k}$ is bounded so we need $(1 + X^2)^{-k} e^{-tH_0'} \in \mathscr{I}_2$ and $e^{-t(K+V_-)}, e^{-tK} \in \mathscr{I}_2$. The former is true by considerations used in the proof of Theorem 21.4. $e^{-tK} \in \mathscr{I}_2$ is proven in Theorem 21.4. Finally, since $H_0 + (1+\delta)V_- \geq a$, we have that

$$K + V_- = \frac{1}{1+\delta}(K + (1+\delta)V_-) + \frac{1}{\gamma} K \geq \frac{1}{\gamma} K + \frac{1}{1+\delta} a$$

so that the eigenvalues of $e^{-t(K+V_-)}$ are dominated by (constant times) eigenvalues of e^{-sK} so $e^{-t(K+V_-)} \in \mathscr{I}_2$. ∎

Theorem 21.8 ([45]) Let V have compact support and let $V = V_+ + V_-$, $V_+ \geq 0$, $V_- \leq 0$ with $V_+ \in L^1_{\text{loc}}(\mathbb{R}^\nu \backslash G)$ and with $H_0 + (1+\delta)V_-$ bounded from below for some $\delta > 0$. Let $H = H_0 + V$. Then $e^{-tH} - e^{-tH_0}$ is trace class for all $t > 0$ and, in particular, $\Omega^\pm(H, H_0)$ exist and are complete.

Proof Consider A, B, C as in Theorem 21.4. A is trace class as it is there. C is trace class as in the last lemma. Moreover, as in Theorem 21.4, we can eliminate the V_+ and need only show that the analogs of P, Q, P', Q' with H_0 replaced by $\tilde{H} = H_0 + V_-$, etc., are Hilbert–Schmidt. This we do by induction, showing they lie in $\mathscr{I}_p$, $p = 2^n$. By hypothesis, $(H_0 + V_- + c)^{-1/2} \leq \alpha(H_0 + 1)^{-1/2}$ so $(1 + X^2)^{-k}(\tilde{H} + c)^{-1/2}(1 + X^2)^{-k} \in \mathscr{I}_{p_0}$ for some k and $p_0 = 2^{n_0}$. It follows that $(1 + X^2)^{-k}(\tilde{H} + c)^{-1/4}$ and thus $(1 + X^2)^{-k}e^{-t\tilde{H}}$ lie in $\mathscr{I}_{2p_0}$. By Lemma 21.6, $(1 + X^2)^l(e^{-t\tilde{H}} - e^{-t\tilde{H}'})(1 + X^2)^m$ is in $\mathscr{I}_{2p_0}$ and by Lemma 21.7, $(1 + X^2)^{-k}e^{-t\tilde{H}'}$ is in $\mathscr{I}_{2p_0}$. Using the P, Q factorization of $e^{-t\tilde{H}} - e^{-t\tilde{H}'}$, we see that this difference lies in $\mathscr{I}_{p_0}$. Since $(1 + X^2)^{-k}e^{-t\tilde{H}'}$ is also in $\mathscr{I}_{p_0}$ (by Lemma 21.7), we see that $(1 + X^2)^{-k}e^{-t\tilde{H}}$ is in $\mathscr{I}_{p_0}$. We can now repeat the above argument $(n_0 - 1)$ times and conclude that all the required operators lie in $\mathscr{I}_2$. ∎

Remark Some condition like $H_0 + (1 + \delta)V_- \geq -a$ is needed for completeness since Pearson [200] has constructed a potential V for which V is C^∞ away from zero, supp V is compact, $H_0 + V$ is bounded below and essentially self-adjoint on $C_0^\infty(\mathbb{R}^3\backslash 0)$, $\Omega^\pm(H, H_0)$ exist, but $\operatorname{Ran}\Omega^+ \neq \operatorname{Ran}\Omega^-$!

22. Crushed Ice and the Wiener Sausage

The title of this section suggests a cocktail party and, indeed, our formal goal will be to understand the first law of cocktail-dynamics: crushed ice is a more effective cooling agent than block ice. That is, n small balls of ice each of volume $4\pi r^3/3$ are a more efficient cooler than one chunk of volume $4\pi n r^3/3$. The folk wisdom is that this is due to surface area; i.e., the n balls are roughly comparable to a chunk with surface area $4\pi n r^2$. In fact this intuition is wrong; the relevant factor is nr: explicitly n coolers of radius r_n will make no difference if $nr_n \to 0$ as $n \to \infty$ and will be infinitely efficient if $nr_n \to \infty$ and the spheres are distributed "throughout" the region to be cooled in a suitable sense. Thus, crushed ice is even better than the folk think!

The model we consider is a bounded, open region Ω in $\mathbb{R}^3$ and n balls, $B_1, \ldots, B_n$, of radius r_n about points $\mathbf{x}_1, \ldots, \mathbf{x}_n \in \Omega$. We consider the lowest eigenvalue $E_1(n)$ of $-\frac{1}{2}\Delta$ on $L^2(\Omega)$ with Dirichlet boundary conditions on $\partial\Omega$ and on $B_1, \ldots, B_n$. The solution of the heat equation $\dot{u} = \frac{1}{2}\Delta u$ with zero boundary conditions behaves asymptotically as $t \to \infty$ as $ce^{-tE_1(n)}$ with $c > 0$ if $u(x, t = 0) > 0$ so that the size of E_1 is a measure of the efficiency of cooling. Putting Dirichlet boundary conditions on $\partial\Omega$ is somewhat unnatural, corresponding to placing the "pitcher," Ω, in an ice bath; Neumann boundary conditions would be more natural but more complicated for both the analytic and probabilistic approaches described below.

This problem (without the picturesque cocktail analogy) was first considered by Kac [141] in relation to some related work by Kac and Luttinger [144]. Kac used Wiener integral methods. Rauch and Taylor [210] considered the problem analytically. We will reverse the historical order and first describe the analytical results following a lecture of Rauch [208] and then describe the probabilistic methods. The latter are more involved but give much more detailed information. In line with the methods, the analytical approach interprets "distributed throughout Ω" as "uniformly distributed" while the probabilistic approach as "randomly distributed"—due to fluctuations these are very different notions.

The crushed ice problem is one of a number of closely related problems, some of them quantum mechanical; these are discussed in [141, 144, 208–211]; here for intuition, we mention one other: Consider optical scattering off n spherical absorbers of radius r_n distributed uniformly through Ω; i.e., the wave equation $\ddot{u} = \Delta u$ where the Laplacian has vanishing boundary conditions on $\partial\Omega$. Then ([209, 210]) as $n \to \infty$, the absorbers become transparent (i.e., the scattering approaches that with no absorbers) if $nr_n \to 0$ and opaque (i.e., the scattering approaches that with zero boundary conditions on $\partial\Omega$) if $nr_n \to \infty$. At first sight this is surprising, for in the limit of ray optics, cross-sections (and so nr_n^2) should count. The point is that ray optics is only good for wavelengths short compared with the dimensions of the problem, so as $r_n \to 0$, for no wavelengths. What is relevant is Brownian paths—the occurrence of nr_n rather than nr_n^2 is an expression of the wigglyness of Brownian paths—indeed it is, in some sense, a restatement of the fact that the hitting probability for a sphere of radius r is proportional to r and not r^2 in $\nu = 3$ dimensions.

We first discuss the analytical approach.

Theorem 22.1 ([208]) Fix Ω and define $E_1(n)$ as above and let $E_1(0)$ be the lowest eigenvalue of $-\Delta^\Omega$ on $L^2(\Omega)$ with Dirichlet conditions on $\partial\Omega$. Suppose that $|\mathbf{x}_i - \mathbf{x}_j| \geq 4r_n$ for all i, j. Then for $nr_n \leq d$,

$$\sqrt{E_1(0)} \leq \sqrt{E_1(n)} \leq \sqrt{E_1(0)} + c\sqrt{nr_n} \tag{22.1}$$

for suitable nonzero constants c, d depending only on Ω (and independent of n, r_n and x_i). In particular, $E_1(n) \to E_1(0)$ as $n \to \infty$ if $nr_n \to 0$.

Remark Using more sophisticated methods, Rauch [208] proves a stronger result.

Proof $E_1(0) \leq E_1(n)$ is easy. Let ψ be a normalized eigenvector for $-\Delta^{\Omega}$ with eigenvalue $E_1(0)$. For any "piecewise C^1" function ϕ with $\phi = 0$ on each ball B_i and $0 \leq \phi \leq 1$, we have that

$$E_1(n) \leq \frac{\int |\nabla(\phi\psi)|^2\,dx}{\int |\phi\psi|^2\,dx}$$

by the variational principle. Now clearly as $\phi \leq 1$

$$\left(\int |\nabla(\phi\psi)|^2\,dx\right)^{1/2} \leq \left(\int |\nabla\psi|^2\,dx\right)^{1/2} + \|\psi\|_\infty \left(\int |\nabla\phi|^2\,dx\right)^{1/2}$$

and

$$\begin{aligned}\int |\phi\psi|^2\,dx &\geq 1 - \int_{\{\mathbf{x} \mid \phi(\mathbf{x}) < 1\}} |\psi|^2\,dx \\ &\geq 1 - \|\psi\|_\infty^2\, |\{\mathbf{x} \mid \phi(\mathbf{x}) < 1\}|\end{aligned}$$

Thus (22.1) follows if we find ϕ with

$$\|\nabla\phi\|_2^2 \leq anr_n$$

$$|\{\mathbf{x} \mid \phi(\mathbf{x}) < 1\}| \leq bnr_n^3$$

with a, b independent of $\mathbf{x}_i$, n, r_n. Let $\mathbf{x}_i$ be the center of ball B_i and write r for r_n. Let

$$\phi_i(\mathbf{x}) = \begin{cases} 0, & |\mathbf{x} - \mathbf{x}_i| \leq r \\ \dfrac{|\mathbf{x} - \mathbf{x}_i| - r}{r}, & r \leq |\mathbf{x} - \mathbf{x}_i| \leq 2r \\ 1, & |\mathbf{x} - \mathbf{x}_i| \geq 2r \end{cases}$$

and take $\phi = \prod_{i=1}^n \phi_i$. Then the supp$(1 - \phi_i)$ are disjoint so

$$\|\nabla\phi\|_2^2 = \sum_i \|\nabla\phi_i\|_2^2 = nr^{-2}\frac{4\pi}{3}(7r^3) = \frac{28\pi}{3}nr$$

and

$$|\{\mathbf{x} \mid \phi(\mathbf{x}) < 1\}| = \sum_i |\{\mathbf{x} \mid \phi_i(\mathbf{x}) < 1\}| = \frac{32\pi}{3}nr^3 \quad \blacksquare$$

Definition We say that the B_i are **uniformly distributed of degree m** if and only if there exist R_n so that if $\tilde{B}_i$ are the balls of radius R_n about x_i, then (a) $\Omega \subset \bigcup \tilde{B}_i$ and (b) each $y \in \Omega$ is contained in at most m different $\tilde{B}_i$.

Theorem 22.2 ([208]) If the B_i are uniformly distributed of degree m, then

$$E_1(n) \geq \frac{1.5 - \varepsilon}{m^2} \frac{nr_n}{|\Omega|}\left(\frac{4\pi}{3}\right) \tag{22.2}$$

where ε is a function of $n^{1/3}r_n$ going to zero as $n^{1/3}r_n \to 0$. In particular, if $nr_n \to \infty$, $n^{1/3}r_n \to 0$, then $E_1(n) \to \infty$.

Remark If $nr_n \to \infty$, we can always decrease r_n [which will only decrease $E_1(n)$] so that $nr_n \to \infty$, $n^{1/3}r_n \to 0$. [Notice that uniform distribution is only a property of the x_i *not* of the r_n's.]

Proof Let $c(R, r)$ be the lowest eigenvalue of $-\frac{1}{2}\Delta$ in $L^2(\{\mathbf{x} \mid r < |\mathbf{x}| < R\})$ with the boundary conditions $\phi(\mathbf{x}) = 0$ if $|\mathbf{x}| = r$, $\partial\phi/\partial n = 0$ if $|\mathbf{x}| = R$. Then $c(R, r) = R^{-2}f(r/R)$ by scaling. We first claim that

$$\lim_{y \downarrow 0} y^{-1}f(y) = 1.5 \tag{22.3}$$

For choose $R = 1$, let $\phi(\mathbf{x})$ be the required eigenfunction and note that, by a partial wave expansion, $\phi(\mathbf{x}) = |\mathbf{x}|^{-1}u(|\mathbf{x}|)$ where $-u'' = k^2u$, $f(r) = \frac{1}{2}k^2$, $u(r) = 0$, and $u' - (1/x)u = 0$ at $x = 1$. Thus $u(x) = \sin k(x - r)$ where k is the smallest solution of

$$k\cos(k(1 - r)) = \sin(k(1 - r))$$

Using $\tan x \sim x + \frac{1}{3}x^3 + O(x^5)$ one easily sees that $k^2 \sim 3r$ as $r \to 0$. This proves (22.3).

Thus, using the fact that Neumann boundary conditions correspond to no condition on the form domain [217], for any trial function vanishing on B_i:

$$\int_{\tilde{B}_i} |\nabla\psi|^2\,dx \geq (1.5 - \varepsilon)\frac{r_n}{R_n^3}\int_{\tilde{B}_i} |\psi|^2\,dx \tag{22.4}$$

where $\varepsilon \downarrow 0$ as $r_n/R_n \to 0$. Using

$$m\int_\Omega |\nabla\psi|^2\,dx \geq \sum_i \int_{\tilde{B}_i} |\nabla\psi|^2\,dx$$

$$\sum_i \int_{\tilde{B}_i} |\psi|^2 \geq \int_\Omega |\psi|^2$$

and $m|\Omega| \geq \frac{4}{3}\pi nR_n^3 \geq |\Omega|$, one obtains (22.2) from (22.4). ∎

Remark While the above uses no Wiener integrals, it implies something about Wiener paths; namely, if $\{B_i^{(n)}\}_{i=1}^n$ is a family of uniformly distributed spheres with $nr_n \to \infty$, then almost every Wiener path $\boldsymbol{\omega}$ has the property that for all large n, $\boldsymbol{\omega}$ hits some $B_i^{(n)}$.

* * *

Next we turn to the probabilistic approach. Let $x = (\mathbf{x}_1, \ldots)$ be an infinite sequence of points in Ω. We put the product measure $d\gamma = \times_{n=1}^{\infty} |\Omega|^{-1}\, dx_n$ on these sequences; i.e., the $\mathbf{x}$'s are independently distributed "uniformly" through Ω (intuitively, they are randomly placed). Fix a sequence $r_1, r_2, \ldots$ of positive numbers once and for all. Given x in Ω^∞ and n we let $H(x, n)$ be $-\frac{1}{2}\Delta$ on $L^2(\Omega)$ with Dirichlet boundary conditions on $\partial\Omega$ and on the n balls of radius r_n about $\mathbf{x}_1, \ldots, \mathbf{x}_n$. Let $E_k(x, n)$ be the kth eigenvalue of $H(x, n)$. We view $E_k(x, n)$ as random variables, $E_k(n)$, on $(\Omega^\infty, d\gamma)$. To state the main result we need a new notion:

Definition Let f_n, f be random variables. We say that f_n **converges to f in probability** if and only if

$$\lim_{n\to\infty} E(|f_n - f| > \varepsilon) = 0$$

for each $\varepsilon > 0$.

Remark It is easy to see that $\|f_n - f\|_p \to 0$ implies convergence in probability and also that $f_n(x) \to f(x)$ for almost every x implies convergence in probability. From the former fact one easily sees that if $E(f_n) \to a$ and $E(f_n^2) \to a^2$, then $f_n \to a$ in probability.

Theorem 22.3 ([141]) Let $nr_n \to \alpha$ (α may be zero or infinity) as $n \to \infty$. Then

$$E_k(n) \to E_k(0) + 2\pi\alpha|\Omega|^{-1} \tag{22.5}$$

in probability for each k.

Thus random impurities produce a constant shift of the spectrum. Since $E_k(n)$ is monotone increasing in r_n, it is easy to see that the $0 < \alpha < \infty$ results imply the $\alpha = 0, \infty$ results also, so henceforth we suppose that $0 < \alpha < \infty$. The proof of Theorem 22.3 requires some elaborate preliminary machinery. First, we need a Feynman–Kac formula for $\mathrm{Tr}(e^{-tH(n)})$, a formula already used in Section 10.

Theorem 22.4 Let Ω be an arbitrary open bounded set in $\mathbb{R}^\nu$ and let H^Ω be the operator $-\frac{1}{2}\Delta$ on $L^2(\Omega)$ with Dirichlet conditions on $\partial\Omega$. Then, for any $t > 0$,

$$\operatorname{Tr}(e^{-tH^\Omega}) = (2\pi t)^{-\nu/2} \int_\Omega d^\nu x\, E(\mathbf{x} + \mathbf{b}(s) \in \Omega, 0 \le s \le t \mid \mathbf{b}(t) = 0) \quad (22.6)$$

Proof Let V be the potential used in the proof of Theorem 21.1 and let $H(\lambda, m, \varepsilon) = -\frac{1}{2}\Delta + \lambda \min(V, m) + \varepsilon x^2$. Then, $e^{-tH(\lambda, m, \varepsilon)}$ is trace class and its trace is given by (9.5). We take $m \to \infty$, $\varepsilon \to 0$ and then $\lambda \to 0$ and recover the required result. ∎

Thus

$$\operatorname{Tr}(e^{-tH(x;n)}) = (2\pi t)^{-3/2} \int d^3y\, E(\mathbf{y} + \mathbf{b}(s) \in \Omega;$$

$$\mathbf{y} + \mathbf{b}(s) \notin B_i(x, n), i = 1, \ldots, n, 0 \le s \le t \mid \mathbf{b}(t) = 0) \quad (22.7)$$

where $B_i(x, n)$ are the balls excluded in the definition of $H(x, n)$. Let $\langle \cdot \rangle$ denote expectation with respect to $d\gamma$ on Ω^∞. Our goal is to show that

$$\langle \operatorname{Tr}(e^{-tH(x;n)}) \rangle \to e^{-2\pi t\alpha|\Omega|^{-1}} \operatorname{Tr}(e^{-tH(0)}) \quad (22.8)$$

$$\langle |\operatorname{Tr}(e^{-tH(x;n)})|^2 \rangle \to |e^{-2\pi t\alpha|\Omega|^{-1}} \operatorname{Tr}(e^{-tH(0)})|^2 \quad (22.9)$$

so that for each t, $\sum e^{-tE_k(n)}$ converges to $\sum e^{-t(E_k(0) + 2\pi\alpha|\Omega|^{-1})}$ in probability.

To prove (22.8), we interchange the $d\gamma$ integration with the y and Db integration and do the $d\gamma$ integration first. Thus we will have a fixed path b and want to know whether a randomly placed sphere of radius δ is hit by b. This clearly depends on the volume of a δ-neighborhood of b. We therefore make the following definition.

Definition $W_\delta(a, c)$ is the set-valued random variable on Brownian paths defined by

$$W_\delta(a, c)(\mathbf{b}) = \{\mathbf{x} \mid |\mathbf{x} - \mathbf{b}(s)| \le \delta \text{ for some } s \in [a, c]\}$$

$W_\delta(t) \equiv W_\delta(0, t)$ is called the **Wiener sausage** for obvious geometric and punnish reasons.

We work in $\nu = 3$ dimensions throughout. We are interested in $|W_\delta(t)|$ as $\delta \to 0$. Because of scaling

$$|W_{\lambda\delta}(\lambda^2 t)| \doteq \lambda^3 |W_\delta(t)|$$

for any $\lambda > 0$ so we can consider $|W_1(t)|$ as $t \to \infty$ and then deduce the behavior of $|W_\delta(t)| \doteq \delta^3 |W_1(\delta^{-2}t)|$ as $\delta \to 0$.

Theorem 22.5 (Spitzer [266])

$$E(|W_\delta(t)|) = 2\pi\delta t + 4\delta^2(2\pi t)^{1/2} + (4\pi/3)\delta^3 \tag{22.10}$$

and, in particular,

$$\lim_{\delta \downarrow 0} \delta^{-1} E(|W_\delta(t)|) = 2\pi t \tag{22.11}$$

$$\lim_{t \to \infty} t^{-1} E(|W_\delta(t)|) = 2\pi\delta \tag{22.12}$$

Proof Without loss take $\delta = 1$. Let

$$f(\mathbf{x}, t) = E(\mathbf{x} + \mathbf{b}(s) \in B_1 \text{ for some } 0 \le s \le t)$$

where B_1 is the ball of radius 1 about 0. Now

$$\begin{aligned} E(|W_1(t)|) &= \int Db \int dx \; \{\mathbf{x} \mid \mathbf{x} \in W_1(t)\} \\ &= \int dx \int Db \; \{\mathbf{x} \mid -\mathbf{x} + \mathbf{b}(s) \in B_1 \text{ for some } 0 \le s \le t\} \\ &= \int dx \; f(\mathbf{x}, t) \end{aligned}$$

Next notice that $f(\mathbf{x}, t)$ has the following properties:

(a) for $|\mathbf{x}| > 1$, $\lim_{t \downarrow 0} f(\mathbf{x}, t) = 0$,
(b) for $t > 0$, $\lim_{|\mathbf{x}| \to 1} f(\mathbf{x}, t) = 1$,
(c) in $|\mathbf{x}| > 1$, $t > 0$, f obeys

$$\frac{\partial f}{\partial t} = \frac{1}{2} \Delta f$$

(d) for $|\mathbf{x}| \le 1$, $t > 0$, $f(\mathbf{x}, t) = 1$.

(d) is trivial and (a), (b) follow from the kind of hitting probability considerations in Section 7 (see especially the Aside following Lemma 7.21). (c) follows from (7.12) which says that

$$1 - f = \int P_D(\mathbf{x}, \mathbf{y}; t) \, dy$$

and the formula

$$\frac{\partial P_{\mathrm{D}}}{\partial t} = \frac{1}{2}\,\Delta P_{\mathrm{D}}(\mathbf{x}, \mathbf{y}; t)$$

(using the convergence of the integral, one gets the differential equation for f a priori in distributional sense).

Since f is spherically symmetric, we take $|x|\,f(\mathbf{x}) = g(|x|)$ and find that

$$\frac{\partial g(r, t)}{\partial t} = \frac{1}{2}\frac{\partial^2}{\partial r^2}\, g(r, t)$$

with the boundary conditions $g(r, t) \to 0$ as $t \to 0$ for $r > 1$ and $g(1, t) = 1$. The equation and first boundary condition suggest we try

$$g(r, t) = \int_{-\infty}^{0} (2\pi t)^{-1/2} e^{-(r-y-1)^2/2t}\, d\rho(y)$$

and the second boundary condition gives $d\rho = 2\,dy$. Thus

$$f(\mathbf{x}, t) = 2r^{-1}(2\pi t)^{-1/2} \int_0^\infty dz \exp[-(2t)^{-1}(z + |\mathbf{x}| - 1)^2]$$

for $|\mathbf{x}| \geq 1$. We thus compute

$$\begin{aligned}
\int_{|\mathbf{x}| \geq 1} f(\mathbf{x}, t)\, d^3x &= (4\pi)2(2\pi t)^{-1/2} \int_1^\infty r\, dr \int_0^\infty e^{-(z+r-1)^2/2t}\, dz \\
&= (4\pi)2(2\pi t)^{-1/2} \int_1^\infty r\, dr \int_r^\infty e^{-(y-1)^2/2t}\, dy \\
&= (4\pi)2(2\pi t)^{-1/2} \int_1^\infty dy\, e^{-(y-1)^2/2t} \int_1^y r\, dr \\
&= (4\pi)(2\pi t)^{-1/2} \int_0^\infty (u^2 + 2u) e^{-u^2/2t}\, du \\
&= 2\pi t + 4(2\pi t)^{1/2}
\end{aligned}$$

Since $\int_{|\mathbf{x}| \leq 1} f(\mathbf{x}, t)\, d^3x = 4\pi/3$, we have (22.10). ∎

Remarks 1. Spitzer [266] considers a more general problem, namely $Q(t) \equiv \int_{\mathbf{x} \notin K} f(\mathbf{x}, t)\, d^3x$ where B_1 is replaced by a general set K in defining $f(\mathbf{x}, t)$. He finds the asymptotics of $Q(t)$ to $O(t^{1/2})$. In particular, the $2\pi\delta$ in (22.10) enters as the Newtonian capacity of the ball. Of course, Spitzer's considerations in general are more sophisticated than those above; (22.10) is mentioned as an aside.

2. (22.11) is a quantitative expression of the "fatness" of Wiener paths. For a smooth curve of length l, the volume V_δ of the neighborhood goes as $\pi\delta^2 l$ (note δ^2, not δ) as $\delta \to 0$. In some kind of sense, (22.11) and its analog in higher dimensions say that the image of the Wiener path has "dimension 2."

It is not quite true that $|W_1(nT)| = \sum_{j=1}^{N} |W_1((j-1)nT, jnT)|$ since the successive pieces of sausage overlap, but if T is large this should be negligible. The pieces in this last sum are independent. It is therefore reasonable that some kind of strong law of large numbers holds and that $|W_1(t)|/|t| \to 2\pi$ for almost every **b**. The first half of the following theorem is due to Kesten, Spitzer, and Whitman (quoted in Spitzer [265]). The second half is stated by Kac [141] without proof. The proof we give is based in part on that of Kesten, Spitzer, and Whitman, and in part on unpublished remarks of Varadhan.

Theorem 22.6 With probability one,

$$\lim_{t\to\infty} \left[\frac{|W_\delta(t)|}{t}\right] = 2\pi\delta \tag{22.13a}$$

for fixed δ and

$$\lim_{\delta\downarrow 0} \left[\frac{|W_\delta(t)|}{\delta}\right] = 2\pi t \tag{22.13b}$$

for fixed t.

Proof Without loss, we can take $\delta = 1$ in (22.13a) and $t = 1$ in (22.13b), since the other results then follow by scaling. Now let $A_\delta \equiv |W_\delta(1)|/\delta$ and $B_\delta \equiv |W_1(\delta^{-2})|/\delta^{-2}$. For each *fixed* δ, A_δ and B_δ have the same probability distribution, so that L^p convergence theorems like $E(A_\delta) \to 2\pi$ for one imply the analogous result for the other. However, the joint distributions as δ varies are not the same: For example, $2A_{2\delta} \geq A_\delta$ with probability one while $B_{2\delta} \leq 4B_\delta$ with probability one but these inequalities (which rely on the fact that $|W_\delta(t)|$ is increasing in t and δ) are not true almost everywhere if A and B are interchanged. Thus (22.13a) and (22.13b) are not equivalent statements.

We will prove below that for $\delta < 1$,

$$E([B_\delta - 2\pi]^2) \leq C\delta^{1/2} \tag{22.14}$$

Assuming (22.14), let us prove (22.13). Fix $\rho \in (0, 1)$. Then, by the standard Borel–Cantelli argument we have used often before (see the proof of the strong law of large numbers),

$$\lim_{n\to\infty} B_{\rho^n} = 2\pi$$

with probability one. Moreover, since $W_1(T)$ is monotone in T, if $\rho^{n+1} \leq \delta \leq \rho^n$, then

$$\rho^2 B_{\rho^n} \leq B_\delta \leq \rho^{-2} B_{\rho^{n+1}}$$

so that almost everywhere

$$2\pi\rho^2 \leq \underline{\lim}\, B_\delta \leq \overline{\lim}\, B_\delta < 2\pi\rho^{-2}$$

Since ρ is arbitrary in (0, 1), (22.13a) follows.

To prove (22.13b), we first note that since (22.14) involves a single δ, the same inequality holds for A_δ so that, as before,

$$\lim_{n\to\infty} A_{\rho^n} = 2\pi$$

with probability one. Using now the inequality

$$\rho A_{\rho^{n+1}} \leq A_\delta \leq \rho^{-1} A_{\rho^n}$$

which comes from the monotonicity of $W_\delta(1)$ in δ, we obtain (22.13b) as above.

This leaves the proof of (22.14). Let $r_{T,n} = |W_1((n-1)T, nT)|$. For fixed T, the $r_{T,n}$ are identically distributed and they are independent by the basic property of Brownian motion of starting afresh. By (22.10):

$$|E(r_{T,n})/T - 2\pi| \leq C_1 T^{-1/2} \tag{22.15a}$$

for all $T \geq 1$. Clearly

$$|W_1(t)| \leq \sum_{j=1}^{[t/T]+1} r_{T,j} \tag{22.15b}$$

which implies, taking $T = 1$ and changing t to T, that

$$E(r_{T,n}^2) \leq (T+1)^2 E(|W_1(1)|^2) \tag{22.15c}$$

In this last inequality, $E(|W_1(1)|^2) < \infty$, for if $\max_{0\leq s\leq 1}|\mathbf{b}(s)| \leq m+1$, then $W_1(1)$ is less than or equal to the volume of a sphere of radius $m+2$, so

$$E(|W_1(1)|^2) \leq \sum_{m=1}^{\infty} \left(\frac{4\pi}{3}(m+2)^3\right)^2 E\left(\max_{0\leq s\leq 1} |b(s)| \geq m\right) < \infty$$

Now use (22.15b) again and the independence to note that

$$E(|W_1(t)|^2) \leq \left(\left[\frac{t}{T}\right]+1\right)E(r_{T,1}^2) + \left(\left[\frac{t}{T}\right]+1\right)^2 E(r_{T,1})^2$$

so using (22.15a, c) we see that (taking $T = t^{1/2}$)

$$E([|W_1(t)| - 2\pi t]^2) \leq Ct^{7/4}$$

for $t > 1$. This is just a restatement of (22.14). ▮

In (22.7) it is not arbitrary Wiener paths that enter but paths with $\boldsymbol{\omega}(0) = \boldsymbol{\omega}(t)$. We therefore need the following corollary.

Corollary 22.7 For almost every path for the (three-dimensional) Brownian bridge

$$\lim_{\delta \downarrow 0} \delta^{-1} |W_\delta(1)| = 2\pi$$

Proof By considerations similar to those in Theorem 5.5, the Brownian bridge α restricted to $[0, t]$ is absolutely continuous with respect to Brownian motion so long as $t < 1$. Thus for any $t < 1$ and almost every Brownian bridge path:

$$\lim_{\delta \downarrow 0} \delta^{-1} |W_\delta(t)| = 2\pi t$$

Since $|W_\delta(1)| \geq |W_\delta(t)|$ we see that

$$\underline{\lim}_{\delta \downarrow 0} \delta^{-1} |\mathrm{W}_\delta(1)| \geq 2\pi$$

Since $\alpha(1 - t) \doteq \alpha(t)$ we see that $|W_\delta(\frac{1}{2})| \doteq |W_\delta(\frac{1}{2}, 1)|$ so that $\delta^{-1}|W_\delta(\frac{1}{2}, 1)| \to \pi$ for almost every α-path. Since

$$|W_\delta(1)| \leq |W_\delta(\tfrac{1}{2})| + |W_\delta(\tfrac{1}{2}, 1)|$$

we have that

$$\overline{\lim}_{\delta \downarrow 0} \delta^{-1} |W_\delta(1)| \leq \pi + \pi \quad \blacksquare$$

We now have the tools for proving (22.8). To prove (22.9) we must consider a somewhat more complicated problem. Namely, fix $\mathbf{x}$ and $\mathbf{y}$ (perhaps equal) and let $\mathbf{b}, \mathbf{b}'$ be independent Brownian motions with sausages W, W'. Then we need to know that

$$\delta^{-1} |(\mathbf{x} + W_\delta(1)) \cup (\mathbf{y} + W'_\delta(1))| \to 4\pi$$

i.e., that the overlap of the two sausages is negligible relative to the total size as $\delta \to 0$. A moment's reflection will convince the reader that this should be true and that stopping times are the right tools for its proof.

Theorem 22.8 Let W and W' be the Wiener sausages for two independent Brownian motions and let $\mathbf{x}, \mathbf{y}$ be fixed. Then, with probability one

$$\delta^{-1} |(\mathbf{x} + W_\delta(t)) \cup (\mathbf{y} + W'_\delta(t))| \to 4\pi t$$

as $\delta \downarrow 0$. The same result is true for the (three-dimensional) Brownian bridge and $t \le 1$.

Proof Without loss take $\mathbf{y} = 0$. Consider first the case of Brownian motion. Clearly, we need to show that

$$\delta^{-1}|(\mathbf{x} + W_\delta(t)) \cap W'_\delta(t)| \to 0 \tag{22.16}$$

Fix the $\mathbf{b}'$ path and ε and define stopping times for the $\mathbf{b}$-path, $\tau_1, \sigma_1, \tau_2, \ldots,$ inductively by

$$\tau_1 = \inf\{s \mid b(s) + x \in W'_{2\varepsilon}(t)\}$$
$$\sigma_i = \inf\{s \ge \tau_i \mid b(s) + x \notin W'_{3\varepsilon}(t)\}$$
$$\tau_{i+1} = \inf\{s \ge \sigma_i \mid b(s) + x \in W'_{2\varepsilon}(t)\}$$

with the proviso that the stopping times are t if no such s exists. Clearly for $\delta \le \varepsilon$, $x + W_\delta$ and W'_ε intersect only for b-times belonging to the intervals (τ_i, σ_i), so that

$$|(x + W_\delta(t)) \cap W'_\varepsilon(t)| \le \sum_{i=1}^{\infty} |W_\delta(\tau_i, \sigma_i)|$$

By the Dynkin–Hunt theorem, each $\mathbf{b}(s + \tau_i) - \mathbf{b}(\tau_i)$ is a Brownian motion, so with probability one

$$\overline{\lim_{\delta \downarrow 0}}\, \delta^{-1}|(\mathbf{x} + W_\delta(t)) \cap W'_\varepsilon(t)| \le 2\pi \sum_{i=1}^{\infty} (\sigma_i - \tau_i)$$
$$\le 2\pi|\{s \mid \mathbf{x} + \mathbf{b}(s) \in W'_{3\varepsilon}(t); s \le t\}| \equiv 2\pi g_\varepsilon$$

Thus, (22.16) follows from

$$\lim_{\varepsilon \downarrow 0} g_\varepsilon = 0$$

with probability one. But g_ε decreases as ε does so, by the monotone convergence theorem, it suffices that

$$\lim_{\varepsilon \downarrow 0} E(g_\varepsilon) = 0 \tag{22.17}$$

This we can compute by Fubini's theorem:

$$E(g_\varepsilon) = \int Db \int_0^t ds\, \{(s, \mathbf{b}) \mid \mathbf{x} + \mathbf{b}(s) \in W'_{3\varepsilon}(t)\}$$
$$= \int_0^t ds\, E(\mathbf{x} + \mathbf{b}(s) \in W'_{3\varepsilon}(t))$$

For each $s \ne 0$, the probability distribution of $\mathbf{x} + \mathbf{b}(s)$ is absolutely continuous so that $E(\mathbf{x} + \mathbf{b}(s) \in W'_{3\varepsilon}(t)) \downarrow 0$ as $\varepsilon \to 0$ since $|W'_{3\varepsilon}| \to 0$ by Theorem

22.6. Thus (22.17) follows by the monotone convergence theorem. This proves the theorem in the Brownian motion case.

For the Brownian bridge, we note first that for $t < 1$, the result follows from the Brownian motion case as in Corollary 22.7. Thus, for $t = 1$

$$\varliminf_{\delta \downarrow 0} \delta^{-1} |(\mathbf{x} + W_\delta) \cup W'_\delta| \geq \sup_{t<1}(4\pi t) = 4\pi$$

On the other hand, by Corollary 22.7

$$\varlimsup_{\delta \downarrow 0} \delta^{-1} |(\mathbf{x} + W_\delta) \cup W'_\delta| \leq \lim \delta^{-1}(|W_\delta| + |W'_\delta|) = 4\pi \quad \blacksquare$$

We now have the tools for proving (22.8) and (22.9). To complete the proof of (22.5) we need one more fact:

Lemma 22.9 Suppose we are given a doubly indexed family of random variables $E_j(n)$ and two sequences a_j and b_j of numbers so that

(i) $$0 \leq E_j(n) \leq E_{j+1}(n)$$

(ii) $$a_j \leq E_j(n)$$

(iii) $$a_j \leq b_j$$

(iv) $$\textstyle\sum_j e^{-ta_j} < \infty \qquad \text{for each} \quad t$$

(v) $$\textstyle\sum_j e^{-tE_j(n)} \to \sum_j e^{-tb_j}$$

in probability as $n \to \infty$ for each $t > 0$. Then, as $n \to \infty$

$$E_j(n) \to b_j$$

in probability.

Proof It suffices to prove that $E_1(n) \to b_1$ in probability, for once we have that, $\sum_{j \geq 2} e^{-tE_j(n)}$ will converge in probability to $\sum_{j \geq 2} e^{-tb_j}$ so by induction we will obtain $E_j(n) \to b_j$. Let us show that the probability that $E_1(n) \geq b_1 + \delta$ goes to zero for each $\delta > 0$. The proof of the other inequality is similar.

Suppose that $E_1(n) \geq b_1 + \delta$. Pick j_0 so that $a_{j_0+1} \geq b_1 + \delta$. Then for $t \geq 1$

$$\begin{aligned}
\sum_j e^{-tE_j(n)} &= \sum_{j=1}^{j_0} e^{-tE_j(n)} + \sum_{j_0+1}^{\infty} e^{-tE_j(n)} \\
&\leq \sum_{j=1}^{j_0} e^{-tE_1(n)} + \sum_{j_0+1}^{\infty} e^{-ta_j} \\
&\leq e^{-t(b_1+\delta)} \left[j_0 + \sum_{j_0+1}^{\infty} e^{-(a_j - b_1 - \delta)} \right]
\end{aligned}$$

Thus, for suitable t large and some $\varepsilon > 0$,

$$\sum e^{-tE_j(n)} \leq e^{-tb_1} - \varepsilon$$

Thus condition (v) implies that $E(E_1(n) \geq b_1 + \delta) \to 0$ as $n \to \infty$. ∎

Proof of Theorem 22.3 Fix $\mathbf{b}$. Then

$$\langle \mathbf{y} + \mathbf{b}(s) \notin B_i(x; n); 0 \leq s \leq t \rangle = 1 - \frac{|(\mathbf{y} + W_{r_n}(t)) \cap \Omega|}{|\Omega|}$$

since the center $\mathbf{x}_i$ must lie outside $W_{r_n}(t)$. Since the $\mathbf{x}_i$'s are independent, we have by (22.7) that

$$\langle \mathrm{Tr}(e^{-tH(x;n)}) \rangle = (2\pi t)^{-3/2} \int d^3y \, E([1 - |\Omega|^{-1} |(W_{r_n}(t) + \mathbf{y}) \cap \Omega|]^n;$$
$$\mathbf{y} + \mathbf{b}(s) \in \Omega \mid \mathbf{b}(t) = 0)$$

For almost every path with $\mathbf{b}(t) = 0$ and $\mathbf{y} + \mathbf{b}(s) \in \Omega$, $\mathbf{y} + W_{r_n}(t) \subset \Omega$ for r_n small and thus we can replace $|(\mathbf{y} + W_{r_n}(t)) \cap \Omega|$ by $|W_{r_n}(t)|$ for n large. For almost every path

$$[1 - |\Omega|^{-1} |W_{r_n}(t)|]^n \to e^{-2\pi t\alpha|\Omega|^{-1}}$$

by the compound interest formula and

$$n|W_{r_n}(t)| = (nr_n)\frac{|W_{r_n}(t)|}{r_n} \to (2\pi t)\alpha$$

on account of Corollary 22.7. Thus, by the dominated convergence theorem and Theorem 22.4,

$$\lim_{n\to\infty} \langle \mathrm{Tr}(e^{-tH(x;n)}) \rangle = (2\pi t)^{-3/2} e^{-2\pi t\alpha|\Omega|^{-1}} \int d^3y \, E(\mathbf{y} + \mathbf{b}(s) \in \Omega \mid \mathbf{b}(t) = 0)$$
$$= e^{-2\pi t\alpha|\Omega|^{-1}} \, \mathrm{Tr}(e^{-tH(0)})$$

Thus, we have (22.8). The proof of (22.9) is similar, except we need two independent Brownian motions to write out $\mathrm{Tr}(\cdot)^2$ and then doing the $\langle \cdot \rangle$-average first, we need information on the volume of two independent sausages. This information is provided by Theorem 22.8 giving (22.9).

By the Remark preceding Theorem 22.3, (22.8) and (22.9) imply that $\mathrm{Tr}(e^{-tH(x;n)})$ converges in probability to $\mathrm{Tr}(e^{-t(H(0)+2\pi\alpha|\Omega|^{-1})})$ whence (22.5) follows from Lemma 22.9. [Choose $a_j = E_j(0)$.] ∎

Remark Rauch and Taylor [210] apply operator theoretic methods to avoid some of the more involved probabilistic considerations above, especially Theorem 22.8. For example, as in the proof of Theorem 22.3, we can use the Feynman–Kac formula and Theorem 22.6 to see that

$$\lim_{n\to\infty} \langle (f, e^{-tH(x;n)}g) \rangle = (f, e^{-tA}g) \tag{22.18}$$

where $A = H(0) + 2\pi\alpha|\Omega|^{-1}$. Since (22.18) holds for t and $2t$ we have that

$$\lim_{n\to\infty} \langle \|(e^{-tH(x;n)} - e^{-tA})f\|^2 \rangle = 0$$

This "strong convergence in probability," which is neither stronger nor weaker than (22.5), is often physically useful. Similarly, we are able to prove (22.5) directly from (22.8) and (22.18) without Lemma 22.9 or (22.9): For (22.18) and (22.8) for t and $2t$ imply that

$$\langle \mathrm{Tr}[(e^{-tH(x;n)} - e^{-tA})^2] \rangle = 0 \tag{22.19}$$

Moreover, a theorem of Lidskii (see Kato [152, Section II.6.5]) implies that

$$\sum (\mu_n(C) - \mu_n(D))^2 \leq \mathrm{Tr}((C - D)^2)$$

for any bounded positive operators C and D with eigenvalues $\mu_1(C) \geq \mu_2(C) \geq \cdots$ so that (22.19) yields (22.5).

23. The Statistical Mechanics of Charged Particles with Positive Definite Interactions

Most of the applications described so far have involved the nonrelativistic quantum mechanics of at most a few particles and Gaussian measures on continuous functions (Db, Dq, and $D\alpha$) or their perturbations. In the next two sections we want to briefly indicate some applications to statistical mechanics and quantum field theory. These applications will require us to utilize some more complicated Gaussian processes than those considered thus far including some which cannot be naturally realized on the space of continuous functions.

In this section, we consider some statistical mechanical models. There has been considerable application of the Wiener process itself to study quantum statistical mechanics. We will not discuss this here but we refer the reader to Edwards–Lenard [73] and Siegert [239] for two of the earliest discussions, and to Ginibre [99, 100] and Brydges–Federbush [27] for some deep applications to the construction of correlation functions at low β and/or z. Here we want to discuss a Gaussian realization of the basic objects of the classical statistical mechanics of a special but very interesting class of interactions. This realization has been described by various authors, e.g., [239, 1, 93]. By essentially passing first to a Wiener integral realization and then to this Gaussian realization inside the Wiener integrals, one can extend this formalism and the results below to quantum statistical mechanics with Boltzmann statistics (i.e., no symmetrization or antisymmetrization) (see [239, 94]). After presenting this Gaussian realization we will describe some beautiful results of Fröhlich and Park [94] controlling the thermodynamic limit for this special class of interactions at all β, z.

We want to consider ν-dimensional particles coming in two charges ± 1 interacting with a *bounded* potential $V(\mathbf{x} - \mathbf{y})$ obeying

$$\sum_{i,j=1}^{n} \bar{z}_i z_j V(\mathbf{x}_i - \mathbf{x}_j) \geq 0 \tag{23.1}$$

for all $z_1, \ldots, z_n \in \mathbb{C}$ and $\mathbf{x}_1, \ldots, \mathbf{x}_n \in \mathbb{R}^\nu$; that is, two particles with equal charges at points x and y have interaction $V(\mathbf{x} - \mathbf{y})$ and with opposite charges $-V(\mathbf{x} - \mathbf{y})$. The **grand canonical partition function** in volume $\Lambda \subset \mathbb{R}^\nu$ is defined by

$$\Xi_\Lambda(z, \beta) = \sum_{N=0}^{\infty} \frac{z^N}{N!} 2^{-N} \sum_{\substack{\varepsilon_j = \pm 1 \\ j=1,\ldots,N}} \int_{\Lambda^N} d^{N\nu}x \, e^{-\beta U_N(\mathbf{x};\, \varepsilon)} \tag{23.2}$$

where

$$U_N(\mathbf{x}_1, \ldots, \mathbf{x}_N; \varepsilon_1, \ldots, \varepsilon_N) = \sum_{i<j} \varepsilon_i \varepsilon_j V(\mathbf{x}_i - \mathbf{x}_j) \tag{23.3}$$

The **pressure** in region Λ is given by

$$P_\Lambda(z, \beta) = |\Lambda|^{-1} \log \Xi_\Lambda(z, \beta) \tag{23.4}$$

and the **correlation functions** (essentially giving the probability density of finding particles of charge ε_i at points $\mathbf{x}_i$) are given by

$$\begin{aligned} &\rho_\Lambda^{(n)}(\mathbf{x}_1, \ldots, \mathbf{x}_n; \varepsilon_1, \ldots, \varepsilon_n; z, \beta) \\ &\quad = \Xi_\Lambda^{-1} z^n \sum_{N=0}^{\infty} \frac{z^N}{N!} 2^{-N} \sum_{\substack{\varepsilon'_j = \pm 1 \\ j=1,\ldots,N}} \int_{\Lambda^N} d^{N\nu}x' \exp(-\beta U_{N+n}(\mathbf{x}, \mathbf{x}'; \varepsilon, \varepsilon')) \end{aligned} \tag{23.5}$$

The results below depend critically on the somewhat artificial restriction to a single fugacity z for both charges. For a discussion of the physics of the grand canonical formalism see Huang [129].

Our goal in this section is to prove the following.

Theorem 23.1 (Fröhlich–Park [94]) Under the hypothesis (23.1) we have that

(a) $\lim_{|\Lambda| \to \infty} P_\Lambda(z, \beta) \equiv P(z, \beta)$ exists for all positive z and β where the $\lim_{|\Lambda| \to \infty}$ is intended in the sense of choosing Λ to run through hypercubes and taking all the sides to infinity.

(b) $\lim_{\Lambda \to \mathbb{R}^\nu} \rho_\Lambda^{(n)}(\mathbf{x}_i, \varepsilon_i; z, \beta) \equiv \rho^{(n)}(\mathbf{x}_i, \varepsilon_i; z, \beta)$ exists for all positive z and β, and $\varepsilon_i = \pm 1$, $\mathbf{x}_i \in \mathbb{R}^\nu$ where the $\lim_{\Lambda \to \mathbb{R}^\nu}$ is intended in the sense of the net of all regions ordered by inclusion. The $\rho^{(n)}$ are translation invariant [i.e., $\rho^{(n)}(\mathbf{x}_i + \mathbf{a}) = \rho^{(n)}(\mathbf{x}_i)$ (same $\mathbf{a}$)], are charge reversal invariant (i.e., under $\varepsilon_j \to -\varepsilon_j$, all j), have any symmetry of V [i.e., if $V(Tx) = V(x)$ for some linear T, then $\rho^{(n)}(T\mathbf{x}_i) = \rho^{(n)}(\mathbf{x}_i)$] and are monotone increasing in z.

Remarks 1. Since V is supposed bounded, this result does not include the Coulomb potential but does include cutoff Coulomb potentials, e.g., $V(\mathbf{x}) = \int |\mathbf{x} - \mathbf{y}|^{-1} \rho(\mathbf{y})\, dy$ with $\rho \geq 0$ in $L^1 \cap L^\infty$ and positive definite. In fact, for Coulomb potentials, the integrals in (23.2) diverge; i.e., matter is only stable for quantum mechanics (needed for the traces which replace the integrals to converge) of fermions [needed for the sum in (23.2) to converge].

2. Essentially for free, the limit in (a) can be replaced by "van Hove convergence" (see [226]). Actually by a little more work, Fröhlich–Park [94] extend the limit to a very general type.

3. With only the small cost of slightly more complicated notation, one can extend all the results to "generalized charges" such as dipole moments; see [94].

4. It is useful to compare this result with those obtained by more "standard" methods [226]. The convergence of the pressure, part (a) of Theorem 23.1, can be proven under much weaker conditions than (23.1) but only if some rather strong falloff ($\sim |\mathbf{x}|^{-\nu-\varepsilon}$) is assumed on V; no falloff is required here. More significantly, convergence of the correlation functions is only known for general V's at small z or β and then only with considerably more effort than we will require! Since it can be absorbed into V, *we henceforth set* $\beta = 1$.

The basic philosophy with which we begin is that any positive definite function is crying out to be the covariance of a Gaussian process. Thus we construct a Gaussian process $\{q(\mathbf{x})\}_{\mathbf{x} \in \mathbb{R}^\nu}$ with covariance $V(\mathbf{x} - \mathbf{y})$; we use $d\mu(q)$ to denote the corresponding measure. Occasionally, it will be useful to think of $\sum_{i=1}^n a_i q(\mathbf{x}_i)$ as $\phi(\sum_{i=1}^n a_i \delta_{\mathbf{x}_i})$ in the notation of Corollary 2.4. If V is Hölder continuous, then one can prove a multidimensional Kolmogorov lemma and realize $d\mu$ on $C(\mathbb{R}^\nu)$ and in any event, by Minlos' theorem, we can realize $d\mu$ on $\mathscr{S}'(\mathbb{R}^\nu)$. In fact, the explicit realization of $d\mu$ will play no serious role. Of course, we have that ($\langle\cdot\rangle = \int \cdot\, d\mu$).

$$\left\langle \exp\left(i\sum_1^n a_i q(\mathbf{x}_i)\right)\right\rangle = \left\langle \exp\left(i\phi\left(\sum_1^n a_i \delta_{\mathbf{x}_i}\right)\right)\right\rangle$$
$$= \exp(-\tfrac{1}{2}\sum a_i a_j V(\mathbf{x}_i - \mathbf{x}_j))$$
$$= \exp\left(-\tfrac{1}{2}V(\mathbf{0})\sum a_i^2 - \sum_{i<j} a_i a_j V(\mathbf{x}_i - \mathbf{x}_j)\right)$$

Therefore

$$\left\langle \prod_{j=1}^n \cos(q(\mathbf{x}_j))\right\rangle = 2^{-n}\sum_{\varepsilon_j=\pm 1}\left\langle \exp\left(i\sum_{j=1}^n \varepsilon_j q(\mathbf{x}_j)\right)\right\rangle$$
$$= 2^{-n}\exp\left(-\frac{n}{2}V(\mathbf{0})\right)\sum_{\varepsilon_j=\pm 1} e^{-U_n(\mathbf{x};\,\varepsilon)}$$

Looking at the definition (23.2) of Ξ we see that we have the first half of the following theorem.

Theorem 23.2 Define

$$C_\Lambda = \int_{\mathbf{x}\in\Lambda} \cos(q(\mathbf{x}))\, dx \tag{23.6}$$

$$\tilde{z} = z\exp(\tfrac{1}{2}V(\mathbf{0}))$$

Then

(a)
$$\Xi_\Lambda(z) = \langle \exp(\tilde{z}C_\Lambda)\rangle \tag{23.7}$$

(b)
$$\rho_\Lambda^{(n)}(\mathbf{x}, \varepsilon; z) = \tilde{z}^n\left\langle \exp\left(i\sum_{j=1}^n \varepsilon_j q(\mathbf{x}_j)\right)\right\rangle_{\text{int}}$$
$$= \tilde{z}^n\left\langle \cos\left(\sum_{j=1}^n \varepsilon_j q(\mathbf{x}_j)\right)\right\rangle_{\text{int}} \tag{23.8}$$

where

$$\langle\cdot\rangle_{\text{int}} = \Xi_\Lambda(z)^{-1}\langle\cdot\exp(\tilde{z}C_\Lambda)\rangle$$

Remarks 1. The effect of keeping the β is to replace $\cos(q(\mathbf{x}))$ by $\cos(\beta^{1/2}q(\mathbf{x}))$ in (23.6).

2. For the reader worried about the existence of the integral (23.6), we can either add the requirement that q has a version on $C(\mathbb{R}^\nu)$, or else we can regard (23.6) as formal and define C_Λ as the unique L^2-function f with

$$\int f(q)e^{\sum a_i q(\mathbf{x}_i)}\,d\mu = \int_\Lambda \langle \cos(q(\mathbf{y}))e^{\sum a_i q(\mathbf{x}_i)}\rangle\,d\mathbf{y}$$

3. The Gaussian realization, Theorem 23.2, is sometimes called the **Sine–Gordon transformation** since it relates the Coulomb gas to the Sine–Gordon field theory; see [93].

Proof (23.7) is already proven. (23.8) is similar if we note that

$$\left\langle \exp\left(i\sum_{j=1}^{n} \varepsilon_j q(\mathbf{x}_j)\right) \prod_{k=1}^{N} \cos(q(\mathbf{x}'_k)) \right\rangle$$
$$= 2^{-n} \exp\left[-\left(\frac{n+N}{2}\right)V(\mathbf{0})\right] \sum_{\varepsilon'_k = \pm 1} e^{-U_{n+N}(\mathbf{x}, \mathbf{x}'; \varepsilon, \varepsilon')}$$

and that ρ is real. ∎

Corollary 23.3

(a) $$\Xi_\Lambda(z) \le \exp(\tilde{z}|\Lambda|)$$
(b) $$P_\Lambda(z) \le \tilde{z}$$
(c) $$\rho_\Lambda^{(n)}(x, \varepsilon; z) \le \tilde{z}^n$$

Proof (a) and (c) follow immediately from the bound $|\cos u| \le 1$ and (23.7), (23.8). (b) is obvious from (a). ∎

The key to the proof of Theorem 23.1 is two correlation inequalities; the first is due to Park [198] and the second is a closely related result of Fröhlich–Park [94]. Both have proofs closely related to Ginibre's proof [101] of his correlation inequalities for "plane rotors." We state them for general Gaussian processes, but use $\sum a_i \phi(v_i)$ rather than the integral needed above. The results for the integrals which follow similarly will be used below without comment.

Theorem 23.4 Let ϕ be the Gaussian process over some Hilbert space, $\mathscr{H}$, and let $\langle\cdot\rangle$ denote expectation with respect to $d\mu$. Fix $v_1, \ldots, v_n \in \mathscr{H}$. For $a_1, \ldots, a_n \ge 0$ define

$$Z(a_1, \ldots, a_n) = \left\langle \exp\left(\sum_{i=1}^{n} a_i \cos\phi(v_i)\right)\right\rangle$$

and

$$\langle \cdot \rangle_{a_1, \ldots, a_n} = Z^{-1} \langle \cdot \exp(\sum a_i \cos(\phi(v_i))) \rangle$$

Then

(a) For any $v, w \in \mathscr{H}$,

$$\langle \cos \phi(v) \cos \phi(w) \rangle_{a_i} \geq \langle \cos \phi(v) \rangle_{a_i} \langle \cos \phi(w) \rangle_{a_i}$$

(b) $Z(a_i + b_i) \geq Z(a_i)Z(b_i)$.

Proof We use the method of duplicate systems exploited already in Section 12 in proving GKS II. Let ϕ' be an independent copy of ϕ and let $\langle \cdot \rangle_{\text{dupl}}$ be $\int \cdot \, d\mu(\phi) \, d\mu(\phi')$. Let ψ, η be defined by

$$\phi(v) = \psi(v) - \eta(v)$$

$$\phi'(v) = \psi(v) + \eta(v)$$

Then ψ, η are independent Gaussian processes with the same covariance, namely, $\frac{1}{2}(v, w)$; i.e., $\int \cdot \, d\mu(\phi) \, d\mu(\phi') = \int \cdot \, d\nu(\psi) \, d\nu(\eta)$ for a suitable $d\nu$. As in Section 12,

$$\begin{aligned} \alpha &\equiv 2Z^2[\langle \cos \phi(v) \cos \phi(w) \rangle_{a_i} - \langle \cos \phi(v) \rangle_{a_i} \langle \cos \phi(w) \rangle_{a_i}] \\ &= \langle [\cos \phi(v) - \cos \phi'(v)][\cos \phi(w) - \cos \phi'(w)] \\ &\quad \times \exp(\sum a_i[\cos \phi(v_i) + \cos \phi'(v_i)]) \rangle_{\text{dupl}} \end{aligned}$$

Using

$$\cos(x + y) + \cos(x - y) = 2 \cos x \cos y$$

$$\cos(x + y) - \cos(x - y) = -2 \sin x \sin y$$

we see that

$$\begin{aligned} \alpha = 4 \langle &[\sin \psi(v)][\sin \eta(v)][\sin \psi(w)][\sin \eta(w)] \\ &\times \exp(2 \sum a_i \cos \psi(v_i) \cos \eta(v_i)) \rangle \end{aligned}$$

Expanding the exponential we have a sum of terms of the form

$$\langle f(\psi) f(\eta) \rangle_{\text{dupl}} = \left[\int f(\psi) \, d\nu(\psi) \right]^2$$

which is obviously positive. This proves (a).

To prove (b) we write

$$\begin{aligned} \beta &\equiv Z(a + b) - Z(a)Z(b) \\ &= \langle \exp(\sum (a_i + b_i) \cos \phi(v_i)) - \exp(\sum a_i \cos \phi(v_i) + b_i \cos \phi'(v_i)) \rangle \end{aligned}$$

and use

$$\cos(x \mp y) = \cos x \cos y \pm \sin x \sin y$$

to write

$$\beta = 2\langle \exp[\textstyle\sum (a_i + b_i)\cos \psi(v_i)\cos \eta(v_i) + \sum a_i \sin \psi(v_i)\sin \eta(v_i)]$$
$$\times \sinh(\textstyle\sum b_i \sin \psi(v_i)\sin \eta(v_i))\rangle_{\text{dupl}}$$

Expanding exp and sinh in power series, we again get a sum of $\langle f(\psi)f(\eta)\rangle_{\text{dupl}} \geq 0$. ∎

Proof of Theorem 23.1 (a) (See [94] for an alternate proof using Jensen's inequality.) By Theorem 23.4(b), $\ln \Xi_\Lambda$ is superadditive; i.e.,

$$\ln \Xi_{\Lambda_1 \cup \Lambda_2} \geq \ln \Xi_{\Lambda_1} + \ln \Xi_{\Lambda_2}$$

for $\Lambda_1 \cap \Lambda_2 = \varnothing$. Since $\sup(|\Lambda|^{-1}\Xi_\Lambda) < \infty$, by Corollary 23.3(b) the limit exists by standard arguments [226].

(b) By Theorem 23.4(a) and the standard formula,

$$\frac{\partial}{\partial z}\left[\frac{\langle fe^{zg}\rangle}{\langle e^{zg}\rangle}\right]\Bigg|_{z=0} = \langle fg\rangle - \langle f\rangle\langle g\rangle$$

we see that $\rho_\Lambda^{(n)}$ is monotone in z. Moreover writing $C_{\Lambda\cup\Lambda'} - C_\Lambda = f(1) - f(0)$ with $f(\lambda) = C_\Lambda + \lambda C_{\Lambda'}$ and taking derivatives with respect to λ, we see that $\rho_\Lambda^{(n)}$ is monotone in Λ so, since $\sup_\Lambda \rho_\Lambda^{(n)} < \infty$ by Corollary 23.3(a), the limit exists. Its invariance properties follow by standard arguments; see, e.g., [120]. ∎

The Gaussian realization theorem, Theorem 23.2, is also the starting point for a number of detailed analyses of systems obeying (23.1) [sometimes without the hypothesis $V \in L^\infty(V(\mathbf{0}) = \infty$ allowed)]: For example, Fröhlich [93] has studied the stability of two-dimensional Yukawa and Coulomb gases (Yukawa; respectively, Coulomb *means* $\hat{V}(k) = (k^2 + m^2)^{-1}$ or $\hat{V}(k) = k^{-2}$), and Brydges [26] has proven under some additional hypotheses that for suitable z, β, the ρ's decrease exponentially even though the V's have a long range tail ("Debye screening"). In this work, the close analogy between $\langle\cdot\rangle_{\Lambda,z}$ and the quantum field theories described in Section 24 is crucial. Indeed, the two-dimensional statistical mechanics Yukawa model is *identical* to the $:\cos \phi:_2$ model so that Fröhlich in [93] relies on his earlier work on that model [92, 95]. Brydges relies heavily on machinery of Glimm *et al.* [110] proving exponential falloff of correlations ("mass gap") in certain quantum field theories with broken symmetry.

24. An Introduction to Euclidean Quantum Field Theory

In the early days of quantum mechanics while all of classical physics was being quantized, it was quite natural to try to quantize the Maxwell electromagnetic field. The quantization of the field in the absence of sources was not hard and, indeed, Dirac's demonstration of the resulting photon–wave duality was one of the early triumphs of the "new" quantum mechanics. It was reasonable to attempt to add interaction via a perturbation series and it was here that the famous infinities occurred that took 20 years to understand even on a formal and perturbation theoretic level. The resulting theory of Dyson, Feynman, Schwinger, and Tomonoga [234] represents one of the great challenges of mathematical physics: To find a well-defined mathematical model for quantum electrodynamics which has the Feynman perturbation series as asymptotic series and which "makes physical sense."

The first critical step towards a solution of this problem was made in the early 1950's by Gårding and Wightman who gave a precise mathematical definition of a quantum field theory by listing the properties they should have; i.e., they gave exact meaning to "makes physical sense." A considerable and beautiful theory developed of the general study of such objects but nontrivial examples were not known even if the dimension of space–time was decreased (the famous infinities of the theory are less severe in lower dimension, as we shall see).

The period since 1964 has seen the development of a discipline called **constructive quantum field theory** which has succeeded in the construction of nontrivial models in two and three *space–time* dimensions. In its earliest phases, this theory was characterized by a combination of C^*-algebraic and operator theoretic methods. A considerable amount of information was obtained by Glimm and Jaffe and their students with significant contributions by Friedrichs, Nelson and Segal. But even in the simplest models, Lorentz invariance and uniqueness of the vacuum eluded proof.

The key to the completion of the verification of the Wightman axioms in the simplest models as well as a basic element in virtually all progress in the field since 1972 has been the exploitation of "Euclidean" functional integration methods. Formal functional integration (but the analog of Feynman integrals rather than Wiener integrals) was an element of Feynman's original work, and Wiener integrals were used by Glimm, Jaffe, and Nelson as a tool to study certain partial differential operators which entered as approximations to their field theoretic Hamiltonians, but it was Nelson's development in 1971 of an Euclidean covariant infinite-dimensional path integral based, in part, on earlier work of Segal and Symanzik that changed the outlook to a fully "Euclidean" one.

Obviously, we cannot give a complete treatment of the subject in one short section or even give a comprehensive overview. Our goal is to introduce the basic formalism and renormalization theory. See [106, 258, 282] for further discussion.

A quantum field is an operator-valued distribution $\Phi(\mathbf{x}, t)$, i.e., a linear map from $\mathscr{S}(\mathbb{R}^\nu)$ (ν is the dimension of space–time; we occasionally write $x = (\mathbf{x}, t)$ to distinguish the space and time components) to the (unbounded) operators on a Hilbert space with some additional properties called Gårding–Wightman axioms; these and some related theory are described, e.g., in [18, 137, 215, 267]. There is a distinguished vector, the vacuum, in the theory denoted by ψ_0. The ordinary distributions

$$W_n(x_1, \ldots, x_n) = (\psi_0, \Phi(x_1) \cdots \Phi(x_n)\psi_0)$$

are particularly important since the theory can be "reconstructed" given them, and the axioms can be translated into properties of W_n. One consequence of the axioms is that W_n is the boundary value of an analytic function $W_n(z_1, \ldots, z_n)$ analytic in a certain region, R. R includes all points of the form $z_j = (\mathbf{x}_j, is_j)$ with $(\mathbf{x}_j, s_j) \in \mathbb{R}^\nu$ and with $z_k \neq z_l$ (all k, l). The functions

$$\begin{aligned} S_n(y_1, \ldots, y_n) &\equiv S(\mathbf{y}_j, s_j) \\ &= W_n(\mathbf{y}_j, is_j) \end{aligned}$$

are called **Schwinger functions**.

Rather than constructing the Wightman field Φ, one constructs a measure $d\mu$ on $\mathscr{S}'(\mathbb{R}^\nu)$ whose moments are the candidates for Schwinger functions; i.e., one lets

$$\left(\int \phi(y) f(y)\, d^\nu y\right)(T) = T(f)$$

as random variables and takes

$$S_n(y_1, \ldots, y_n) = \int \phi(y_1) \cdots \phi(y_n)\, d\mu \tag{24.1}$$

that is,

$$\int S_n(y_1, \ldots, y_n) f_1(y_1) \cdots f_n(y_n)\, d^{n\nu} y = \int T(f_1) \cdots T(f_n)\, d\mu(T)$$

One thus needs some conditions on the measure $d\mu$ or on S_n which allow us to reconstruct Φ. The earliest such conditions are due to Nelson [193], but they turn out to be difficult to verify in practice. Osterwalder–Schrader [197] gave a set of conditions on the S_n's which modulo growth conditions with n are equivalent to the Gårding–Wightman axioms (no effective axioms strictly equivalent to the Gårding–Wightman axioms are known) and, in particular, they isolated a condition now known as OS (Osterwalder–Schrader) positivity. Fröhlich [91] found a special case of the Osterwalder–Schrader reconstruction theorem with a more direct proof which suffices in almost all applications. His result, which we now quote, is further discussed in [89, 90, 106].

Theorem 24.1 (Fröhlich's reconstruction theorem) Let $d\mu$ be a cylinder measure on $\mathscr{S}'(\mathbb{R}^\nu)$ (ν is the number of space–time dimensions) obeying:

(i) Proper Euclidean motions [i.e., $T(x) \mapsto T(Ax + b)$, $b \in \mathbb{R}^\nu$, $A \in SO(\nu)$] leave $d\mu$ invariant.

(ii) OS positivity; i.e., given a real-valued $f \in \mathscr{S}(\mathbb{R}^\nu)$ with

$$\operatorname{supp} f \subset \{(\mathbf{x}, s),\, s > 0\},$$

let $(\theta f)(\mathbf{x}, s) = f(\mathbf{x}, -s)$. Then for real-valued $f_1, \ldots, f_n$ with the above support and $z_1, \ldots, z_n \in \mathbb{C}$:

$$\sum_{j,k=1}^{n} \bar{z}_k z_j \int \exp(i[\phi(f_k) - \phi(\theta f_j)])\, d\mu \geq 0$$

(iii) For any $f \in \mathscr{S}(\mathbb{R}^\nu)$,

$$\int \exp(\phi(f))\, d\mu < \infty$$

(iv) The action of the translations $(\mathbf{x}, s) \to (\mathbf{x}, s + t)$ is ergodic.

Then, there is a unique (scalar) field theory obeying the Gårding–Wightman axioms whose Schwinger functions are given by (24.1).

The most natural way to construct measures is to try Gaussian $d\mu$'s. These turn out to describe "trivial" field theories in that they describe particles without interactions—they are the analog of the harmonic oscillator. Despite their triviality, they are significant for there is a reasonable way to try to construct nontrivial $d\mu$'s, namely, as perturbations of the Gaussian $d\mu$'s analogous to the $P(\phi)_1$ construction. We therefore begin by analyzing Gaussian $d\mu$'s.

Theorem 24.2 A Gaussian measure $d\mu$ on $\mathscr{S}'(\mathbb{R}^\nu)$ obeys conditions (i)–(iv) of Theorem 24.1 if and only if the covariance $S_2(x, y) = \int \phi(x)\phi(y)\, d\mu$ is of the form:

$$S_2(x, y) = (2\pi)^{-\nu} \int e^{ik(x-y)} \hat{S}_2(k)\, d^\nu k$$

with

$$\hat{S}_2(k) = \int d\rho\,(m^2)(k^2 + m^2)^{-1} + P(k^2) \tag{24.2}$$

where $d\rho$ is a polynomially bounded positive measure on $[0, \infty)$, so that the integral lies in $\mathscr{S}'$ and P is a polynomial which is positive on $[0, \infty)$.

Sketch of proof (i) is equivalent to $S_2(x, y) = f(x - y)$ with f rotation invariant, (iii) is automatic for Gaussian process, and (iv) can be seen to be equivalent to $f(x) \to 0$ as $x \to \infty$. That leaves the analysis of (ii). This is clearly equivalent to

$$\sum_{j,k=1}^{n} \bar{z}_k z_j \exp[-\tfrac{1}{2}S_2(f_k - \theta f_j, f_k - \theta f_j)] \geq 0$$

It can be shown that this, in turn, is equivalent to

$$S_2(\theta f, f) \geq 0$$

for all f with the proper support property. This is an analog to the usual positive definiteness condition that leads to Bochner's theorem. There is a proof of that theorem (see, e.g., [215]) that constructs an auxiliary Hilbert space and uses Stone's theorem on that space. A similar analysis works here, except that now the unitary group of Bochner's theorem is a unitary group in the $\mathbf{x}$ variables and a self-adjoint semigroup in the s-variables. The temperedness of S_2 leads to the boundedness of the semigroup. Thus for $s > 0$ and $d\mu$ a tempered measure:

$$S_2(\mathbf{x}, s, \mathbf{0}, 0) = \frac{2}{(2\pi)^{\nu-1}} \int_{E \geq 0} e^{i\mathbf{k}\cdot\mathbf{x}} e^{-sE}\, d\mu(E, \mathbf{k})$$

Since S_2 is rotationally invariant,

$$\left[x_i \frac{\partial}{\partial s} - s \frac{\partial}{\partial x_i}\right] S_2 = 0$$

This translates into *Lorentz*-invariance of $d\mu$. Since $d\mu$ is supported in the region $E \geq 0$, it follows (see, e.g., [199]) that

$$d\mu(E, \mathbf{k}) = C_0 \delta(E, \mathbf{k}) + \int d\rho(m^2)[E^{-1}\delta(E - \sqrt{\mathbf{k}^2 + m^2})\, d^{\nu-1}\mathbf{k}]\, dE$$

Using ($a > 0$)

$$\frac{1}{2\pi} \int \frac{e^{iak_0}\, dk_0}{k_0^2 + a^2} = \frac{e^{-as}}{2a}$$

we find that for $s > 0$

$$S_2(\mathbf{x}, s; \mathbf{0}, 0) = \frac{2C_0}{(2\pi)^{\nu-1}} + \int \frac{e^{ikx}}{(2\pi)^\nu}\left[\int \frac{d\rho(m^2)}{k^2+m^2}\right] d^\nu k$$

Since $S_2 \to 0$ as $s \to \infty$, $C_0 = 0$.

Rotation invariance yields the applicability of the formula for all $(\mathbf{x}, s) \neq 0$. All that remains is an ambiguity of a positive-definite rotation invariant distribution supported at zero. This gives the polynomial $P(k^2)$ for the Fourier transform. ▮

Remark There is, for $\nu = 2$, a possible infrared ($k = 0$) singularity in (24.2) depending on the behavior of $d\rho$ at $m = 0$. This singularity cannot be canceled by a local in x-space singularity so we need the condition on the integral defining a distribution.

The $P(k^2)$ term in (24.2) will only make the infinities described below more severe so the possibility of such a term is not seriously discussed (it will not effect the W_n's for the Gaussian theory, but should effect the perturbed theory). We therefore take $P = 0$ and $d\rho(m^2) = \delta(m^2 - m_0^2)\, d(m^2)$. The resulting Gaussian process is called **the free Euclidean field of mass m_0**. How can we perturb this process and keep axioms (i)–(iv) of Theorem 24.1? Let us proceed formally at first. Suppose one can define a function $F(\phi(x))$ of $\phi(x)$. Let $U_\Lambda = \int_{x\in\Lambda} F(\phi(x))\, d^\nu x$ and define

$$d\nu_\Lambda = \frac{e^{-U_\Lambda}\, d\mu_0}{\int e^{-U_\Lambda}\, d\mu_0}$$

Let Λ be a square symmetric about the $s = 0$ axis, so $U_\Lambda = U_{\Lambda_1} + \theta(U_{\Lambda_1})$ where $\Lambda_1 = \Lambda \cap \{s > 0\}$. Thus ($Z_\Lambda = \int e^{-U_\Lambda}\, d\mu_0$),

$$Z_\Lambda \int (\theta F)F\, d\nu_\Lambda = \int (Fe^{-U_{\Lambda_1}})\theta(Fe^{-U_{\Lambda_1}})\, d\mu_0 \geq 0$$

so OS positivity for $d\mu_0$ implies the same for $d\mu_\Lambda$. If we can somehow take $\Lambda \to \infty$ and obtain $\lim_{\Lambda\to\infty} d\nu_\Lambda$, then Euclidean invariance is obtained. The regularity conditions (iii) hopefully can still hold. Finally, the ergodicity (iv) can be investigated. In fact, for certain special F's, it should fail for good physical reasons and this has been demonstrated in some cases [97, 109]. The obstructions to the above program are twofold: (a) **Ultraviolet problems** are those that are connected with the fact that $\phi(x)$ is not meaningful as a random variable and thus neither is $F(\phi(x))$. We discuss this further below. (b) **Infinite volume problems** are those controlling $\lim_{\Lambda\to\infty} d\nu_\Lambda$. For suitable theories in $\nu = 2, 3$ dimensions (formally, at least, the problems in $\nu = 4$ are ultraviolet and the methods used for $\nu = 2, 3$ to control infinite volume should work for $\nu = 4$ once the ultraviolet problems are controlled), these problems have been overcome using certain expansions [108, 71] which are the analog of the high temperature expansions of statistical mechanics, and by using the correlation inequality methods of Section 12 [120, 258]. We do not discuss this problem further.

One can see the ultraviolet problem quite explicitly by looking at the support of the measure $d\mu_0$ for the free Euclidean field. Collella and Lanford [36] have shown that for $\nu \geq 2$, the measure of those distributions T which are equal to a signed measure on some open set is zero. Thus $\phi(x)$ is really not meaningful so that $F(\phi(x))$ is not well defined for any F. The way out is to consider objects $\phi_n(x) = \int f_n(x - y)\phi(y)\, d^\nu y$ for a family f_n in $\mathscr{S}$ approaching $\delta(x)$. Then ϕ_n is continuous for each n since $\mathscr{S} * \mathscr{S}' \subset \mathscr{O}_M$, the C^∞ polynomially bounded functions. Thus we can form $F_n(\phi_n(x))$; if we choose F_n suitably, it might happen that $F_n(\phi_n(x))$ has a limit. Of course, the limit $\psi(x)$ cannot be a nonconstant random variable with finite moments if ψ is to be Euclidean covariant and OS positive [for $F(x - y) = E(\psi(x)\psi(y)) - E(\psi(x))E(\psi(y))$ will have the form of (24.2)], but one can hope that the limit exists in the sense of "generalized processes"; i.e., $\int f(x)F_n(\phi_n(x))\, dx$ converges for each nice enough f.

To realize this strategy, we return to the perturbation theory of Section 20. Suppose that we consider replacing q^4 by $q^4 + aq^2 + b$. For the current purpose, give the values of graphs with a general function $g(t)$ replacing $\frac{1}{2}e^{-|t|}$ in the contribution of lines. The $O(\beta)$ term is exactly $3g(0)^2 + ag(0) + b$. Other than this, no b-terms occur. To describe all the other terms, we consider all graphs with n vertices, but each vertex can have either two lines coming out or four. For each vertex with only two lines, we multiply the graph by a. Notice that except for labeling, there is a one–one correspondence between graphs with only two lines coming out of vertices $i_1, \ldots, i_k$ and those graphs with four lines out of each vertex but with a single self-pairing at vertices $i_1, \ldots, i_k$. We can therefore describe all connected graphs with more than two vertices by considering only the q^4-graphs, but modifying the rules so that a self-loop has the value $g(0) + \frac{1}{6}a$ rather than $g(0)$ (the $\frac{1}{6}$ comes from the labeling possibilities). The special choice $a = -6g(0)$, $b = 3g(0)^2$ results in the following graphical rules: *Follow the rules for q^4 but allow no diagrams with lines coming from the same vertex joined together.* The formal rules for a $\phi(x)^4$ perturbation of the free field of mass m_0 follow those of Section 20, but we use two-dimensional integrals and use

$$g(x) = (2\pi)^{-\nu} \int e^{ip\cdot x}(p^2 + m_0^2)^{-1}\, d^\nu p$$

For $\nu \geq 2$, $g(0) = \infty$ but when $\nu = 2$, the singularity at $x = 0$ is only logarithmic and there is exponential falloff at infinity. Thus, the only "divergent" graphs are those with self-loops. Therefore, when $\nu = 2$, we take (:–: agrees with earlier definition of Wick ordering for Gaussian random variables!)

$$:\phi_n^4: = \phi_n^4 - 6\langle\phi_n^2\rangle\phi_n^2 + 3\langle\phi_n^2\rangle^2$$

One expects on the basis of this perturbation theory that (for $\nu = 2$)

$$U(f) \equiv \lim_{n\to\infty} \int f(x):\phi_n^4(x): d^4x$$

exists and this can be proven (see, e.g., [258]). More subtle is the result of Nelson that $\int e^{-U(f)}\, d\mu_0 < \infty$ for f's which are nonnegative and sufficiently nice (actually, $f \in L^2$ will suffice if $f \geq 0$; see [258]). This solves the ultraviolet problem in $\nu = 2$ dimensions at least for ϕ^4 theories.

For $\nu = 3$, all divergent Wick-ordered graphs contain one of three especially simple graphs (of order β^2, β^2, and β^3, respectively) as a subgraph. Formally, one can cancel the infinities by taking

$$F_n(\phi_n) = \beta:\phi_n^4: + a_n\beta^2:\phi_n^2: + b_n\beta^2 + c_n\beta^3$$

with a_n, b_n, c_n diverging as $n \to \infty$. F_n does not have a limit but

$$\frac{e^{-U(f)}\, d\mu_0}{\int e^{-U(f)}\, d\mu_0}$$

does for suitable f's (the basic result on $d\mu$ is due to Glimm–Jaffe [105]; the last sentence includes refinements of [81, 235]). In this way, $\nu = 3$ dimensional theories have been constructed.

For $\nu = 4$, there are an infinite number of divergent graphs which are "primitively divergent" and no definitive progress has been made on nonperturbative control of the ultraviolet problem.

25. Properties of Eigenfunctions, Wave Packets, and Green's Functions

Two questions concerning the properties of eigenfunctions of $H = -\frac{1}{2}\Delta + V$ have been extensively studied. Are they smooth or at least bounded? Do they fall off exponentially? Two related questions which one can ask concern smoothness and boundedness of the integral kernel $e^{-tH}(\mathbf{x}, \mathbf{y})$ for e^{-tH} (often called the Green's function) and of functions ϕ in $C^\infty(H) = \bigcap_n D(H^n)$. We have touched upon continuity of $e^{-tH}(\mathbf{x}, \mathbf{y})$ in Section 6, but only with rather strong hypotheses.

Detailed history and results associated with the "conventional" approach to these problems may be found in [217, Section XIII.12]. Here we want to discuss the attack on these problems using path integrals. Our approach to the L^∞ and smoothness problems is motivated by that of Herbst and Sloan [126]. These authors do not explicitly use path integrals, but as they exploit positivity of $e^{t\Delta}$ (indeed, they consider $H = F(-i\nabla) + V$ for functions F with $\exp(-tF(-i\nabla))$ an operator with a positive integral kernel) and the Trotter product formula, they are "essentially path theoretic"; moreover, they are certainly motivated by a path integral intuition. Our presentation of this aspect is simplified by relying on the Portenko [204a] results described in Theorem 11.2; independently, Carmona [31] noticed this simplifying idea. Our discussion of the exponential falloff behavior is patterned directly on that of Carmona [32]. Devinatz [46a] has used some of these ideas to study self-adjointness problems.

In comparing the results of the conventional approach with those of the path integral method, one finds that generally the latter are somewhat superior. For example, the conditions on the potential in Corollary 25.7 below are somewhat weaker than those of Kato [150] and Simon [243]. Moreover, if $H = -\Delta + V$ and $V \in L^p(\mathbb{R}^\nu)$ with $p > \nu/2$, then Kato and Simon find $D(H^N) \supset L^\infty$ where $N \to \infty$ as $p \to \nu/2$; while we, following Herbst–Sloan [126], find an N independent of p (recently, using non-path-integral methods, Brezis–Kato [25] have found a result of a similar genre which is, in some ways, stronger). Moreover, the conditions on the potential required in the various exponential falloff results obtained by Carmona are weaker than those of Schnol [231] or Simon [244, 245]. However, the constants in the rate of exponential falloff obtained with path integrals are presently far from optimal although that may change with further development. Throughout our discussion, we make no attempt to find good overall constants in front of bounds.

We begin by noting that the proof of Theorem 11.2 actually shows somewhat more than stated there, namely, the following.

Proposition 25.1 Let $W \geq 0$ be a measurable function on $\mathbb{R}^\nu$. If

$$\sup_{\mathbf{x}} E\left(\int_0^t W(\mathbf{x} + \mathbf{b}(s))\, ds\right) = \gamma < 1$$

then, for any $\mathbf{x}$,

$$E\left(\exp\left(\int_0^t W(\mathbf{x} + \mathbf{b}(s))\, ds\right)\right) \leq (1 - \gamma)^{-1} < \infty$$

To estimate $E(\int_0^t W(\mathbf{x} + \mathbf{b}(s))\,ds)$, we note that it is the same as

$$2\int_0^{t/2} (e^{s\Delta}W)(\mathbf{x})\,ds$$

so that the following is useful.

Lemma 25.2 For any $W \geq 0$ and any t,

$$e^{-1}\int_0^t (e^{s\Delta}W)(\mathbf{x})\,ds \leq [(-\Delta + t^{-1})^{-1}W](\mathbf{x}) \tag{25.1}$$

for almost every $\mathbf{x}$.

Proof Since $e^{s\Delta}$ has a positive integral kernel and $W \geq 0$,

$$e^{-1}\int_0^t (e^{s\Delta}W)(\mathbf{x})\,ds \leq \int_0^\infty e^{-s/t}(e^{s\Delta}W)(\mathbf{x})\,ds$$

which equals $[(-\Delta + t^{-1})^{-1}W](\mathbf{x})$. ∎

Definition Let Δ be the cube with unit side in $\mathbb{R}^\nu$, centered about $\mathbf{0}$. We say that f is **uniformly locally** L^p (written $f \in L_u^p(\mathbb{R}^\nu)$) if and only if

$$\|f\|_{L_u^p(\mathbb{R}^\nu)} \equiv \sup_{\mathbf{x}}\left[\int_\Delta |f(\mathbf{x} + \mathbf{y})|^p\,dy\right]^{1/p} < \infty$$

Lemma 25.3

(a) If $W \in L_u^p(\mathbb{R}^\nu)$ where $p > \nu/2$ (for $\nu \geq 2$) or $p \geq 1$ (for $\nu = 1$), then

$$\lim_{\alpha\to\infty} \|(-\Delta + \alpha)^{-1}W\|_\infty = 0 \tag{25.2}$$

(b) If $\mathbf{x} \in \mathbb{R}^\nu$ is written, $\mathbf{x} = (\mathbf{y}, \mathbf{z})$ with $\mathbf{y} \in \mathbb{R}^\mu$, $\mathbf{z} \in \mathbb{R}^{\nu-\mu}$ and if $p > \mu/2$ and $\sup_{\mathbf{z}} \|W(\cdot, \mathbf{z})\|_{L_u^p(\mathbb{R}^\mu)} < \infty$, then (25.2) still holds.

(c) If $W \in L^p(\mathbb{R}^\nu) \cap L^r(\mathbb{R}^\nu)$ and $p > \nu/2 > r$, $\nu \geq 3$, then

$$\|\Delta^{-1}W\|_\infty \leq c_{p,r}[\|W\|_p + \|W\|_r]$$

Remarks 1. The naturalness of uniformly locally L^p-spaces for various contexts related to those here was noted by Strichartz [268]. They are useful, for example, in consideration of periodic media.

2. The point of (b) is the following: To treat N particles in three dimensions, one takes $\nu = 3N$ and $V(\mathbf{x}) = \sum_{i<j} V_{ij}(\mathbf{x}_i - \mathbf{x}_j)$ where $\mathbf{x} = (\mathbf{x}_1, \ldots, \mathbf{x}_N)$ ($\mathbf{x}_i \in \mathbb{R}^3$) and V_{ij} is a function on $\mathbb{R}^3$. Since we can take $\mathbf{y} = \mathbf{x}_i - \mathbf{x}_j$, we see that to control $(-\Delta + \alpha)^{-1}V_{ij}$ we only need $V_{ij} \in L_u^p$ with $p > \frac{3}{2}$, whereas (a) alone would lead one to suspect that $p > 3N/2$ is needed.

Proof (a) Let $f_\alpha(\mathbf{x} - \mathbf{y})$ be the integral kernel for $(-\Delta + \alpha)^{-1}$. Then, letting $q = (1 - p^{-1})^{-1}$:

$$\begin{aligned}
&\int f_\alpha(\mathbf{x} - \mathbf{y})|W(\mathbf{y})|\,dy \\
&\quad = \sum_{\gamma \in \mathbb{Z}^\nu} \int_{\mathbf{y} \in \Delta} f_\alpha(\mathbf{y} + \gamma)|W(\mathbf{x} - \mathbf{y} - \gamma)|\,dy \\
&\quad \le \sum_{\gamma \in \mathbb{Z}^\nu} \left(\int_{\mathbf{y} \in \Delta} |f_\alpha(\mathbf{y} + \gamma)|^q\right)^{1/q} \left(\int_{\mathbf{y} \in \Delta} |W(\mathbf{x} - \mathbf{y} - \gamma)|^p\right)^{1/p} \\
&\quad \le c(\alpha, q)\|W\|_{L_u{}^p(\mathbb{R}^\nu)}
\end{aligned}$$

where

$$c(\alpha, q) = \sum_\gamma c(\alpha, \gamma, q)$$

$$c(\alpha, \gamma, q) = \left(\int_{\mathbf{y} \in \Delta} |f_\alpha(\mathbf{y} + \gamma)|^q\right)^{1/q}$$

From the fact that $f_1(\mathbf{x}) \sim |\mathbf{x}|^{-(\nu-2)}$ $(\nu \ge 3)$, $f_1(\mathbf{x}) \sim \ln|\mathbf{x}|^{-1}$ $(\nu = 2)$, $f_1(\mathbf{x}) \sim 1$ $(\nu = 1)$ near $|\mathbf{x}| = 0$, and $f_1(\mathbf{x}) \sim e^{-|\mathbf{x}|}$ near $|\mathbf{x}| = \infty$, we see that $c(\alpha, q) < \infty$ for $\alpha = 1$. Moreover, $f_\alpha(\mathbf{x}) = \int_0^\infty e^{-\alpha t}(e^{+t\Delta})(\mathbf{x}, 0)\,dt$ shows that $f_\alpha(\mathbf{x})$ is monotone decreasing to zero as $\alpha \to \infty$ for $\mathbf{x} \ne 0$. Thus using the monotone convergence theorem, first on $\int_\Delta \cdot$ and then on $\sum_\gamma$, we see that $c(\alpha, q) \to 0$ as $\alpha \to \infty$.

(b) Write $-\Delta_{\mathbf{x}} = -\Delta_{\mathbf{y}} - \Delta_{\mathbf{z}} = A_1 + A_2$. The method of proof of (a) shows that

$$\int \|(e^{-tA_1}e^{-t\alpha}W)\|_\infty\,dt \le d(\alpha, p)\sup_{\mathbf{z}}\|W(\cdot, \mathbf{z})\|_{L_u^p}$$

where $d \to 0$ as $\alpha \to \infty$. Since e^{tA_2} is a contraction on L^∞, the result is proven.

(c) Part (c) follows from Young's inequality and $|\mathbf{x}|^{-(\nu-2)} \in L^{q-\varepsilon} + L^{q+\varepsilon}$ with $q = \nu/(\nu - 2)$. ∎

We summarize the last three results in the following theorem.

Theorem 25.4

(a) If W obeys the hypotheses of Lemma 25.3(a) or (b), then for all sufficiently small t, and all $\mathbf{x}$:

$$E\left[\exp\left(\int_0^t W(\mathbf{x} + \mathbf{b}(s))\,ds\right)\right] \le Q < \infty \qquad (25.3)$$

where how small t must be and how large Q may be only depend on $\|W\|_{L_u^p(\mathbb{R}^\nu)}$ (or $\sup_{\mathbf{z}}\|W(\cdot, \mathbf{z})\|_{L_u^p(\mathbb{R}^\mu)}$).

(b) Let $\nu \geq 3$ and let $\varepsilon > 0$. Then, there is a $C(\nu, \varepsilon)$ so that

$$\sup_{\mathbf{x}}\left[E\left[\exp\left(\int_0^\infty W(\mathbf{x} + \mathbf{b}(s))\, ds\right)\right]\right] < \infty \tag{25.4}$$

if $\| W \|_{\nu/2+\varepsilon} + \| W \|_{\nu/2-\varepsilon} \leq C(\nu, \varepsilon)$.

Remark As we shall see, the semigroup property implies that (25.3) for all $\mathbf{x}$, and t small yields it for all $\mathbf{x}$ and t if Q is allowed to be t-dependent.

The basic result in the theory of boundedness of Green's functions, eigenfunctions, and wave packets is the following one modeled on results of Herbst and Sloan [126].

Theorem 25.5 Suppose that $V = V_+ - W$, where $W, V_+ \geq 0$, V_+ is in $L^1_{\text{loc}}(\mathbb{R}^\nu)$ and W is a sum of functions W_j each obeying

$$\sup_{\mathbf{z}_j} \| W_j(\cdot, \mathbf{z}_j) \|_{L^{p_j}_u(\mathbb{R}^{\mu_j})} < \infty$$

for some breakup, $\mathbf{x} = (\mathbf{y}_j, \mathbf{z}_j)$ of $\mathbb{R}^\nu = \mathbb{R}^{\mu_j} \times \mathbb{R}^{\nu-\mu_j}$ and $p_j > \mu_j/2$ (if $\mu_j \geq 2$) or $p_j \geq 1$ (if $\mu_j = 1$). Let H be the form sum $-\frac{1}{2}\Delta + V$. Then

(a)
$$\| e^{-tH} \phi \|_p \leq C e^{tA} \| \phi \|_p \tag{25.5}$$

with C, A independent of p for all $p \in [1, \infty]$.

(b) For any r and v with $r \geq v$:

$$\| e^{-tH} \phi \|_r \leq C(t, r, v) \| \phi \|_v \tag{25.6}$$

for all $t > 0$, where for r, v fixed

$$C(t) \leq D e^{At} (t \geq 1); \qquad C(t) \leq D t^{-\mu} (t \leq 1) \tag{25.7}$$

so long as $\mu > \frac{1}{2}\nu(v^{-1} - r^{-1})$.

(c)
$$\| (H + E)^{-L} \phi \|_r \leq C \| \phi \|_v \tag{25.8}$$

so long as E is sufficiently large and $L > \frac{1}{2}\nu(v^{-1} - r^{-1})$.

Remark (25.8) says that H obeys the same inhomogeneous L^p Sobolev smoothing as $-\Delta$ [215].

Proof (a) By the semigroup property, it suffices to show that $\| e^{-tH} \phi \|_p \leq C \| \phi \|_p$ for all small t. By Theorem 25.4, consider t so that (25.3) holds. Then since $V_+ \geq 0$

$$E\left(\exp\left(-\int_0^t V(\mathbf{x} + \mathbf{b}(s))\, ds\right)\right) \leq Q$$

and thus, for any $f \in L^\infty$:

$$E\left(\exp\left(-\int_0^t V(\mathbf{x} + \mathbf{b}(s))\, ds\right) f(\mathbf{x} + \mathbf{b}(t))\right) \leq Q \| f \|_\infty$$

It follows from the Feynman–Kac formula, that

$$\| e^{-tH} f \|_\infty \leq Q \| f \|_\infty$$

for all small t. The general result now follows by duality (which says that the dual of $e^{-tH} \upharpoonright L^\infty$ which is $e^{-tH} \upharpoonright L^1$ is bounded by Q) and interpolation [215].

(b) By the Feynman–Kac formula,

$$\int \exp\left(-\lambda \int_0^t V(\boldsymbol{\omega}(s))\, ds\right) d\mu_{0, \mathbf{x}, \mathbf{y}; t} \equiv e^{-tH(\lambda)}(\mathbf{x}, \mathbf{y})$$

is the integral kernel of $\exp[-t(-\frac{1}{2}\Delta + \lambda V)] \equiv e^{-tH(\lambda)}$, in the sense that

$$\int \overline{f(\mathbf{x})} e^{-tH(\lambda)}(\mathbf{x}, \mathbf{y}) g(\mathbf{x})\, dx\, dy = (f, e^{-tH(\lambda)} g)$$

at least for nice f and g. By Hölder's inequality on the path integral

$$e^{-tH}(\mathbf{x}, \mathbf{y}) \leq [e^{-tH(p)}(\mathbf{x}, \mathbf{y})]^{1/p} [e^{t\Delta}(\mathbf{x}, \mathbf{y})]^{1/q} \tag{25.9}$$

if $p^{-1} + q^{-1} = 1$ and $p, q > 1$. Using Hölder's inequality again:

$$|(e^{-tH} f)(\mathbf{x})| \leq \| e^{-tH(p)} 1 \|_\infty^{1/p} (e^{t\Delta} | f |^q)^{1/q}(\mathbf{x})$$

so using (25.5), we see that (noting that the p in (25.5) can be ∞; C_p, A_p are the constants for $H(p)$)

$$\| e^{-tH} f \|_r \leq [C_p e^{tA_p}]^{1/p} \| e^{t\Delta} | f |^q \|_{r/q}^{1/q} \tag{25.10}$$

But $e^{t\Delta}$ is a convolution operator with a convolution kernel in all L^s-spaces with L^s-norm $D_s t^{-(1/2)\nu(1-s^{-1})}$. It follows by Young's inequality [215] that

$$\| e^{-tH} f \|_r \leq \tilde{C}_p D_s e^{t\tilde{A}_p} t^{-(1/2)\nu(1-s^{-1})} \| f \|_v$$

where

$$(1 - s^{-1}) = q(v^{-1} - r^{-1})$$

Since we can take p arbitrarily close to infinity, and thus q arbitrarily close to one, (25.7) results.

(c) Part (c) follows from (25.7) and the estimate

$$\| (H + E)^{-L} \phi \|_r \leq \| \phi \|_v \int_0^\infty t^{L-1} e^{-tE} C(t, r, v)\, dt \quad ∎$$

Remarks 1. (25.5) implies that e^{-tH} defined a priori on L^2 is a strongly continuous semigroup on each L^p. Its infinitesimal generator thus gives a "natural" meaning to $-\frac{1}{2}\Delta + V$ as an operator on L^p. A rather different approach to this question has been developed by Schechter [228] and Weder [283a]; these authors exploit ellipticity of $-\Delta$ rather than the fact that $e^{+\Delta}$ is a positivity-preserving operator.

2. (25.9) was first exploited independently by [45] (see Section 21) and [126]. In the above context, it was used by [126].

3. For further discussion of the L^p-norms of e^{-tH}, see [260a] where, in particular, it is proven that $\lim_{t\to\infty} t^{-1} \ln\|e^{-tH}\|_{p,p}$ is independent of p. Also considered is the question of when

$$\sup_t \|e^{-tH}\|_{2,2} < \infty \quad \text{implies that} \quad \sup_t \|e^{-tH}\|_{\infty,\infty} < \infty$$

Corollary 25.6 For any $L > \nu/4$, $D(|H|^L) \subset L^\infty$. In particular, any L^2-eigenfunction of H lies in L^∞.

Proof Take $r = \infty$, $v = 2$ in (25.8). ▌

Definition If $0 < \theta \leq 1$, we define $\mathbf{C}_\theta$ to be the set of functions in L^∞ with $\sup_{x,y}|f(x) - f(y)|/|x - y|^\theta < \infty$. If $1 < \theta \leq 2$, we define $\mathbf{C}_\theta$ to be the set of C^1-functions with ∇f in $C_{\theta-1}$.

Corollary 25.7 Suppose that V_+ is a sum of $V_{+,j} > 0$ obeying the same L_u^p-conditions as W_j, that $\theta < 2 - \mu_j/p_j$ for each j that $\theta < 2$, and that $L > \nu/4 + 1$. Then any $\phi \in D(|H|^L)$ (and, in particular, any eigenfunction) lies in C_θ.

Proof By hypothesis and the method of proof of Lemma 25.3, we can find α so that $\frac{1}{2}\theta < \alpha < 1$ and $\|(H_0 + E)^{-(1-\alpha)}Vf\|_\infty \leq C\|f\|_\infty$. It follows, by (25.8), Corollary 25.6, and

$$(H + E)^{-L} = (H_0 + E)^{-1}(H + E)^{-(L-1)} - (H_0 + E)^{-1}V(H + E)^{-L}$$

that any $\phi \in D(H^L)$ is of the form $\phi = (H_0 + E)^{-\alpha}\psi$ with ψ in L^∞. Thus, we need only show that $(H_0 + E)^{-\alpha}$ maps L^∞ into C_θ. We consider the case $\theta < 1$ and $\nu \geq 2$. The others are similar.

Let f be the function on $\mathbb{R}^\nu$ with $f(\mathbf{x} - \mathbf{y})$ the integral kernel of $(H_0 + E)^{-\alpha}$. Since $f(\mathbf{x} - \mathbf{y})$ is less than or equal to the integral kernel of

$$H_0^{-\alpha} \equiv c|\mathbf{x} - \mathbf{y}|^{-\nu+2\alpha}$$

and since f falls exponentially at infinity (since $\hat{f} = \text{const}(p^2 + E)^{-\alpha}$ is analytic in a strip [215]), we have that

$$|f(\mathbf{x})| \leq C|\mathbf{x}|^{-\nu+2\alpha}e^{-\beta|\mathbf{x}|} \tag{25.11}$$

for some $\beta > 0$. Since f is spherically symmetric, and $(\mathbf{x} \cdot \nabla)f$ has a Fourier transform $C[(\nu - 2\alpha)(p^2 + E)^{-\alpha} + (2E)(p^2 + E)^{-\alpha+1}]$, we see that

$$|\nabla f(\mathbf{x})| \leq C|\mathbf{x}|^{-\nu+2\alpha-1}e^{-\beta|\mathbf{x}|} \tag{25.12}$$

From (25.11), we see that for $|\mathbf{x}| < |\mathbf{y}|$,

$$|f(\mathbf{x}) - f(\mathbf{y})| \leq 2C|\mathbf{x}|^{-(\nu-2\alpha)}e^{-\beta|\mathbf{x}|}$$

and from (25.12) that

$$|f(\mathbf{x}) - f(\mathbf{y})| \leq 2C|\mathbf{x}|^{-\nu+2\alpha-1}e^{-\beta|\mathbf{x}|}|\mathbf{x} - \mathbf{y}|$$

Thus for all $\mathbf{x}, \mathbf{y}$,

$$|f(\mathbf{x}) - f(\mathbf{y})| \leq 2C[|\mathbf{x}|^{-\nu+2\alpha-\theta}e^{-\beta|\mathbf{x}|} + |\mathbf{y}|^{-\nu+2\alpha-\theta}e^{-\beta|\mathbf{y}|}]|\mathbf{x} - \mathbf{y}|^\theta$$

From this last equation, we see that

$$\int |f(\mathbf{x} - \mathbf{y}) - f(\mathbf{y})|\, dy \leq D|\mathbf{x}|^\theta \tag{25.13}$$

since $2\alpha > \theta$. (25.13) together with $f \in L^1$ easily imply that convolution with f is bounded from L^∞ to C_θ. ∎

Corollary 25.8 Suppose that the hypotheses of Corollary 25.7 hold with the exception that the $V_{+,j}$ are only assumed to lie in L^p_{loc} (same conditions on p). Then any eigenfunction is continuous.

Proof Let $H\phi = E\phi$ and let V_n be the function obtained by replacing $V_+(\mathbf{x})$ by zero if $|\mathbf{x}| \geq n$. Let $H_n = -\frac{1}{2}\Delta + V_n$ and $\phi_n(\mathbf{x}) = (e^{-H_n}\phi)(\mathbf{x})$. By the monotone convergence theorem, and the Feynman–Kac formula, $\phi_n(\mathbf{x}) \to e^{-E}\phi(\mathbf{x})$ almost everywhere in x. By Corollary 25.7, each $\phi_n(\mathbf{x})$ is continuous, so it suffices to prove uniform convergence of ϕ_n on compact subsets. Let $|\mathbf{x}| \leq R$ and $n, m \geq 2R$. Let $A = \{\mathbf{b} \mid \sup_{0 \leq s \leq 1} |\mathbf{b}(s)| \geq R\}$. Then

$$V_n(\mathbf{x} + \mathbf{b}(s)) = V_m(\mathbf{x} + \mathbf{b}(s))$$

for $b \notin A$, so

$$|\phi_n(\mathbf{x}) - \phi_m(\mathbf{x})| \leq E\left(\chi_A \exp\left(+ \int_0^1 W(\mathbf{x} + \mathbf{b}(s))\right) |\phi(\mathbf{x} + \mathbf{b}(1))|\right)$$

since $|e^{-a} - e^{-b}| \le 1$ for $a, b \ge 0$. Thus, by the Schwarz inequality,

$$|\phi_n(\mathbf{x}) - \phi_m(\mathbf{x})| \le [\|e^{-\tilde{H}}|\phi|^2\|_\infty E(\chi_A)]^{1/2}$$

where $\tilde{H} = -\frac{1}{2}\Delta - 2W$. Since $\phi \in L^\infty$ (by Corollary 25.6) $e^{-\tilde{H}}$ is bounded on L^∞ and $E(\chi_A) \to 0$ as $R \to \infty$, we have the desired uniform convergence on compact subsets. ∎

Corollary 25.9 ([126]) If V obeys the hypotheses of Theorem 25.5, then for any $t > 0$, e^{-tH} has an integral kernel in $L^\infty(\mathbb{R}^{2\nu})$.

Proof e^{-tH} maps L^1 to L^∞ and so defines a continuous bilinear form on $L^1(\mathbb{R}^\nu)$. Every such form has an L^∞ integral kernel by a general theorem; see, e.g., Trèves [278]. ∎

Corollary 25.10A Let V obey the hypotheses of Theorem 25.5. Let $f \in L^2(\mathbb{R}^\nu)$. Then fe^{-tH} (where we use f to also stand for a multiplication operator) is Hilbert–Schmidt.

Proof By (25.9) and Corollary 25.9,

$$|f(\mathbf{x})e^{-tH}(\mathbf{x}, \mathbf{y})| \le C_t|f(\mathbf{x})|[e^{t\Delta}(\mathbf{x}, \mathbf{y})]^{1/2}$$

which is easily seen to lie in $L^2(\mathbb{R}^{2\nu})$. ∎

This improves results of [126]. It is useful in certain scattering theory contexts; see Avron *et al.* [4] and Davies–Simon [41]. There is also a trace class version of this corollary. It is not true that if $|A(\mathbf{x}, \mathbf{y})| \le B(\mathbf{x}, \mathbf{y})$ and B is trace class, then A is trace class so that the above proof does not extend but a trick of Avron *et al.* [4] is available. The best results use an idea and a space introduced by Birman and Solomjak (see [259] for references and extended discussion).

Definition $l_1(L^2) = \{f \in L^2(\mathbb{R}^\nu) \mid \sum_{\alpha \in \mathbb{Z}^\nu} \|f\chi_\alpha\|_2 < \infty\}$ where χ_α is the characteristic function of Δ_α, the unit cube with center at $\alpha \in \mathbb{Z}^\nu \subset \mathbb{R}^\nu$.

Corollary 25.10B Let V obey the hypotheses of Theorem 25.5. Let $f \in l_1(L^2)$. Then fe^{-tH} is trace class.

Proof Suppose that we show that for all g's supported in Δ_0, $\|ge^{-tH}\|_1 \le C\|g\|_2$ (where $\|\cdot\|_1$ is trace class norm) and C is only dependent on t and on various translation invariant norms of V. Then, by translation covariance for any $f \in l_1(L^2)$,

$$\| f\chi_\alpha e^{-tH} \|_1 \leq C \| f\chi_\alpha \|_2$$

so that $\sum_\alpha f\chi_\alpha e^{-tH}$ converges in trace norm. Since it converges in Hilbert–Schmidt norm to fe^{-tH}, the result will be proven.

Thus suppose $g \in L^2$ has support in Δ_0. Write $ge^{-tH} = AB$, where $A = ge^{-tH/2}(1 + x^2)^\nu$ and $B = (1 + x^2)^{-\nu} e^{-tH/2}$. By Corollary 25.10A, B is Hilbert–Schmidt. By the proof of that corollary and the fact that

$$\|g(1 + x^2)^\nu\|_2 \leq C\|g\|_2$$

A is Hilbert–Schmidt with a norm bounded by $\|g\|_2$. ∎

* * *

Next, we discuss the problem of exponential falloff.

Theorem 25.11 (Carmona's estimate [32]) Let $V = W + U$ with $W \in L^1_{\text{loc}}$, $W_\infty = \inf W(\mathbf{x}) > -\infty$, and $U \in L^p + L^\infty$ with $p > \nu/2$ $(\nu \geq 2)$ or $p = 1$ $(\nu = 1)$. Let $H = -\frac{1}{2}\Delta + V$ and suppose that $H\phi = E\phi$; $\phi \in L^2$. Then, for any a, t and almost all $\mathbf{x}$:

$$|\phi(\mathbf{x})| \leq C(t)(\exp[-Dt^{-1}a^2 - tW_\infty + tE] + \exp[t(E - W_a(\mathbf{x}))]) \tag{25.14}$$

where $D > 0$ and $W_a(\mathbf{x}) = \inf\{W(\mathbf{y}) \mid |\mathbf{x} - \mathbf{y}| \leq a\}$ and $C(t) \leq Ae^{Bt}$. If $\nu \geq 3$ and $U \in L^{(1/2)\nu+\varepsilon} \cap L^{(1/2)\nu-\varepsilon}$ has small enough L^p-norms, or if $U = 0$ we can take $B = 0$.

Proof By the Feynman–Kac formula:

$$\begin{aligned}
|\phi(\mathbf{x})| &= e^{tE}|(e^{-tH}\phi)(\mathbf{x})| \\
&\leq e^{tE} E\left(\exp\left(-\int_0^t V(\mathbf{x} + \mathbf{b}(s))\, ds\right)|\phi(\mathbf{x} + \mathbf{b}(t))|\right) \\
&\leq e^{tE}\|\phi\|_\infty E\left(\exp\left[-2\int_0^t U(\mathbf{x} + \mathbf{b}(s))\, ds\right]\right)^{1/2} \\
&\quad \times E\left(\exp\left[-2\int_0^t W(\mathbf{x} + \mathbf{b}(s))\, ds\right]\right)^{1/2}
\end{aligned} \tag{25.15}$$

where we have used the Schwarz inequality in the last step. Clearly

$$E\left(\exp\left[-2\int_0^t W(\mathbf{x} + \mathbf{b}(s))\, ds\right]\right) \leq e^{-2tW_\infty} E\left(\sup_{0 \leq s \leq t} |\mathbf{b}(s)| > a\right) + e^{-2tW_a(\mathbf{x})}$$

since either the path leaves $\{\mathbf{y} \mid |\mathbf{x} - \mathbf{y}| \leq a\}$ or it does not. By (7.6) and (3.4), $E(\sup_{0 \leq s \leq t} |\mathbf{b}(s)| > a) \leq C_0 \exp(-2Dt^{-1}a^2)$. Thus by (25.15) we can conclude that (25.14) holds with

$$C(t) = C_0 E\left(\exp\left[-2\int_0^t U(\mathbf{x} + \mathbf{b}(s))\, ds\right]\right)^{1/2}$$
$$\leq C_0 \|\exp(-t[-\tfrac{1}{2}\Delta + 2U])1\|_\infty^{1/2}$$

$C(t) \leq Ae^{Bt}$ follows from (25.5) and the result for $U \in L^{(1/2)\nu+\varepsilon} \cap L^{(1/2)\nu-\varepsilon}$ from Theorem 11.2. ∎

The following two corollaries are essentially from [32].

Corollary 25.12 Under the hypotheses of Theorem 25.11, suppose that $E < 0$ and $\underline{\lim}_{\mathbf{x}\to\infty} W(\mathbf{x}) \geq 0$. Suppose, moreover, that either $U = 0$ or $\nu \geq 3$ and $U \in L^{(1/2)\nu+\varepsilon} \cap L^{(1/2)\nu-\varepsilon}$ with small norms. Then

$$|\phi(\mathbf{x})| \leq Ce^{-\delta|\mathbf{x}|}$$

for some $\delta > 0$.

Proof By hypothesis and Theorem 25.11, (25.14) holds with $C(t)$ bounded in t. Choose $a = \frac{1}{2}|\mathbf{x}|$ and $t = \varepsilon|\mathbf{x}|$. Then, for $|\mathbf{x}|$ large:

$$|\phi(\mathbf{x})| \leq C[\exp(-\alpha|\mathbf{x}|) + \exp(-\beta|\mathbf{x}|)]$$

where $\alpha = \frac{1}{4}D\varepsilon^{-1} + \varepsilon W_\infty - \varepsilon E$ can be arranged to be positive by choosing ε small, and β can be made arbitrarily close to $-\varepsilon E$ by taking $|\mathbf{x}|$ large [since $\underline{\lim}\, W_a(\mathbf{x}) \geq 0$ by hypothesis]. ∎

Corollary 25.13 Under the hypotheses of Theorem 25.11, suppose that $\lim_{\mathbf{x}\to\infty} W(\mathbf{x}) = \infty$. Then, for some C, D:

$$|\phi(\mathbf{x})| \leq C\exp(-D|\mathbf{x}|\tilde{W}(\mathbf{x})^{1/2}) \tag{25.16}$$

with $\tilde{W}(\mathbf{x}) = W_{(1/2)|\mathbf{x}|}(\mathbf{x})$.

Proof Take $a = \frac{1}{2}|\mathbf{x}|$ and $t^{-1} = \tilde{W}(\mathbf{x})^{1/2}|\mathbf{x}|^{-1}$ in (25.14). ∎

Remarks 1. Corollary 25.12 can be extended to allow $U \in L^p + L_\varepsilon^\infty$, since any such U can be written $U = \tilde{U} + R$ with $\tilde{U} \in L^{(1/2)\nu+\varepsilon} \cap L^{(1/2)\nu-\varepsilon}$ with small norms and $R \in L^\infty$ and $R \to 0$ at infinity. We can then take $\tilde{W} = R + W$.

2. Corollary 25.13 includes a result of Schnol [231] that $|\phi(\mathbf{x})|$ falls off faster than any exponential if W is bounded below and $W \to \infty$ at infinity. It is stronger in that local singularities are allowed.

3. The bound (25.16) is of "WKB type," in that the WKB behavior of ϕ is $\exp(-|\mathbf{x}| \int_0^1 (2W(s\mathbf{x}))^{1/2}\, ds)$, which is qualitatively similar to (25.16); e.g., if $W(x) \sim a|\mathbf{x}|^{2n}$, then (25.16) says $|\phi(\mathbf{x})| \le C \exp(-D|\mathbf{x}|^{n+1})$. However, we note that with stronger hypotheses and different methods, one can show [245] that D can be taken near $(2a)^{1/2}(n+1)^{-1}$, a result not currently available by path integral methods.

4. The last two corollaries can be proven without the hypothesis $U \in L^{(1/2)\nu-\varepsilon}$; see [32].

Finally, we discuss lower bounds on the lowest eigenfunction of $-\frac{1}{2}\Delta + V$. We aim for a result which captures the proper qualitative behavior for potentials which look like $|\mathbf{x}|^{2m}$ at infinity (but which are not assumed central). Afterward, we will describe in some remarks how one can deal with more general potentials. Here, too, we are following the basic scheme of Carmona [32].

Lemma 25.14 Fix ε and let $\mathbf{b}$ be ν-dimensional Brownian motion. Then, there exists a D, so that for any t, $\mathbf{x}$, R, δ:

$$\begin{aligned} E\Big(|\mathbf{b}(t) - \mathbf{x}| \le \delta,\ \sup_{0\le s\le t} |\mathbf{b}(s)| \le R\Big) &\ge \tau_\nu \delta^\nu (2\pi t)^{-\nu/2} \exp(-(2t)^{-1}[|\mathbf{x}| + \delta]^2) \\ &\quad - D[1 + (t^{-1/2}R)^{\nu-1}] \exp(-R^2/2t) \end{aligned} \tag{25.17}$$

where $\tau_\nu = |\{\mathbf{x} \,|\, |\mathbf{x}| \le 1\}|$.

Proof Clearly the left-hand side of (25.17) is larger than $e_1 - e_2$ where $E(|\mathbf{b}(t) - \mathbf{x}| \le \delta) = e_1$, $E(\sup_{0\le s\le t}|\mathbf{b}(s)| \ge R) = e_2$. Given the probability distribution for $\mathbf{b}(t)$, the first term on the right-hand side of (25.16) is obviously a lower bound on e_1. To see that the term subtracted is an upper bound on e_2, we use (7.6′). ∎

Theorem 25.15 Let V obey the hypotheses of Corollary 25.8 (in particular, suppose that V is globally bounded from below and locally bounded from above). Suppose that $E = \inf \operatorname{spec}(H)$ and that $H\phi = E\phi$, for some $\phi \ne 0$ in L^2. Then ϕ (after multiplication by some overall phase) is everywhere strictly positive and thus bounded away from zero on every compact set.

Proof Since the positive functions are total, we can find $\eta \ge 0$ pointwise so that $(\eta, \phi) \ne 0$. Since

$$\phi = (\phi, \eta)^{-1} \lim_{t\to\infty} e^{tE} e^{-tH} \eta$$

the Feynman–Kac formula shows that ϕ is nonnegative up to an overall phase [namely, the phase of (η, ϕ)]. Since ϕ is continuous, we can find an open set C, and a $\lambda > 0$ so that $\phi(\mathbf{x}) \geq \lambda$ on C. For simplicity of notation, we suppose that $C = \{\mathbf{y} \mid |\mathbf{y}| \leq \delta\}$. Fix r and $|\mathbf{x}| \leq r$ and let $W_R(\mathbf{y}) = W(\mathbf{y})$ if $|\mathbf{y}| \geq r + R$ and zero otherwise. Let

$$A_R = \left\{\mathbf{b} \,\middle|\, |\mathbf{b}(1) + \mathbf{x}| \leq \delta,\ \sup_{0 \leq s \leq 1} |\mathbf{b}(s)| \leq R\right\}$$

Then, by the Feynman–Kac formula and $V_- \leq 0$ and the fact that ϕ is nonnegative:

$$\begin{aligned}\phi(\mathbf{x}) &= e^E(e^{-H}\phi)(\mathbf{x}) \\ &\geq e^E E\left(\chi_A \exp\left(-\int_0^1 W_R(\mathbf{x} + \mathbf{b}(s))\, ds\right)\phi(\mathbf{x} + \mathbf{b}(1))\right) \\ &\geq \lambda e^E E(\chi_A)^2 E\left(\exp\left(+\int_0^1 W_R(\mathbf{x} + \mathbf{b}(s))\, ds\right)\right)^{-1} \qquad (25.18)\end{aligned}$$

where we use $\phi(\mathbf{y}) \geq \lambda$ if $|y| \leq \delta$ and the Schwarz inequality on

$$\chi_A = \left[\chi_A \exp\left(-\frac{1}{2}\int W\right)\right]\left[\exp\left(+\frac{1}{2}\int W\right)\right]$$

in the last step. Since W_R is in L^p we get an upper bound B_R on

$$\sup_{\mathbf{x}} E\left(\exp\left(\int_0^1 W_R(\mathbf{x} + \mathbf{b}(s))\, ds\right)\right) = \|\exp(\tfrac{1}{2}\Delta + W_R)1\|_\infty$$

by (25.5). By Lemma 25.14, $E(\chi_A) > 0$ for R sufficiently large. ∎

Remark Carmona [31] has remarked that if one just wants strict positivity without explicit lower bounds, then there is a simpler proof as follows: Fix $\mathbf{x}$. For every R,

$$E\left(\exp\left(\int_0^1 W_R(\mathbf{x} + \mathbf{b}(s))\, ds\right)\right) < \infty$$

so $\int_0^1 W_R(\mathbf{x} + \mathbf{b}(s))\, ds < \infty$ for all R and almost every $\mathbf{b}$. Thus

$$\int_0^1 W(\mathbf{x} + \mathbf{b}(s))\, ds < \infty$$

for almost every $\mathbf{b}$ by continuity of paths. It follows that

$$\exp\left(-\int_0^1 V(\mathbf{x} + \mathbf{b}(s))\, ds\right) > 0$$

for almost every $\mathbf{b}$ and thus $\phi \geq 0$ and $(e^{-tH}\phi)(\mathbf{x}) = 0$ implies that $(e^{-tH_0}\phi)(\mathbf{x}) = 0$ so that $\phi \equiv 0$.

Example Let $H = -\Delta_1 - \Delta_2 - 2|\mathbf{x}_1|^{-1} - 2|\mathbf{x}_2|^{-1} + |\mathbf{x}_1 - \mathbf{x}_2|^{-1}$ be the helium atom Hamiltonian on $L^2(\mathbb{R}^6)$ where $\mathbf{x} = (\mathbf{x}_1, \mathbf{x}_2)$ with $\mathbf{x}_i \in \mathbb{R}^3$. The eigenfunction ψ corresponding to the lowest eigenvalue (which is an acceptable physical state, even though it is symmetric in the electron coordinates since the spin variables can be used to accommodate the Pauli principle demand of total antisymmetry) is pointwise strictly positive since $W(\mathbf{x}) = |\mathbf{x}|^{-1}$ is in $L_u^p(\mathbb{R}^3)$ for any $p < 3$ and, in particular, for some $p > \frac{3}{2}$. This is a new result, for Simon [245] needs local regularity for V and thus cannot assert strict positivity at points with $\mathbf{x}_1 = \mathbf{x}_2$ while Carmona [32] does not allow localization or for functions of the projection of $\mathbb{R}^6$ to $\mathbb{R}^3$ and so requires $V \in L^p + L^\infty$ for $p > 3 = \frac{1}{2}(6)$.

Lemma 25.14 is also the key to proving lower bounds of WKB type:

Theorem 25.16 Let $V(\mathbf{x})$ be a function which is bounded on any compact set of $\mathbb{R}^\nu$ and let $V^{(\infty)}(\mathbf{x}) = \sup_{|\mathbf{y}| \leq 3|\mathbf{x}|} V(\mathbf{y})$. Suppose that $V(\mathbf{x}) \to \infty$ as $|\mathbf{x}| \to \infty$ so that $E = \inf \operatorname{spec}(-\frac{1}{2}\Delta + V)$ is an eigenvalue. Let ϕ be the corresponding eigenfunction normalized so that $\phi(\mathbf{0}) > 0$. Then

$$\phi(\mathbf{x}) \geq C \exp(-D|\mathbf{x}|[V^{(\infty)}(\mathbf{x})]^{1/2}) \tag{25.19}$$

Proof The argument is similar to the proof of (25.18). Let

$$A(\mathbf{x}, t) = \left\{\mathbf{b} \,\middle|\, |\mathbf{b}(t) + \mathbf{x}| \leq 1, \sup_{0 \leq s \leq t} |\mathbf{b}(s)| \leq 2|\mathbf{x}|\right\}$$

Then, by Lemma 25.14, there exists R, c_1, and c_2 so that

$$E(\chi_A) \geq c_1 t^{-\nu/2} \exp(-(2t)^{-1}(|\mathbf{x}| + 1)^2) \tag{25.20a}$$

for all $\mathbf{x}$, t with $|\mathbf{x}| \geq R$ and

$$|\mathbf{x}|^2/t \geq c_2|\ln[t + 1]|^2 \tag{25.20b}$$

By the Feynman–Kac formula:

$$\begin{aligned}\phi(\mathbf{x}) &= e^{tE}(e^{-tH}\phi)(\mathbf{x}) \\ &\geq e^{tE}E\left(\chi_A \exp\left(-\int_0^t V(\mathbf{b}(s) + \mathbf{x})\, ds\right)\phi(\mathbf{x} + \mathbf{b}(t))\right) \\ &\geq d e^{tE} e^{-tV^{(\infty)}(\mathbf{x})} E(\chi_A)\end{aligned} \tag{25.21}$$

where $d = \inf_{|\mathbf{y}| \leq 1} \phi(\mathbf{y}) > 0$ by Theorem 25.15. (25.19) now holds for suffiently large $\mathbf{x}$ by (25.20) and (25.21) with the choice $t = [V^{(\infty)}(\mathbf{x})]^{-1/2}|\mathbf{x}|$. Knowing it for large $\mathbf{x}$, implies it for all $\mathbf{x}$ by Theorem 25.15 again. ∎

Remarks 1. For example, if $V(\mathbf{x}) \geq C|\mathbf{x}|^{2n}$ for n large, then we obtain an $\exp(-D|\mathbf{x}|^{n+1})$ lower bound. This has "supercontractive" consequences [222, 245, 31].

2. By using the ideas which lead to (25.19), one can easily accommodate positive local singularities in L^p with $p > \nu/2$. Negative singularities always have no adverse effect on lower bounds for ϕ.

3. It is somewhat disturbing that (25.19) depends on a supremum over a sphere and does not allow consideration of different growth of V in different directions. However, when $V(x) \geq C|\mathbf{x}|^{2+\varepsilon}$, it is clear that $V^{(\infty)}(\mathbf{x})$ can be replaced by the supremum over a small cigar-shaped region about a straight line from zero to $\mathbf{x}$. For, in the above proof, $t \to 0$ as $|\mathbf{x}| \to \infty$ so that the paths will stay in such a region with overwhelming probability.

4. The ideas in the first part of this section can also be used in the study of exponential falloff; one shows under suitable circumstances that $H\phi = E\phi$ and $e^f\phi \in L^2$ implies automatically that $e^f\phi \in L^\infty$. See [44a].

26. Inverse Problems and the Feynman–Kac Formula

In this final section, we want to describe certain aspects of the solution by Trubowitz [280] of the inverse problem for periodic potentials. Similar considerations are involved in the solution of the inverse scattering problem on the line by Deift and Trubowitz [46]. "Inverse problems" concern the determination of a potential V given suitable "spectral" or "scattering" data for $-\Delta + V$.

Here we will consider a function V on $(-\infty, \infty)$ which is C^∞ and periodic with period one. One labels the eigenvalues of $-d^2/dx^2 + V(x)$ on $L^2(0, 1)$ with the boundary conditions $u(0) = u(1)$, $u'(0) = u'(1)$, by $\lambda_0 < \lambda_3 \leq \lambda_4 < \lambda_7 \leq \lambda_8 \cdots < \lambda_{4n-1} \leq \lambda_{4n} < \cdots$ and with the boundary conditions $u(0) = -u(1)$, $u'(0) = -u'(1)$ by $\lambda_1 \leq \lambda_2 < \lambda_5 \leq \lambda_6 < \cdots < \lambda_{4n+1} \leq \lambda_{4n+2} < \cdots$. As the notation suggests, one can show [69, 166, 179, 217] that

$$\lambda_0 < \lambda_1 \leq \lambda_2 < \lambda_3 \leq \lambda_4 < \cdots \tag{26.1}$$

Moreover, [128], since V is C^∞, one has that

$$\lim_{n\to\infty} |\lambda_{2n} - \lambda_{2n-1}|(n+1)^{-m} = 0 \tag{26.2}$$

for any m (one can show [185] that if V is a priori only L^2_{loc} and (26.2) holds, then V is C^∞). Moreover,

$$\lambda_{2n}, \lambda_{2n-1} = n^2\pi^2 + \int_0^1 V(x)\, dx + O(n^{-2}) \tag{26.3}$$

Next, let $\mu_n(s)$ be the nth eigenvalue of $-d^2/dx^2 + V(x)$ on $L^2(s, 1 + s)$ with the boundary condition $u(s) = u(1 + s) = 0$. Then [69, 166],

$$\lambda_{2n-1} \leq \mu_n(s) \leq \lambda_{2n} \tag{26.4}$$

for all s. Using (26.1–4), we wish to prove the following which goes back at least to McKean–van Moerbeke [184].

Theorem 26.1 Let V be C^∞ of period one. For any s:

$$V(s) = \lambda_0 + \sum_{n\geq 1} [\lambda_{2n-1} + \lambda_{2n} - 2\mu_n(s)] \tag{26.5}$$

where the sum is absolutely convergent.

(26.5) says that $V(s)$ can be recovered from the λ_m's and the $\mu_m(s)$'s. This is only part of the story. For the $\mu_n(s)$'s can be found once one knows the λ_m's and the values $\mu_n(s_0)$ and $(d\mu_n/dt)(s_0)$ for a single s_0. This makes it sound as if the μ's obey a second order differential equation. Actually, they obey a first order equation, but one which is quadratic in $d\mu/ds$. Thus, the λ_m's and the $\mu_k(s_0)$'s determine $|(d\mu_n/ds)(s_0)|$, and only a choice of each sign is needed to determine μ for all s and thus V. One can further show that the $\mu_n(s_0)$ and a set of positive numbers a_n called **norming constants** determine the λ_m and the sign of $(d\mu_n/ds)(s_0)$ and therefore this data determines V. This yields a result first obtained by Gel'fand and Levitan [98] using very different methods. The results just described are contained in Trubowitz' paper [280]; here we will concentrate only on proving (26.5). We note that the C^∞-condition is replaced by a C^3-condition in [280] and it is likely that mere continuity of V will suffice especially if one is satisfied with conditional convergence of the sum in (26.5). We also remark that a number of features of the inverse problem are made particularly transparant by the above solution; e.g., $V(x)$ has period $1/m$ if and only if $\lambda_{2n} = \lambda_{2n-1}$ if m does not divide n.

We will prove (26.5) in two steps:

Definition A_p, A_a, and A_d will stand for the operators $-d^2/dx^2 + V(x)$ on $L^2(0, 1)$ with periodic $[u(0) = u(1);\ u'(0) = u'(1)]$, antiperiodic $[u(0) = -u(1);\ u'(0) = -u'(1)]$, and Dirichlet $[u(0) = u(1) = 0]$ boundary conditions, respectively.

Theorem 26.2 For each $t > 0$, e^{-tA_p}, e^{-tA_a} and e^{-tA_d} are trace class. Moreover, the sum on the right-hand side of (26.5) is absolutely convergent and

$$\begin{aligned}\operatorname{Tr}(e^{-tA_p} + e^{-tA_a} - 2e^{-tA_d})\\ = 1 - t\left[\lambda_0 + \sum_{n\geq 1}(\lambda_{2n-1} + \lambda_{2n} - 2\mu_n(0))\right] + o(t) \qquad (26.6)\end{aligned}$$

as $t \downarrow 0$.

Theorem 26.3 As $t \downarrow 0$

$$\operatorname{Tr}(e^{-tA_p} - e^{-tA_d}) = \tfrac{1}{2} - \tfrac{1}{2}tV(0) + o(t) \qquad (26.7)$$

$$\operatorname{Tr}(e^{-tA_a} - e^{-tA_d}) = \tfrac{1}{2} - \tfrac{1}{2}tV(0) + o(t) \qquad (26.8)$$

Proof of Theorem 26.2 This follows fairly directly from (26.1)–(26.4). By (26.3) and (26.4), we see that e^{-tA_y} is trace class for $y = $ p, a, d. Moreover:

$$\begin{aligned}\operatorname{Tr}(e^{-tA_p} + e^{-tA_a} - 2e^{-tA_d}) - 1\\ = (e^{-t\lambda_0} - 1) + \sum_{n\geq 1}[e^{-t\lambda_{2n}} + e^{-t\lambda_{2n-1}} - 2e^{-t\mu_n(0)}] \qquad (26.9)\end{aligned}$$

where we have used the trace class properties to freely arrange the sums. Dividing the right-hand side of (26.9) by t and taking t to zero, the first term converges to $-\lambda_0$ and each term in the sum to $-\lambda_{2n} - \lambda_{2n-1} + 2\mu_n(0)$. Moreover, from the estimate

$$|e^{-x} - e^{-y}| \leq e^{-\min(x,y)}|x - y|$$

and (26.4), we see that

$$|e^{-t\lambda_{2n}} + e^{-t\lambda_{2n-1}} - 2e^{-t\mu_n(0)}| \leq 2e^{-t\lambda_{2n-1}}|\lambda_{2n} - \lambda_{2n-1}|$$

so that we can use (26.2) and the dominated convergence theorem to justify taking a term-by-term limit inside the sum in (26.9). ∎

The proof of Theorem 26.3 depends on the same mechanism that allowed us to control the classical limit (Section 10) and the falloff of eigenfunctions (Section 25), namely, that Brownian paths do not go very far in short times; explicitly, we need the following slight strengthening of an argument used already in the proof of Theorem 10.1.

Lemma 26.4 Fix ε in $(0, \frac{1}{2})$. Then for $t \leq 1$,

$$E\left(\sup_{0 \leq s \leq t} |b(s)| \geq t^{1/2-\varepsilon} \mid b(t) = 0\right) \leq C_1 \exp(-C_2 t^{-2\varepsilon}) \tag{26.10}$$

for suitable $C_1, C_2 > 0$.

Proof As in the proof of Lemma 7.10, the left-hand side of (26.10) is equal to

$$1 - P(0, 0; t)^{-1} P_{\mathrm{D};t}(0, 0; t)$$

where P is the kernel of $\exp(+t(d^2/dx^2))$ and $P_{\mathrm{D};t}$ is the analogous kernel with Dirichlet boundary conditions at $x = \pm t^{1/2-\varepsilon}$. By the method of images:

$$P_{\mathrm{D};t}(0, 0; t) = \sum_{n=-\infty}^{\infty} (-1)^n P(0, 2nt^{1/2-\varepsilon}; t)$$

so that

$$|P_{\mathrm{D};t}(0, 0; t) - P(0, 0; t)| \leq Ct^{-1/2} \exp(-2t^{-2\varepsilon})$$

from which (26.10) follows. ▮

Proof of Theorem 26.3 We begin by writing e^{tA_y} for $y = \mathrm{p}, \mathrm{a}$ in terms of Feynman–Kac formulas. Let $H = -\frac{1}{2}(d^2/dx^2) + \frac{1}{2}V(x)$ as an operator on $L^2(-\infty, \infty)$. By the method of images:

$$(e^{-tA_y})(x, x') = \sum_{n=-\infty}^{\infty} \varepsilon_y^n (e^{-2tH})(x, x' + n) \tag{26.11}$$

where $\varepsilon_p = 1$ and $\varepsilon_a = -1$. We will not give the details of the proof of (26.11) leaving that to the reader but we note that the kind of estimates on $e^{-tH}(x, x')$ which follow from the Feynman–Kac formula, e.g.,

$$|e^{-tH}(x, x')| \leq e^{t\|V\|_\infty} e^{-tH_0}(x, x')$$

are useful and that there are two possible approaches: One can verify (26.11) for $V = 0$ and use a Trotter product formula or one can directly show that the right-hand side of (26.11) when smeared in x' obeys the correct differential equation in t and boundary condition at $t = 0$. Either proof establishes that the right-hand side of (26.11) is continuous in x, x' so that (for $y = \mathrm{p}, \mathrm{a}$)

$$\mathrm{Tr}(e^{-tA_y}) = \sum_{n=-\infty}^{\infty} (\varepsilon_y)^n \int_0^1 dx \int \exp\left(-\frac{1}{2}\int_0^{2t} V(\omega(s))\, ds\right) d\mu_{0,x,x+n;\,2t}(\omega)$$

Since the nth term in this sum is bounded by

$$(4\pi t)^{-1/2} \exp(-n^2/4t) e^{t\|V\|_\infty}$$

References

0. M. Aizenman and E. Lieb, On semi-classical bounds for eigenvalues of Schrödinger operators, *Phys. Lett.* **66** (1978), 427–429.
1. S. Albeverio and R. Hoegh-Krohn, Uniqueness of the physical vacuum and the Wightman functions in the infinite volume limit for some non-polynomial interactions, *Comm. Math. Phys.* **30** (1973), 171–200.
2. S. Albeverio and R. Hoegh-Krohn, *Mathematical Theory of Feynman Path Integrals*, Lecture Notes in Math. **523**, Springer-Verlag, Berlin and New York, 1976.

2a. T. W. Anderson, The integral of a symmetric unimodular function over a symmetric convex set and some probability inequalities, *Proc. Amer. Math. Soc.* **6**, (1955), 170–176.

3. A. M. Arthurs, ed., *Functional Integration and Its Applications*, Oxford Univ. Press (Clarendon), London and New York, 1975.
4. J. Avron, I. Herbst, and B. Simon, Schrödinger operators with magnetic fields, I. General interactions, *Duke Math. J.* **45** (1978), 847–883.
5. J. Avron, I. Herbst, and B. Simon, Schrödinger operators with magnetic fields, II. Separation of the center of mass in homogeneous magnetic fields, *Ann. Physics* **114** (1978), 431–451.
6. J. Avron, I. Herbst, and B. Simon, Schrödinger operators with magnetic fields, III. Atoms in constant magnetic field, to be submitted to *Comm. Math. Phys.*

6a. D. Babbitt, Wiener integral representations for certain semi-groups which have infinitesmal generators with matrix coefficients, *J. Math. Mech.* **19** (1970), 1057–1067.

7. C. Bender and T. T. Wu, Anharmonic oscillator, *Phys. Rev.* **184** (1969), 1231–1260.
8. C. Bender and T. T. Wu, Large order behavior of perturbation theory, *Phys. Rev. Lett.* **16** (1971), 461–465.
9. C. Bender and T. T. Wu, Anharmonic oscillator, II. A study of perturbation theory in large order, *Phys. Rev. D* **7** (1973), 1620–1636.
10. C. Bender and T. T. Wu, Statistical analysis of Feynman diagrams, *Phys. Rev. Lett.* **37** (1976), 117–120.
11. S. Berberian, *Measure and Integration*, Macmillan, New York, 1965.
12. F. A. Berezin, Wick and anti-Wick operator symbols, *Mat. Sb.* **B6** (1971) [Transl.: *Math. U.S.S.R. Sb.* **15** (1971), 577–606].
13. A. Berthier and B. Gaveau, Critère de convergence des fonctionelles de Kac et application en mécanique quantique et en géométrie, *J. Functional Analysis*, **29** (1978), 416–424.

14. P. Billingsley, *Convergence of Probability Measures*, Wiley, New York, 1968.
15. M. Birman, On the number of eigenvalues in a quantum scattering problem, *Vestnik LSU* **16** (No. 13) (1961), 163–166.
16. M. Birman and V. V. Borzov, On the asymptotics of the discrete spectrum of some singular differential operators, *Topics in Math. Phys.* **5** (1972), 19–30.
17. R. Blankenbecler, M. L. Goldberger, and B. Simon, The bound state of weakly coupled, long-range, one-dimensional quantum Hamiltonians, *Ann. Physics* **108** (1977), 69–78.
18. N. N. Bogoliubov, A. A. Logunov, and I. T. Todorov, *Introduction to Axiomatic Quantum Field Theory*, Benjamin, New York, 1975 (Russian original, 1969).
18a. C. Borell, Convex measures on locally convex spaces, *Ark. Mat.* **12** (1974), 239–252.
18b. C. Borell, Convex set functions in *d*-space, *Period. Math. Hungar.* **6** (1975), 111–136.
18c. C. Borell, The Brunn–Minkowski inequality in Gauss space, *Invent. Math.* **30** (1975), 207–216.
19. H. Brascamp and E. H. Lieb, Some inequalities for Gaussian measures, pp. 1–14 in [3].
20. H. Brascamp and E. H. Lieb, On extensions of the Brunn–Minkowski and Prékopa–Leindler theorems, including inequalities for log concave functions, and with an application to the diffusion equation, *J. Functional Analysis* **22** (1976), 366–389.
21. H. Brascamp and E. H. Lieb, Best constants in Young's inequality, its converse, and its generalization to more than three functions, *Advances in Math.* **20** (1976), 151–173.
22. H. Brascamp, E. H. Lieb, and J. M. Luttinger, A general rearrangement inequality for multiple integrals, *J. Functional Analysis* **17** (1974), 227–237.
23. L. Breiman, *Probability*, Addison-Wesley, Reading, Massachusetts, 1968.
24. E. Brezin, J. C. LeGuillou, and J. Zinn-Justin, Perturbation theory at large order, I. The ϕ^{2N} interaction, *Phys. Rev. D* **15** (1971), 1544–1557.
25. H. Brezis and T. Kato, Remarks on the Schrödinger operator with singular complex potential, *J. Math. Pures et Appl.*, to be published.
26. D. Brydges, A rigorous approach to Debye screening in dilute classical Coulomb screening, *Comm. Math. Phys.* **58** (1978), 315–350.
27. D. Brydges and P. Federbush, The cluster expansion for potentials with exponential falloff, *Comm. Math. Phys.* **53** (1977), 19–30.
28. D. Burkholder, R. Gundy, and M. Silverstein, A maximal function characterization of the class H^p, *Trans. Amer. Math. Soc.* **157** (1971), 137–153.
29. R. Cameron, The Ilstow and Feynman integrals, *J. Analyse Math.* **10** (1962/63), 287–361.
30. R. H. Cameron and W. T. Martin, Transformations of Wiener integrals by non-linear transformations, *Trans. Amer. Math. Soc.* **66** (1949), 252–283.
31. R. Carmona, Regularity properties of Schrödinger and Dirichlet semigroups, *J. Functional Analysis*, to be published.
32. R. Carmona, Pointwise bounds for Schrödinger eigenstates, *Comm. Math. Phys.* **62** (1978), 97–106.
33. P. Cartier, unpublished.
34. P. Chernoff, Note on product formulas for operator semigroups, *J. Functional Analysis* **2** (1968), 238–242.
35. Z. Ciesielski, Hölder conditions for realizations of Gaussian processes, *Trans. Amer. Math. Soc.* **99** (1961), 403–413.
36. P. Collella and O. Lanford, III, Sample field behavior for free Markov random fields, pp. 44–77 in [282].
37. J. M. Combes, R. Schrader, and R. Seiler, Classical bounds and limits for energy distributions of Hamiltonian operators in electromagnetic fields, *Ann. Physics* **111** (1978), 1–18.
38. M. Combescure and J. Ginibre, Scattering and local absorptions for the Schrödinger operator, *J. Functional Analysis* **29** (1978), 54–73.

39. H. Cramer and M. R. Leadbetter, *Stationary and Related Stochastic Processes: Sample Function Properties and Their Application*, Wiley, New York, 1967.
39a. P. Cvitanovic, B. Lautrup, and R. Pearson, The number and weights of Feynman diagrams, *Phys. Rev.* **D18** (1978), 1939–1949.
40. M. Cwickel, Weak type estimates and the number of bound states of Schrödinger operators, *Ann. of Math.* **106** (1977), 93–102.
40a. Yu. Daleckii, Continual integrals associated to certain differential equations and systems, *Soviet Math. Dokl.* **2** (1961), 259–263.
41. E. B. Davies and B. Simon, Scattering theory for systems with different spatial asymptotics on the left and right, *Comm. Math. Phys.* **63** (1978), 277–301.
42. B. Davis, Picard's theorem and Brownian motion, *Trans. Amer. Math. Soc.* **213** (1975), 353–362.
43. B. DeFacio and C. L. Hammer, Remarks on the Klauder phenomenon, *J. Mathematical Phys.* **15** (1974), 1071–1077.
44. P. Deift, *Classical Scattering Theory with a Trace Condition*, Princeton Univ. Press, Princeton, New Jersey, to appear.
44a. P. Deift, W. Hunziker, B. Simon, and E. Vock, Pointwise bounds on eigenfunctions and wave packets in N-body quantum systems, IV, *Comm. Math. Phys.* **64** (1979), 7–34.
45. P. Deift and B. Simon, On the decoupling of finite singularities from the question of asymptotic completeness in the two-body quantum systems, *J. Functional Analysis* **23** (1976), 218–238.
46. P. Deift and E. Turbowitz, Inverse scattering on the line, *Comm. Pure Appl. Math.* **32** (1979), 121–251.
46a. A. Devinatz, Schrödinger operators with singular potentials, *J. Op. Th.*, to be published.
47. C. DeWitt-Morette, Feynman's path integral. Definition without limiting procedure, *Comm. Math. Phys.* **28** (1972), 47–67.
48. C. DeWitt-Morette, Feynman path integrals, I. Linear and affine transformations, II. The Feynman Green's function, *Comm. Math. Phys.* **37** (1974), 63–81.
49. C. DeWitt-Morette, The semiclassical expansion, *Ann. Physics* **97** (1976), 367–399.
50. C. DeWitt-Morette, A. Maheshwari, and B. Nelson, Path integrals in phase space, *General Relativity and Gravitation*, **8** (1977), 581–593.
50a. C. DeWitt-Morette, A. Maheshwari, and B. Nelson, Path integration in non-relativistic quantum mechanics, *Physics Reports*, **50** (1979), 255–372.
51. C. DeWitt-Morette and R. Stora, eds. *Statistical Mechanics and Quantum Field Theory, Les Houches, 1970*, Gordon & Breach, New York, 1971.
52. J. Dimock, Perturbation series asymptotic to Schwinger functions in $P(\phi)_2$, *Comm. Math. Phys.* **35** (1974), 347–356.
53. M. Donsker, An invariance principle for certain probability limit theorems, *Mem. Amer. Math. Soc.* **6** (1951).
54. M. Donsker and S. R. S. Varadhan, Asymptotic evaluation of certain Wiener integrals for large time, pp. 15–33 in [3].
55. M. Donsker and S. R. S. Varadhan, Asymptotic evaluation of certain Markov expectations for large time, I, *Comm. Pure Appl. Math.* **28** (1975), 1–47.
56. M. Donsker and S. R. S. Varadhan, Asymptotic evaluation of certain Markov expectations for large time, II, *Comm. Pure Appl. Math.* **28** (1975), 279–301.
57. M. Donsker and S. R. S. Varadhan, Asymptotic evaluation of certain Markov expectations for large time, III, *Comm. Pure Appl. Math.* **29** (1976), 389–461.
58. M. Donsker and S. R. S. Varadhan, Asymptotics for the Wiener sausage, *Comm. Pure Appl. Math.* **27** (1975), 525–565.
59. M. Donsker and S. R. S. Varadhan, On a variational formula for the principal eigenvalue for operators with maximum principle, *Proc. Nat. Acad. Sci. U.S.A.* **72** (1975), 780–783.

60. M. Donsker and S. R. S. Varadhan, On the principal eigenvalue of second order elliptic differential operators, *Comm. Pure Appl. Math.* **29** (1976), 595–621.
61. J. Doob, *Stochastic Processes*, Wiley, New York, 1953.
61a. R. M. Dudley, Sample functions of the Gaussian Process, *Ann. Probability* **1** (1973), 66–103.
62. F. Dunlop and C. Newman, Multicomponent field theories and classical rotators, *Comm. Math. Phys.* **44** (1975), 223–235.
63. A. Dvoretsky, P. Erdös, and S. Kakutani, Double Points of Brownian motion in n-space, *Acta Sci. Math.* (*Szeged*) **12** (1950), 75–81.
64. A. Dvoretsky, P. Erdös, and S. Kakutani, Multiple points of paths of Brownian motion in the plane, *Bulletin of the Research Council of Israel*, **3** (1954), 364–371.
65. A. Dvoretsky, P. Erdös, and S. Kakutani, Nonincreasing everywhere of the Brownian motion process, *Proc. 4th Berk. Symp. Math. Statist. Probability* **3** (1961), 103–116.
66. A. Dvoretsky, P. Erdös, S. Kakutani, and S. J. Taylor, Triple points of Brownian paths in three space, *Proc. Camb. Phil. Soc.* **53** (1957) 856–862.
67. E. Dynkin, *Markov Processes*, Springer-Verlag, Berlin and New York, 1965.
68. F. Dyson and A. Lenard, Stability of Matter, I, II; *J. Mathematical Phys.* **8** (1967), 423–434; **9** (1968), 698–711.
69. M. S. P. Eastham, *The Spectral Theory of Periodic Differential Equations*, Scottish Academic Press, 1973.
70. J. P. Eckmann, Hypercontractivity for anharmonic oscillators, *J. Functional Analysis* **16** (1974), 388–404.
71. J. P. Eckmann, *Relativistic Bose Quantum Field Theories in Two Space–Time Dimensions*, Rome Lecture Notes, 1976.
72. J. P. Eckmann, J. Magnen, and R. Séneor, Decay properties and Borel summability for Schwinger functions in $P(\phi)_2$ Theories, *Comm. Math. Phys.* **39** (1974), 251–271.
73. S. F. Edwards and A. Lenard, Exact statistical mechanics of a one-dimensional system with Coulomb forces, II. The method of functional integration, *J. Mathematical Phys.* **3** (1962), 778–792.
74. R. Ellis, J. Monroe, and C. Newman, The GHS and other correlation inequalities for a class of even ferromagnets, *Comm. Math. Phys.* **46** (1976), 167–182.
75. R. Ellis and C. Newman, Necessary and sufficient conditions for the GHS inequality with applications to analysis and probability, *Trans. Amer. Math. Soc.* **237** (1978), 83–99.
76. R. Ellis and C. Newman, Limit theorems for sums of two dependent random variables occurring in statistical mechanics, *Z. Wahrscheinlichkeitstheorie und Verw. Gebiete*, **44** (1978), 117–139.
77. P. Erdös and M. Kac, On certain limit theorems in the theory of probability, *Bull. Amer. Math. Soc.* **52** (1946), 292–302.
78. P. Erdös and A. Rennyi, On Cantor's series with convergent $1/q_n$, *Ann. Univ. Sci. Budapest Eötvös Sect. Math.* **2** (1959), 93–109.
79. H. Ezawa, J. R. Klauder, and L. A. Shepp, Vestigial effects of singular potentials in diffusion theory and quantum mechanics, *J. Mathematical Phys.* **16** (1975), 783–799.
80. W. Faris and B. Simon, Degenerate and nondegenerate ground states for Schrödinger operators, *Duke Math. J.* **42** (1975), 559–567.
80a. J. Feldman, Equivalence and perpendicularity of Gaussian processes, *Pacific J. Math.* **8** (1958), 699–708.
81. J. Feldman, The $\lambda\phi_3^4$ field theory in a finite volume, *Comm. Math. Phys.* **37** (1974), 93–120.
82. W. Feller, *An Introduction to Probability Theory and Its Applications I, II*, Wiley, New York, 1951, 1961.
82a. X. Fernique, Regularité de processus gaussiens, *Invent. Math.* **12** (1971), 304–320.

82b. X. Fernique, Regularité des trajectoires des fonctions aleatoires gaussiennes, pp. 2–96, in *Ecole d'Eté de Probabilitiés de Saint-Flour IV—1974*, Springer-Verlag, Berlin and New York, Lecture Notes in Math. **480**, 1975.

82c. X. Fernique, Evaluations de processus gaussiens composés, pp. 67–83 in *Probability in Banach Spaces—Oberwolfach 1975*, Lecture Notes in Math. **526**, Springer-Verlag, Berlin and New York, 1976.

83. R. P. Feynman, Space–time approach to nonrelativistic quantum mechanics. *Rev. Modern Phys.* **20** (1948), 367–387.

84. R. P. Feynman and A. Hibbs, *Quantum Mechanics and Path Integrals*, McGraw-Hill, New York, 1965.

85. C. Fortuin, P. Kasteleyn, and J. Ginibre, Correlation inequalities on some partially ordered sets, *Comm. Math. Phys.* **22** (1971), 89–103.

85a. D. Freedman, *Brownian Motion and Diffusion*, Holden-Day, San Francisco, California, 1971.

86. A. Friedman, *Stochastic Differential Equations and Applications*, Academic Press, New York, 1975.

87. C. Friedman, Perturbations of the Schrödinger equation by potentials with small support, *J. Functional Analysis* **10** (1972), 346–360.

88. K. O. Friedrichs, *Perturbation of Spectra in Hilbert Space*, Amer. Math. Soc., Providence, Rhode Island, 1965.

89. J. Fröhlich, Schwinger functions and their generating functionals, I, *Helv. Phys. Acta* **47** (1974), 265–306.

90. J. Fröhlich, Schwinger functions and their generating functionals, II, *Advances in Math.* **23** (1977), 119–180.

91. J. Fröhlich, Verification of axioms for Euclidean and relativistic fields and Haag's theorem in a class of $P(\phi)_2$ models, *Ann. Inst. H. Poincaré Sect. A* **21** (1974), 271–317.

92. J. Fröhlich, Quantized "Sine–Gordon" equation with a nonvanishing mass term in two space–time dimensions, *Phys. Rev. Lett.* **34** (1975), 833–836.

93. J. Fröhlich, Classical and quantum statistical mechanics in one and two dimensions: Two component Yukawa and Coulomb systems, *Comm. Math. Phys.* **47** (1976), 233–268.

94. J. Fröhlich and Y. M. Park, Correlation inequalities and the thermodynamic limit for classical and quantum continuous systems, *Comm. Math. Phys.* **59** (1978), 235–266.

95. J. Fröhlich and E. Seiler, The massive Thirring–Schwinger model (QED_2): Convergence of perturbation theory and particle structure, *Helv. Phys. Acta* **49** (1976), 889–924.

96. J. Fröhlich and B. Simon, Pure states for general $P(\phi)_2$ theories: Construction, regularity and variational equality, *Ann. of Math.* **105** (1977), 493–526.

97. J. Fröhlich, B. Simon, and T. Spencer, Infrared bounds, phase transitions and continuous symmetry breaking, *Comm. Math. Phys.* **50** (1976), 79–95.

97a. D. Fujiwara, Fundamental solution of partial differential operators of Schrödinger's type, I, II, III, *Proc. Japan Acad.*, **50**, 566–569; 699–701 (1974); **54** (1978), 62–66.

97b. D. Fujiwara, A construction of the fundamental solution for Schrödinger equation, Preprint.

98. I. M. Gel'fand and B. M. Levitan, On the determination of a differential equation from its spectral function, *Izv, Akad. Nauk.* **15** (1951), 309–360.

99. J. Ginibre, Reduced density matrices of quantum gases, I. Limit of infinite volume, II. Cluster properties, III. Hard core potentials, *J. Mathematical Phys.* **6** (1965), 238–251, 256–262, 1432–1446.

100. J. Ginibre, Some applications of functional integration in statistical mechanics, pp. 327–428 in [51].

101. J. Ginibre, General formulation of Griffiths' inequality, *Comm. Math. Phys.* **16** (1970), 310–328.

102. V. Glaser, H. Grosse, and A. Martin, Bounds on the number of eigenvalues of the Schrödinger operator, *Comm. Math. Phys.* **59** (1978), 197–212.

103. V. Glaser, H. Grösse, A. Martin, and W. Thirring, A family of optimal conditions for the absence of bound states in a potential, pp. 169–194 in [174].

104. J. Glimm, Boson fields with nonlinear self-interaction in two dimensions, *Comm. Math. Phys.* **8** (1968), 12–25.

105. J. Glimm and A. Jaffe, Positivity of the $(\phi^4)_3$ Hamiltonian, *Fortschr. Physik* **21** (1973), 327–376.

106. J. Glimm and A. Jaffe, Functional integration methods in *New Developments in Quantum Mechanics and Statistical Mechanics; Cargèse, 1976*, (M. Lévy and P. Mitter, eds.) Plenum, New York, 1977.

107. J. Glimm, A. Jaffe, and T. Spencer, The particle structure of the weakly coupled $P(\phi)_2$ model and other applications of high temperature expansions, Part I. Physics of quantum field models, pp. 133–198 in [282].

108. J. Glimm, A. Jaffe, and T. Spencer, The particle structure of the weakly coupled $P(\phi)_2$ model and other applications of high temperature expansions, Part II. The cluster expansion, pp. 199–242 in [282].

109. J. Glimm, A. Jaffe, and T. Spencer, Phase transitions for ϕ_2^4 quantum field models, *Comm. Math. Phys.* **45** (1975), 203–216.

110. J. Glimm, A. Jaffe, and T. Spencer, A convergent expansion about mean field theory, I, II, *Ann. Physics* **101** (1976), 610–630, 631–689.

111. S. Golden, Lower bounds for the Helmholtz function, *Phys. Rev. B* **137** (1965), 1127–1128.

112. J. Goldstein, ed., *Partial Differential Equations and Related Topics*, Lecture Notes in Math. **446**, Springer-Verlag, Berlin and New York, 1975.

113. S. Graffi, V. Grecchi, and B. Simon, Borel summability: Application to the anharmonic oscillator, *Phys. Lett. B* **32** (1970), 631–634.

114. W. M. Greenlee, Singular perturbation theory for semibounded operators, *Bull. Amer. Math. Soc.* **82** (1976), 341–343.

115. R. Griffiths, Correlation inequalities in Ising ferromagnets, I, II, *J. Mathematical Phys.* **8** (1967), 478–483, 484–489.

116. R. Griffiths, Phase transitions, pp. 241–280 in [51].

117. R. Griffiths, C. Hurst, and S. Sherman, Concavity of the magnetization of an Ising ferromagnet in positive external field, *J. Mathematical Phys.* **11** (1970), 790–795.

118. L. Gross. Logarithmic Sobolev inequalities, *Amer. J. Math.* **97** (1976), 1061–1083.

118a. L. Gross, Measurable functions on Hilbert space, *Trans. Amer. Math. Soc.* **105** (1962), 372–390.

119. C. Gruber, A. Hinterman, and D. Merlini, *Group Analysis of Classical Lattice Systems*, Lecture Notes in Physics **60**, Springer-Verlag, Berlin and New York, 1977.

120. F. Guerra, L. Rosen, and B. Simon, The $P(\phi)_2$ Euclidean quantum field theory as classical statistical mechanics, *Ann. of Math.* **101** (1975), 111–259.

121. F. Guerra, L. Rosen, and B. Simon, Boundary conditions for the $P(\phi)_2$ Euclidean quantum field theory, *Ann. Inst. H. Poincaré Sect. A* **25** (1976), 231–334.

121a. J. Hajek, On a property of the normal distribution of any stochastic process, *Czechoslovak Math. J.* **8** (1958), 610–618.

122. P. Halmos, *Lectures in Ergodic Theory*, Chelsea, Bronx, New York, 1956.

123. E. Harrell, II, Singular perturbation potentials, *Ann. Physics* **105** (1977), 379–406.

123a. E. Harrell and B. Simon, The mathematical theory of resonances whose widths are exponentially small, to appear.

124. W. K. Hayman and P. B. Kennedy, *Subharmonic Functions*, Vol. I, Academic Press, New York, 1976.

125. K. Hepp, Renormalization theory, pp. 429–500 in [51].
126. I. Herbst and A. Sloan, Perturbation of translation invariant positivity preserving semigroups in $L^2(\mathbb{R}^N)$, *Trans. Amer. Math. Soc.* **236** (1978), 325–360.
127. H. Hess, P. Schrader, and D. Uhlenbrock, Domination of semigroups and generalizations of Kato's inequality, *Duke Math. J.* **44** (1977), 893–904.
128. H. Hochstadt, Estimates on the stability intervals for Hill's equation, *Proc. Amer. Math. Soc.* **14** (1963), 930–932.
129. K. Huang, *Statistical Mechanics*, Wiley, New York, 1963.
130. G. Hunt, Some theorems concerning Brownian motion, *Trans. Amer. Math. Soc.* **81** (1956), 294–319.
131. R. Hwa and V. Teplitz, *Homology and Feynman Integrals*, Benjamin, New York, 1966.
132. T. Ikebe and T. Kato, Uniqueness of the self-adjoint extension of singular elliptic differential operators, *Arch. Rational Mech. Anal.* **9** (1962), 77–92.
133. R. Israel, *Convexity and the Theory of Lattice Gases*, Princeton Univ. Press, Princeton, New Jersey, 1978.
134. K. Itô, Stochastic integrals, *Proc. Imp. Acad. Tokyo* **20** (1944), 519–524.
135. K. Itô, On a formula concerning stochastic differentials, *Nagoya Math. J.* **3** (1951), 55–65.
136. K. Itô and H. McKean, *Diffusion Processes and Their Sample Paths*, Springer-Verlag, Berlin and New York, 1965.
136a. A. Jensen and T. Kato, Asymptotic behavior of the scattering phase for exterior domains, *Commun. P.D.E.*, to be published.
137. R. Jost, *The General Theory of Quantized Fields*, Amer. Math. Soc., Providence, Rhode Island, 1965.
138. M. Kac, On some connections between probability theory and differential equations, *Proc. 2nd Berk. Symp. Math. Statist. Probability* (1950), 189–215.
139. M. Kac, *Probability and Related Topics in the Physical Sciences*, Wiley (Interscience), New York, 1959.
140. M. Kac, Can you hear the shape of a drum. *Amer. Math. Monthly* **73** (1966), 1–23 (Slaught Mem. Papers, No. 11).
141. M. Kac, Probabilistic methods in some problems of scattering theory, *Rocky Mountain J. Math.* **4** (1974), 511–537.
142. M. Kac, On the asymptotic number of bound states for certain attractive potentials, pp. 159–167, in *Topics in Functional Analysis* (I. Gohberg and M. Kac, eds.), Academic Press, New York, 1978.
143. M. Kac, unpublished.
144. M. Kac and J. M. Luttinger, Bose Einstein condensation in the presence of impurities, II, *J. Mathematical Phys.* **15** (1974), 183–186.
145. S. Kakutani, On Brownian motion in n-space, *Proc. Japan Acad.* **20** (1944), 648–652.
146. S. Kakutani, Two-dimensional Brownian motion and harmonic functions, *Proc. Japan Acad.* **20** (1944), 706–714.
147. Y. Kannai, Off-diagonal short-time asymptotics for fundamental solutions of differential equations, *Comm. Partial Differential Equations* **2** (1977), 781–830.
148. J. Karamata, Neuer Beweis und Verallgemeinerung der Tauberschen Sätze, welche die Laplacesche und Stieltjes Transformation betreffen, *J. Reine Angew. Math.* **164** (1931), 27–39.
149. T. Kato, On the convergence of the perturbation method, I, II, *Progr. Theoret. Phys.* **4** (1949), 514–523; **5** (1950), 95–101, 207–212.
150. T. Kato, On the eigenfunctions of many particle systems in quantum mechanics, *Comm. Pure Appl. Math.* **10** (1957), 151–171.
151. T. Kato, Wave operators and similarity for non-self-adjoint operators, *Math. Ann.* **162** (1966), 258–279.

152. T. Kato, *Perturbation Theory for Linear Operators*, Springer-Verlag, Berlin and New York, 1966; 2nd ed., 1976.

153. T. Kato, Schrödinger Operators with Singular Potentials, *Israel J. Math.* **13** (1973), 135–148.

154. T. Kato, Trotter's product formula for an arbitrary pair of self-adjoint contraction semigroups, pp. 185–195, in *Topics in Functional Analysis* (I. Gohberg and M. Kac, eds.), Academic Press, New York, 1978.

155. T. Kato, Remarks on Schrödinger operators with vector potentials, *Integral Equations and Operator Theory*, **1** (1978), 103–113.

155a. T. Kato and K. Masuda, Trotter's product formula for nonlinear semigroups generated by the subdifferentials of convex functionals. *J. Math. Soc. Japan* **30** (1978), 169–178.

156. D. Kelley and S. Sherman, General Griffiths' inequalities on correlations in Ising ferromagnets, *J. Mathematical Phys.* **9** (1968), 466–484.

157. A. Ya. Khintchine, *Asymptotische Gesetze der Wahrscheinlichkeitsrechnung*, Ergebnis Math. **2** (No. 4), Springer-Verlag, Berlin and New York, 1933.

158. J. F. C. Kingman and S. J. Taylor, *Introduction to Measure and Probability*, Cambridge Univ. Press, London and New York, 1966.

159. J. Klauder, Field structure through model studies: Aspects of non-renormalizable theories, *Acta Phys. Austriaca Suppl.* **11** (1973), 341–387.

160. J. Klauder and L. Detweiler, Supersingular quantum perturbations, *Phys. Rev. D* **11** (1975), 1436–1441.

161. M. Klaus, On the bound state of Schrödinger operators in one dimension, *Ann. Physics* 108 (1977), 288–300.

162. A. N. Kolmogorov and Yu. V. Prohorov, *Zufällige Funktionen und Grenzverteilungssätze, Bericht über die Tagung Wahrscheinlichkeiterschnung und mathematische Statistik*, (1954) 113–126.

163. J. Kupsch and W. Sandhas, Møller operators for scattering on singular potentials, *Comm. Math. Phys.* **2** (1966), 147–154.

164. N. S. Landhof, *Foundations of Modern Potential Theory*, Springer-Verlag, Berlin and New York, 1972.

165. J. Lebowitz, GHS and other inequalities, *Comm. Math. Phys.* **35** (1974), 87–92.

166. B. M. Levitan and I. Sargsten, *Introduction to Spectral Theory*, Amer. Math. Soc. Monograph Translations, No. 39, Providence, Rhode Island, 1975.

167. P. Lévy, *Theorie de l'Addition des Variables Aléatoires*, Gauthier-Villars, Paris, 1937.

168. P. Lévy, *Processus Stochastique et Mouvement Brownien*, Gauthier-Villars, Paris, 1948.

169. P. Lévy, Le Mouvement Brownien plan, *Amer. J. Math.* **62** (1940), 487–550.

170. E. Lieb, The classical limit of quantum spin systems, *Comm. Math. Phys.* **31** (1973), 327–340.

171. E. Lieb, Bounds on the eigenvalues of the Laplace and Schrödinger operators, *Bull. Amer. Math. Soc.* **82** (1976), 751–753, and Proc. 1979 AMS Honolulu Conference.

172. E. Lieb, The stability of matter, *Rev. Modern. Phys.* **48** (1976), 553–569.

172a. E. Lieb, The $N^{5/3}$ Law for bosons, *Phys. Lett.* **70A** (1979), 71.

173. E. Lieb and B. Simon, The Thomas–Fermi theory of atoms, molecules, and solids, *Advances in Math.* **23** (1977), 22–116.

173a. E. Lieb and B. Simon, Monotonicity of the electronic contribution to the Born–Oppenheimer energy, *J. Phys. B* **11** (1978), L537–L542.

174. E. Lieb, B. Simon, and A. S. Wightman, eds., *Studies in Mathematical Physics, Essays in Honor of Valentine Bargmann*, Princeton Univ. Press, Princeton, New Jersey, 1976.

175. E. Lieb and W. Thirring, Bound for the kinetic energy of fermions which proves the stability of matters, *Phys. Rev. Lett.* **35** (1975), 687–689.

176. E. Lieb and W. Thirring, Inequalities for the moments of the eigenvalues of the Schrödinger Hamiltonian and their relation to Sobolev inequalities, pp. 269–304 in [174].

177. L. N. Lipatov, Divergence of the perturbation theory series and pseudoparticles, *Soviet Phys. JETP Lett.* **25** (1977), 116–119 [Eng. transl.: *Soviet Phys. JETP Lett.* **25** (1977), 104–107].

178. J. M. Luttinger, Generalized Isoperimetric inequalities, I, II, III, *J. Mathematical Phys.* **14** (1973), 586–593, 1444–1447, 1448–1450.

179. W. Magnus and W. Winkler, *Hill's Equation*, Wiley (Interscience), New York, 1966.

179a. A. Majda and J. Ralston, An analogue of Weyl's theorem for unbounded domains, I, *Duke Math. J.* **45** (1978), 183–196.

180. M. Marcus and L. Shepp, Sample behavior of Gaussian processes, *Proc. 6th Berk. Symp. Math. Statist. Probability*, **II** (1972) 423–439.

181. A. Martin, Bound states in the strong coupling limit, *Helv. Phys. Acta.* **45** (1972), 140–148.

182. A. Martin, unpublished, quoted in [6].

183. H. McKean, *Stochastic Integrals*, Academic Press, New York, 1969.

184. H. McKean and P. van Moerbeke, The spectrum of Hill's equation, *Invent. Math.* **30** (1975), 217–274.

185. H. McKean and E. Trubowitz, Hill's operator and hypoelliptic function theory in the presence of infinitely many branch points, *Comm. Pure Appl. Math.* **29** (1976), 143–226.

186. P. Meyer, *Probability and Potentials*, Ginn (Blaisdell), Boston, Massachusetts, 1966.

187. R. A. Minlos, Generalized random processes and their extension to a measure, *Trudy Moskov. Mat. Obšč.* **8** (1959), 497–518.

188. E. Nelson, Regular probability measures on function space, *Ann. of Math.* **67** (1954), 630–643.

189. E. Nelson, Feynman integrals and the Schrödinger equation, *J. Mathematical Phys.* **5** (1964), 332–343.

190. E. Nelson, Derivation of the Schrödinger equation from Newtonian mechanics, *Phys. Rev.* **150** (1966), 1079–1085.

191. E. Nelson, *Dynamical Theories of Brownian Motion*, Princeton Univ. Press, Princeton, New Jersey, 1967.

192. E. Nelson, A quartic interaction in two dimensions, pp. 69–73, in *Mathematical Theory of Elementary Particles* (R. Goodman, and I. Segal, eds.), MIT Press, Cambridge, Massachusetts, 1966.

193. E. Nelson, Construction of quantum fields from Markoff Fields, *J. Functional Analysis* **12** (1973), 97–112.

194. E. Nelson, The free Markoff field, *J. Functional Analysis* **12** (1973), 211–227.

195. E. Nelson, unpublished.

195a. J. Neveu, *Processus Aléatoires Gaussiens*, Univ. de Montréal Lecture Notes, 1968.

195b. J. Neveu, Sur l'espérance conditionelle par un rapport à un mouvement Brownien, *Ann. Inst. H. Poincaré Sect. B* **12** (1976), 105–109.

196. C. Newman, Gaussian correlation inequalities for ferromagnets, *Z. Wahrscheinlichkeitstheorie und Verw. Gebiete* **33** (1975), 75–93.

196a. M. Nisio, On the extreme values of Gaussian processes, *Osaka J. Math.* **4** (1967), 313–326.

197. K. Osterwalder and R. Schrader, Axioms for Euclidian Green's functions I, II, *Comm. Math. Phys.* **31** (1973), 83–112; **42** (1975), 281–305.

198. Y. M. Park, Massless quantum sine–Gordon equation in two space–time dimensions: Correlation inequalities and infinite volume limit, *J. Mathematical Phys.* **18** (1977), 2423–2426.

198a. K. R. Parthasarathy, *Probability on Measure Spaces*, Academic Press, New York, 1967.

199. R. Payley, N. Wiener, and A. Zygmund, Note on random functions, *Math. Z.* **37** (1933), 647–668.

200. D. Pearson, An example illustrating the breakdown of asymptotic completeness, *Comm. Math. Phys.* **40** (1975), 125–146.

201. D. Pearson, General theory of potential scattering with absorption at local singularities, *Helv. Phys. Acta* **48** (1974), 249–264.

202. J. Percus, Correlation inequalities for Ising spin lattices, *Comm. Math. Phys.* **40** (1972) 283–308.

203. K. E. Peterson, *Brownian Motion, Hardy Spaces and Bounded Mean Oscillation*, Cambridge Univ. Press, London and New York, 1977.

204. M. Pincus, Gaussian processes and Hammerstein integral equations, *Trans. Amer. Math. Soc.* **134** (1968), 193–216.

204a. N. I. Portenko, Diffusion processes with unbounded drift coefficient, *Theor. Probability Appl.* **20** (1976), 27–37.

205. A. Prékopa, Logarithmic concave measures with application to stochastic programming, *Acta Sci. Math. (Szeged)* **32** (1971), 301–315.

206. Yu. V. Prohorov, Probability distributions in functional spaces, *Uspehi Mat. Nauk.* **8** (1953), 165–167.

207. Yu. V. Prohorov, Convergence of random processes and limit theorems in probability theory, *Theor. Probability Appl.* **1** (1956), 157–214.

208. J. Rauch, The mathematical theory of crushed ice, pp. 370–379 in [112].

209. J. Rauch, Scattering by many tiny obstacles, pp. 380–389 in [112].

210. J. Rauch and M. Taylor, Potential and scattering theory on wildly perturbed domains, *J. Functional Analysis* **18** (1975), 27–59.

211. J. Rauch and M. Taylor, Electrostatic screening, *J. Mathematical Phys.* **16** (1975), 284–288.

212. D. B. Ray, On spectra of second order differential operators, *Trans. Amer. Math. Soc.* **77** (1954), 299–321.

213. J. W. S. Rayleigh, *The Theory of Sound*, Macmillan, New York, 1894.

214. M. Reed and B. Simon, *Methods of Modern Mathematical Physics, I. Functional Analysis*, Academic Press, New York, 1972.

215. M. Reed and B. Simon, *Methods of Modern Mathematical Physics, II. Fourier Analysis, Self-Adjointness*, Academic Press, New York, 1975.

216. M. Reed and B. Simon, *Methods of Modern Mathematical Physics, III. Scattering Theory*, Academic Press, New York, 1979.

217. M. Reed and B. Simon, *Methods of Modern Mathematical Physics, IV. Analysis of Operators*, Academic Press, New York, 1978.

218. F. Rellich, Störungstheorie der Spektralzerlegung, I–IV, *Math. Ann.* **113** (1937), 600–619; **113** (1937), 677–685; **116** (1939), 555–570; **117** (1940), 356–382; **118** (1942), 462–484.

219. E. Rellich, *Perturbation Theory of Eigenvalue Problems*, Gordon & Breach, New York, 1969.

220. Y. Rinott, On convexity of measures, Thesis, Weizmann Institute, Rehovot, Israel, 1973.

221. D. Robinson, *The Thermodynamic Pressure in Quantum Statistical Mechanics*, Springer-Verlag, Berlin and New York, 1971.

222. J. Rosen, Sobolev inequalities for weight spaces and supercontractivity, *Trans. Amer. Math. Soc.* **222** (1976), 367–376.

223. J. Rosen and B. Simon, Fluctuations in the $P(\phi)_1$ process, *Ann. Probability* **4** (1976), 155–174.

224. G. Rosenbljum, The distribution of the discrete spectrum for singular differential

operators, *Dokl. Akad. Nauk SSSR* **202** (1972) [Transl.: *Soviet Math. Dokl.* **13** (1972), 245–249].

225. D. Ruelle, A variational formulation of equilibrium statistical mechanics and the Gibbs phase rule, *Comm. Math. Phys.* **5** (1967), 324–239.
226. D. Ruelle, *Statistical Mechanics*, Benjamin, New York, 1969.
227. D. Ruelle, *Thermodynamic Formalism: The Mathematical Structure of Classical Equilibrium Statistical Mechanics*, Addison-Wesley, Reading, Massachusetts, 1978.
228. M. Schechter, *Spectra of Partial Differential Operators*, North-Holland Publ., Amsterdam, 1971.
229. M. Schechter, Essential self-adjointness of the Schrödinger operator with magnetic vector potential, *J. Functional Analysis* **20** (1975), 93–104.
230. M. Schilder, Some asymptotic formulas for Wiener integrals, *Trans. Amer. Math. Soc.* **125** (1965), 63–85.
231. J. E. Schnol, On the behavior of the eigenfunctions of Schrödinger's equations, *Mat. Sb.* **46** (1957), 275–286.
232. E. Schrödinger, Quantisierung als Eigenwertproblem, IV. Störungstheorie mit Anwendung auf den Starkeffekt der Balmerlinien, *Ann. Physik.* **80** (1926), 437–490.
233. J. Schwinger, On the bound states of a given potential, *Proc. Nat. Acad. Sci. U.S.A.* **47** (1967), 122–129.
234. J. Schwinger, *Selected Papers on Quantum Electrodynamics*, Dover, New York, 1958.
235. E. Seiler and B. Simon, Nelson's symmetry and all that in the Yukawa$_2$ and ϕ_3^4 field theories, *Ann. Physics* **97** (1976), 470–518.
236. Yu. Semenov, Wave operators for the Schrödinger equation with strongly singular short-range potentials, *Lett. Math. Phys.* **1** (1977), 457–461.
237. D. Shale, Linear symmetries of the free boson field, *Trans. Amer. Math. Soc.* **103** (1962), 149–167.
238. P. Shields, *The Theory of Bernoulli Shifts*, Chicago Lecture Notes in Math., Univ. of Chicago Press, 1973.
239. A. J. S. Siegert, Partition functions as averages of functionals of Gaussian random functions, *Physica* **26** (1960), S30–S35.
240. B. Simon, Distributions and their Hermite expansions, *J. Mathematical Phys.* **12** (1971), 140–148.
241. B. Simon, Coupling constant analyticity for the anharmonic oscillator, *Ann. Physics* **58** (1970), 149–167.
242. B. Simon, Schrödinger operators with singular magnetic vector potentials, *Math. Z.* **131** (1973), 361–370.
243. B. Simon, Pointwise bounds on eigenfunctions and wave packets in N-body quantum systems, I, *Proc. Amer. Math. Soc.* **42** (1974), 395–401.
244. B. Simon, Pointwise bounds on eigenfunctions and wave packets in N-body quantum systems, II, *Proc. Amer. Math. Soc.* **45** (1974), 454–456.
245. B. Simon, Pointwise bounds on eigenfunctions and wave packets in N-body quantum systems, III, *Trans. Amer. Math. Soc.* **208** (1975), 317–329.
246. B. Simon, Quadratic forms and Klauder's phenomenon: A remark on very singular perturbations, *J. Functional Analysis* **14** (1973), 295–298.
247. B. Simon, Existence of the scattering matrix for the linearized Boltzmann equation, *Comm. Math. Phys.* **41** (1975), 99–108.
248. B. Simon, The bound state of weakly coupled Schrödinger operators in one and two dimensions, *Ann. Physics* (1976). 276–288.
249. B. Simon, Analysis with weak trace ideals and the number of bound states of Schrödinger operators, *Trans. Amer. Math. Soc.* **224** (1976), 367–380.

250. B. Simon, Universal diamagnetism of spinless boson systems, *Phys. Rev. Lett.* **36** (1976), 804–806.
251. B. Simon, On the number of bound states of two body Schrödinger operators—a review, pp. 305–326, in [174].
252. B. Simon, Classical boundary conditions as a technical tool in modern mathematical physics, *Advances in Math.* **30** (1978), 268–281.
253. B. Simon, A canonical decomposition for quadratic forms with application to monotone convergence theorems, J. Functional Analysis **28** (1978), 377–385.
254. B. Simon, Lower semicontinuity of quadratic forms, *Proc. Roy. Soc. Edinburgh* **79** (1977), 267–273.
255. B. Simon, An abstract Kato's inequality for generators of positivity preserving semigroups, *Indiana Univ. Math. J.* **26** (1977), 1067–1073.
256. B. Simon, Kato's inequality and the comparison of semigroups, *J. Functional Analysis*, to appear.
257. B. Simon, *Quantum Mechanics for Hamiltonians Defined as Quadratic Forms*, Princeton Univ. Press, Princeton, New Jersey, 1971.
258. B. Simon, *The $P(\phi)_2$ Euclidian (Quantum) Field Theory*, Princeton Univ. Press, Princeton, New Jersey, 1974.
259. B. Simon, *Trace Ideal Methods*, Cambridge Univ. Press, London and New York, 1979.
260. B. Simon, Maximal and minimal Schrödinger operators and forms, *J. Operator Theory Appl.* **1** (1979), 37–47.
260a. B. Simon, Brownian motion, L^p properties of Schrödinger operators and the localization of binding, *J. Functional Analysis* to be published.
260b. B. Simon, The Classical Limit of Quantum Partition Functions, to appear.
261. B. Simon and R. Griffiths, The ϕ_2^4 field theory as a classical Ising model, *Comm. Math. Phys.* **33** (1973), 145–164.
262. E. Speer, *Generalized Feynman Amplitudes*, Princeton Univ. Press, Princeton, New Jersey, 1969.
263. T. Spencer, The mass gap for the $P(\phi)_2$ quantum field model with strong external field, *Comm. Math. Phys.* **39** (1974), 63–76.
264. T. Spencer, The absence of even bound states for $\lambda(\phi^4)_2$, *Comm. Math. Phys.* **39** (1974), 77–99.
265. F. Spitzer, *Principles of Random Walk*, Van Nostrand–Reinhold, New York, 1964.
266. F. Spitzer, Electrostatic capacity, heat flow, and Brownian motion, *Z. Wahrscheinlichkeitstheorie und Verw. Gebiete* **3** (1964), 110–121.
267. R. Streater and A. S. Wightman, *PCT, Spin and Statistics and All That*, Benjamin, New York, 1964.
268. R. Strichartz, Multipliers on fractional Sobolev spaces, *J. Math. Mechanics* **16** (1967), 1031–1060.
268a. D. Stroock, On certain systems of parabolic equations, *Comm. Pure Appl. Math.* **23** (1970) 447–457.
269. G. Sylvester, Representations and inequalities for Ising Model Ursell functions, *Comm. Math. Phys.* **42** (1975), 209–220.
270. K. Symanzik, Proof and refinements of an inequality of Feynman, *J. Mathematical Phys.* **6** (1965), 1155–1156.
271. B. Sz-Nagy, Perturbations des transformations autoadjointes dans l'espace de Hilbert, *Comment. Math. Helv.* **19** (1946/47), 347–366.
272. L. Takas, *Combinatorial Methods in the Theory of Stochastic Processes*, Wiley, New York, 1967.

273. H. Tamura, The asymptotic eigenvalue distribution for nonsmooth elliptic operators, *Proc. Japan Acad.* **50** (1974), 19–22.

274. E. Teller, On the stability of molecules in the Thomas–Fermi theory, *Rev. Modern Phys.* **34** (1962), 627–631.

275. C. T. Thompson, Inequality with applications in statistical mechanics, *J. Mathematical Phys.* **6** (1965), 1812–1813.

276. E. C. Titchmarsh, *Eigenfunction Expansions*, I, II, Oxford Univ. Press, London and New York, Part I, 1962; Part II, 1958.

277. E. C. Titchmarsh, Some theorems on perturbation theory, I–V, *Proc. Roy. Soc. London Ser. A* **200** (1949), 34–46; **201** (1950), 473–479; **207** (1951), 321–328; **210** (1951), 30–47; *J. Analyse Math.* **4** (1954/56), 187–208.

278. F. Trèves, *Topological Vector Spaces, Distributions, and Kernels*, Academic Press, New York, 1967.

279. H. Trotter, On the product of semigroups of operators, *Proc. Amer. Math. Soc.* **10** (1959), 545–551.

280. E. Trubowitz, The inverse problem for periodic potentials, *Comm. Pure Appl. Math.* **30** (1977), 321–337.

280a. A. Truman, Feynman path integrals and quantum mechanics as $\hbar \to 0$, *J. Mathematical Phys.* **17** (1976), 1852–1862.

280b. A. Truman, The classical action in non-relativistic quantum mechanics, *J. Mathematical Phys.* **18** (1977), 1499–1509.

280c. A. Truman, Classical mechanics, the diffusion (heat) equation, and the Schrödinger equation, *J. Mathematical Phys.* **18** (1977), 2308–2315.

280d. A. Truman, The Feynman maps and the Wiener integral, *J. Mathematical Phys.* **19** (1978), 1742–175.

280e. A. Truman, Some applications of vector space measure to non-relativistic quantum mechanics, pp. 418–441, in *Vector Space Measures and Applications*, I, (R. Aron and S. Dineen, eds.) Lecture Notes in Math., **644**, Springer-Verlag, New York and Berlin, 1978.

280f. A. Truman, The polygonal path formulation of the Feynman path integral, to appear in Lecture Notes in Math, Volume on proceedings 1978 Marseille conference on path integrals, Springer-Verlag, Berlin and New York.

281. G. E. Uhlenbeck and L. S. Ornstein, On the theory of Brownian motion, I, *Phys. Rev.* **36** (1930), 823–841.

282. G. Velo and A. S. Wightman, eds., *Constructive Quantum Field Theory*, Springer-Verlag, Berlin and New York, 1973.

283. L. van Hemmen, unpublished.

283a. R. Weder, The unified approach to spectral analysis, *Comm. Math. Phys.* **60** (1978), 291–299.

284. S. Weinberg, High energy behavior in quantum field theory, *Phys. Rev.* **118** (1960), 838–849.

285. H. Weyl, Das asymptotische Verteilungsgesetz der Eigenwerte linearer partieller Differentialgleichungen, *Math. Ann.* **71** (1911), 441–469.

286. N. Wiener, Differential space, *J. Mathematical and Physical Sci.* **2** (1923), 132–174.

287. A. S. Wightman, Hilbert's sixth problem: Mathematical treatment of the axioms of physics, pp. 147–240, in *Mathematical Developments Arising from Hilbert Problems*, (F. Browder, ed.), American Mathematic Society, New York, 1976.

288. K. Yajima, The quasi-classical limit of quantum scattering theory, *Commun. Math. Phys.*, to be published.

289. J. Rosner, C. Quigg, and H. B. Thacker, Determining the fifth quark's charge: The role of the leptonic widths, *Phys. Lett.* **74B** (1978), 350–352.
290. C. Leung and J. Rosner, Mass dependence of Schrödinger wave functions, *J. Math. Phys.* to be published.
291. G. Hagedorn, Semiclassical quantum mechanics. I. The $\hbar \to 0$ limit for coherent states, *Comm. Math. Phys.* to be published.

Index

N

Newman's inequality, 129
Newtonian capacity, 84–87
n-fold points of Brownian motion, 81–87
Nonanticipatory functional, 153
Nondifferentiability of paths (Th. 5.4), 46

O

Ornstein–Uhlenbeck velocity process, 35
Oscillator process, 34

P

$P(\phi)_1$ process, 57
Park's correlation inequality (Th. 23.4a), 250
Path's properties of
  arcsin law, 58–60
  hitting probabilities, 70–72, 82–84
  Hölder continuity, 45
  iterated logarithm, 60–64
  n-fold points, 81–87
  nondifferentiability, 46
  nonrectifiable, 148
  recurrence, 73–81
  Wiener sausage, 236
  zeros of, 61
Percus' lemma (Prop. 12.11), 130
Portenko's lemma (Th. 11.2), 117
Positive definite, 10
Pressure, 200–201, 246
Probability distribution, joint, 8
Probability distributions of Brownian motion
  $E(\max_{0 \leq s \leq t} b(s) \geq \lambda)$, 64–67, 76
  $E(|(R, 0, \ldots, 0) + b(s)| \leq r$; some $s)$, 70–72, 83–84
  $E(\inf\{s \mid b(s) = 1\} \geq s_0)$, 76–79
  $E(|\{s \leq 1 \mid b(s) \geq 0\}|)$, 59–60
Probability measure space, 8
Prohorov's theorem (Th. 17.4), 177

Q

Quasi-classical limit, *see* Semiclassical limit

R

Random variable, 8
Recurrence properties of Brownian motion, 73–81
Reflection principle, 25, 64
Regular set, 69

S

Schwinger functions, 253
Self-intersection, *see* n-fold points, of Brownian motion,
Semiclassical limit, 1, 7, 105–114, 164–167, 195–198
Shale's theorem (Th. 2.5), 17, 189
Sine–Gordon transformation, 249
Stability of matter (Ths. 9.6, 15.12), 98–105, 168
Statistical mechanical analog, 58, 198
Stochastic integral, 170
Stochastic process, 13
Stopping time, 65
  discretization of, 66
Strong law of large numbers, 75
Submartinaggale, 22–24
Symanzik's inequality (Th. 9.2), 94
Symmetric decreasing function, 142
  rearrangement, 143
Symmetry numbers, 219

T

Tauberian theorem (Th. 10.3), 108
Teller's lemma (Lemma 9.10), 105
Teller's theorem (Th. 9.11), 105
Thomas–Fermi theory, 98, 102–105
Tight family of measures, 177
Trotter product formula (Th. 1.1), 4–6

U

Uniformly distributed of degree m, 233
Uniformly locally L^p, 260
Unlabeled graph, 219
Ursell functions, 129

V

Versions, 13–14, 44

W

Wave operators, 226

Pure and Applied Mathematics

A Series of Monographs and Textbooks

RECENT TITLES

D. V. WIDDER. The Heat Equation

IRVING EZRA SEGAL. Mathematical Cosmology and Extragalactic Astronomy

J. DIEUDONNÉ. Treatise on Analysis: Volume II, enlarged and corrected printing; Volume IV; Volume V; Volume VI

WERNER GREUB, STEPHEN HALPERIN, AND RAY VANSTONE. Connections, Curvature, and Cohomology: Volume III, Cohomology of Principal Bundles and Homogeneous Spaces

I. MARTIN ISAACS. Character Theory of Finite Groups

JAMES R. BROWN. Ergodic Theory and Topological Dynamics

C. TRUESDELL. A First Course in Rational Continuum Mechanics: Volume 1, General Concepts

GEORGE GRATZER. General Lattice Theory

K. D. STROYAN AND W. A. J. LUXEMBURG. Introduction to the Theory of Infinitesimals

B. M. PUTTASWAMAIAH AND JOHN D. DIXON. Modular Representations of Finite Groups

MELVYN BERGER. Nonlinearity and Functional Analysis: Lectures on Nonlinear Problems in Mathematical Analysis

CHARALAMBOS D. ALIPRANTIS AND OWEN BURKINSHAW. Locally Solid Riesz Spaces

JAN MIKUSINSKI. The Bochner Integral

THOMAS JECH. Set Theory

CARL L. DEVITO. Functional Analysis

MICHIEL HAZEWINKEL. Formal Groups and Applications

SIGURDUR HELGASON. Differential Geometry, Lie Groups, and Symmetric Spaces

ROBERT B. BURCKEL. An Introduction to Classical Complex Analysis: Volume 1

JOSEPH J. ROTMAN. An Introduction to Homological Algebra

LOUIS HALLE ROWEN. Polynominal Identities in Ring Theory

BARRY SIMON. Functional Integration and Quantum Physics

IN PREPARATION

C. TRUESDELL AND R. G. MUNCASTER. Fundamentals of Maxwell's Kinetic Theory of a Simple Monatomic Gas: Treated as a Branch of Rational Mechanics

ROBERT B. BURCKEL. An Introduction to Classical Complex Analysis: Volume 2

DRAGOS M. CVETLOVIC, MICHAEL DOOB, AND HORST SACHS. Spectra of Graphs

DAVID KINDERLEHRER and GUIDO STAMPACCHIA. Introduction to Variational Inequalities and Their Applications.